Models for Multi-State Survival Data

Multi-state models provide a statistical framework for studying longitudinal data on subjects when focus is on the occurrence of events that the subjects may experience over time. They find application particularly in biostatistics, medicine, and public health. The book includes mathematical detail which can be skipped by readers more interested in the practical examples. It is aimed at biostatisticians and at readers with an interest in the topic having a more applied background, such as epidemiology. This book builds on several courses the authors have taught on the subject.

Key Features:

- Intensity-based and marginal models.
- Survival data, competing risks, illness-death models, recurrent events.
- Full chapter on pseudo-values.
- Intuitive introductions and mathematical details.
- Practical examples of event history data.
- Exercises.

Software code in R and SAS and the data used in the book can be found on the book's webpage.

Henrik Ravn is senior statistical director at Novo Nordisk A/S, Denmark. He graduated with an MSc in theoretical statistics in 1992 from University of Aarhus, Denmark and completed a PhD in Biostatistics in 2002 from the University of Copenhagen, Denmark. He joined Novo Nordisk in late 2015 after more than 22 years of experience doing biostatistical and epidemiological research, at Statens Serum Institut, Denmark and in Guinea-Bissau, West Africa. He has co-authored more than 160 papers, mainly within epidemiology and application of survival analysis and has taught several courses as external lecturer at Section of Biostatistics, University of Copenhagen.

Per Kragh Andersen is professor of Biostatistics at the Department of Public Health, University of Copenhagen, Denmark since 1998. He earned a mathematical statistics degree from the University of Copenhagen in 1978, a PhD in 1982, and a DMSc degree in 1997. From 1993 to 2002 he worked as chief statistician at Danish Epidemiology Science. He is author or co-author of more than 125 papers on statistical methodology and more than 250 papers in the medical literature. His research has concentrated on survival analysis, and he is co-author of the 1993 book *Statistical Models Based on Counting Processes*. He has taught several courses both nationally and internationally both for students with a mathematical background and for students in medicine or public health.

CHAPMAN & HALL/CRC
Texts in Statistical Science Series

Recently Published Titles

Sampling
Design and Analysis, Third Edition
Sharon L. Lohr

Theory of Statistical Inference
Anthony Almudevar

Probability, Statistics, and Data
A Fresh Approach Using R
Darrin Speegle and Brain Claire

Bayesian Modeling and Computation in Python
Osvaldo A. Martin, Raviv Kumar and Junpeng Lao

Bayes Rules!
An Introduction to Applied Bayesian Modeling
Alicia Johnson, Miles Ott and Mine Dogucu

Stochastic Processes with R
An Introduction
Olga Korosteleva

Design and Analysis of Experiments and Observational Studies using R
Nathan Taback

Time Series for Data Science: Analysis and Forecasting
Wayne A. Woodward, Bivin Philip Sadler and Stephen Robertson

Statistical Theory
A Concise Introduction, Second Edition
Felix Abramovich and Ya'acov Ritov

Applied Linear Regression for Longitudinal Data
With an Emphasis on Missing Observations
Frans E.S. Tan and Shahab Jolani

Fundamentals of Mathematical Statistics
Steffen Lauritzen

Modelling Survival Data in Medical Research, Fourth Edition
David Collett

Applied Categorical and Count Data Analysis, Second Edition
Wan Tang, Hua He and Xin M. Tu

Geographic Data Science with Python
Sergio Rey, Dani Arribas-Bel and Levi John Wolf

Models for Multi-State Survival Data
Rates, Risks, and Pseudo-Values
Per Kragh Andersen and Henrik Ravn

For more information about this series, please visit: https://www.routledge.com/Chapman--HallCRC-Texts-in-Statistical-Science/book-series/CHTEXSTASCI

Models for Multi-State Survival Data

Rates, Risks, and Pseudo-Values

Per Kragh Andersen and Henrik Ravn

Figures by Julie Kjærulff Furberg

CRC Press is an imprint of the
Taylor & Francis Group, an **informa** business
A CHAPMAN & HALL BOOK

Designed cover image: © Gustav Ravn

First edition published 2024
by CRC Press
6000 Broken Sound Parkway NW, Suite 300, Boca Raton, FL 33487-2742

and by CRC Press
4 Park Square, Milton Park, Abingdon, Oxon, OX14 4RN

CRC Press is an imprint of Taylor & Francis Group, LLC

ISBN: 978-0-367-14002-1 (hbk)
ISBN: 978-1-032-56869-0 (pbk)
ISBN: 978-0-429-02968-4 (ebk)

DOI: 10.1201/9780429029684

Typeset in Nimbus Roman font
by KnowledgeWorks Global Ltd.

Publisher's note: This book has been prepared from camera-ready copy provided by the authors.

Access the Support Material: https://multi-state-book.github.io/companion/

Contents

Preface

Multi-state models provide a statistical framework for studying longitudinal data on subjects when focus is on the occurrence of events that the subjects may experience over time. The most simple situation is when only a single event, 'death' is of interest – a situation known as *survival analysis*. We shall use the phrase *multi-state survival data* for the data that arise in the general case when observing subjects over time and several events may be of interest during their life spans.

As indicated in the sub-title of the book, models for multi-state survival data can either be specified via *rates* of transition between states or by directly addressing the *risk* of occupying a given state at given points in time. A general approach for addressing risks and other *marginal* parameters is via *pseudo-values* about which a whole chapter is provided.

The background for writing this book is our teaching of several courses on various aspects of multi-state survival data – either for participants with a basic training in statistics or with a clinical/epidemiological background. Several texts on multi-state models already exist. These include the books by Beyersmann et al. (2012), Broström (2012), Geskus (2016), and more recently, Cook and Lawless (2018). The book by Kalbfleisch and Prentice (1980, 2002) focuses on survival analysis but also discusses general multi-state models. Books with more emphasis on mathematics are those by Andersen et al. (1993), Hougaard (2000), Martinussen and Scheike (2006), and Aalen et al. (2008). In addition, several review papers have appeared (e.g., Hougaard, 1999; Andersen and Keiding, 2002; Putter et al., 2007; Andersen and Pohar Perme, 2008; Bühler et al., 2023). In spite of the existence of these texts, we were unable to identify a suitable common book for these different types of participants and, importantly, none of the cited texts provide detailed discussions on the development of methods based on *pseudo-values*.

With this book we aim at filling this gap (at least for ourselves) and to provide a text that is applicable as the basis for courses for mixed groups of participants. By addressing at least two types of readers, we deliberately run the risk of falling between two chairs; however, we believe that readers with different mathematical backgrounds and interests should all benefit from the book. Those readers who appreciate some mathematical details can read the book from the beginning to the end, thereby first getting a (hopefully) more intuitive introduction to the topics, including practical examples and, subsequently, in sections marked with '(*)' get more details. This will, unavoidably, entail some repetitions. On the other hand, readers with less interest in mathematical details can read all sections that are not marked with '(*)' without losing the flow of the book. It should be emphasized that we will from time to time refer to (*)-marked sections and to more technical publications from those more intuitive sections. The text includes several *summary boxes* that emphasize highlights from recent sections.

The book discusses a number of practical examples of event history data that are meant to illustrate the methods discussed. Sometimes, different statistical models that are fitted to the same data are mathematically incompatible and we will make remarks to this effect along the way. Software code for the calculations in the examples is *not* documented in the book. Rather, code in `R` and `SAS` and data can be found on the book's webpage. The webpage also includes code for solving the practical exercises found at the end of each chapter and solutions to theoretical exercises marked with (*).

The cover drawing probably calls for an explanation, as follows. When analyzing recurrent events data and ignoring competing risks (which is quite common in applications), then a curve like the top one on the figure may be obtained – a curve that is upwards biased. However, then comes the book by the crow (Per's middle name) and the raven (Henrik's last name) as a rescue and forces the curve downwards to avoid the bias. We wish to thank Gustav Ravn for the cover drawing.

There are a number of other people to whom we wish to address our thanks and without whose involvement the writing of this book would not have been possible.

First and foremost, our sincere thanks go to Julie K. Furberg who has carefully created all figures and validated all analyses quoted in the book. She also gave valuable feedback on the text. Eva N.S. Wandall thoroughly created solutions to practical exercises and contributed to some of the examples.

Several earlier drafts of chapters were read and commented upon by Anne Katrine Duun-Henriksen, Niels Keiding, Thomas H. Scheike, and Henrik F. Thomsen. Torben Martinussen provided important input for Chapter 6.

A special thank you goes to those who have provided data for the practical examples: Peter Aaby, Jules Angst, Flemming Bendtsen, John P. Klein, Bjørn S. Larsen, Thorkild I.A. Sørensen, Niels Tygstrup, and Tine Westergaard. Permission to present analyses of LEADER data was given by Novo Nordisk A/S.

We thank our employers: University of Copenhagen and Novo Nordisk A/S for letting us work on this book project during working hours for several years. A special thank to Novo Nordisk A/S for granting a stay at Favrholm Campus to finalize the book. Communication with the publishers has been smooth and we are grateful for their patience.

Bagsværd
March 2023

Per Kragh Andersen
Henrik Ravn

List of symbols and abbreviations

The following list describes main symbols and abbreviations used in the book:

$\alpha(t)$	Hazard (intensity, rate) function
β	Regression coefficient
$\varepsilon_h(\tau)$	Expected length of stay in state h in $[0, \tau]$
$\lambda(t)$	Intensity process for counting process
$\mu(t)$	Mean number of recurrent events until time t; $\mu(t) = E(N(t))$
$A(t)$	Cumulative hazard function; $A(t) = \int_0^t \alpha(u)du$
A_i	Frailty for cluster/subject i
B_i	Delayed entry time for subject i
C_i	Right-censoring time for subject i
D_i	Event indicator for subject i: $\{0, 1\}$; with competing risks $\{0, 1, \ldots, k\}$
DF	Degrees of freedom
$E(T)$	Expectation of the random variable T
$F(t)$	Cumulative distribution function for the random variable T
$F_h(t)$	Cumulative incidence function for cause h
$G(t)$	Survival distribution function for C (censoring)
GEE	Generalized estimating equation
h	State in multi-state model
HR	Hazard ratio
$I(\cdots)$	Indicator function; $I(\cdots)$ is 1 if $\cdots$ is true and 0 otherwise
J	Inspection time giving rise to interval-censoring
$K(t)$	Weight function in logrank test and other non-parametric tests
L_i	Likelihood contribution from subject i
LP	Linear predictor $= \beta_1 Z_1 + \beta_2 Z_2 + \cdots + \beta_p Z_p$

LRT	Likelihood ratio test
$M(t)$	Martingale process
$N(t)$	Counting process; $N_i(t)$ for subject i
$P(\cdot)$	Probability
PL	Partial likelihood
$Q_h(t)$	Probability to be in state h at time t (state occupation probability)
$R(t)$	Risk set at time t
$S(t)$	Survival distribution function for the random variable T; $1-F(t)$
SD	Standard deviation
T_i	Event time for individual i
$U(\cdot)$	Score function or other estimating function
$V(t)$	Multi-state process
W_i	Weight for subject i
X_i	Observation time for subject i; $X_i = \min(T_i, C_i)$
$Y(t)$	Number at risk at time t
$Y_i(t)$	At-risk indicator for subject i
Z_i	Covariate for subject i, may be time-dependent: $Z_i(t)$

Chapter 1

Introduction

In many fields of quantitative science, subjects are followed over *time* for the occurrence of certain *events*. Examples include clinical studies where cancer patients are followed from time of surgery until time of death from any cause (*survival analysis*), epidemiological cohort studies, e.g., registry-based, where disease-free individuals ('exposed' or 'unexposed') are followed from a given calendar time until diagnosis of a certain disease, or demographic studies where women are followed through child-bearing ages with the focus on ages at which they give birth to live-born children. Data from such studies may be represented as events occurring in *continuous time* and a mathematical framework in which to study such phenomena is that of *multi-state models* where an event is considered a *transition* between certain (discrete) *states*. We will denote the resulting data as *multi-state survival data* or *event history data.*

Possible scientific questions that may be addressed in *event history analyses* include how the mortality of the cancer patients is associated with individual *prognostic variables* such as age, disease stage or histological features of the tumor, or what is the probability that exposed or unexposed subjects are diagnosed with the disease within a certain time interval, or what is the expected time spent for women as nulliparous depending on the socio-economic status of her family.

An important feature of event history data is that of *incomplete observation.* This means that observation of the event(s) of interest is precluded by the occurrence of another event, such as end-of-study, drop-out of study, or death of the individual (in case the event of interest is non-fatal). Here, as we shall discuss in more detail in Section 1.3, an important distinction is between *avoidable* events (*right-censoring*) representing practical restrictions in data collection that prevent further observation of the subject (e.g., end-of-study or drop-out) and *non-avoidable events* (*competing risks*), such as the death of a patient. For the former class of avoidable events, it is an important question whether the incomplete data that are available to the investigator after censoring still suitably represent the population for which inference is intended. This is the notion of *independent censoring* that will also be further discussed in Section 1.3.

In this book we will discuss two classes of statistical models for multi-state survival data: Intensity-based models and marginal models. Briefly, *intensities* or *rates* are parameters that describe the immediate future development of the process conditionally on past information on how the process has developed, while *marginal parameters*, such as the *risk* of being in a given state at a particular time, do not involve such a conditioning. Both classes of models often involve *explanatory variables* (or *covariates/prognostic variables/risk factors* – terms that we will use interchangeably in the book).

The first model class targets intensities and is inspired by standard hazard models for survival data, and we shall see that models such as the Cox (1972) proportional hazards model also play an important role for more general multi-state survival data. Throughout, we will use the terms *intensity, hazard,* and *rate* interchangeably. Models for intensities are discussed in Chapters 2 and 3.

The second model class targets marginal parameters (e.g., risks) and, here, one approach is *plug-in* methods where the marginal parameter is estimated using intensity-based models. Thus, the results from these models are either inserted ('plugged') into an equation giving the relationship between the marginal parameter and the intensities, or they are used as the basis for simulating a large number of realizations of the multi-state process, whereby the marginal parameter may be estimated, a technique known as *micro-simulation*. Another approach is models that *directly* target marginal parameters, and a number of such models will also be presented. Marginal models are discussed in Chapters 4 and 5. For direct marginal models (or simply direct models), as we shall see in Chapter 6, *pseudo-values* (or *pseudo-observations*) are useful. In the final Chapter 7, a number of further topics are briefly discussed.

Sections marked with '(*)' contain, as indicated in the Preface, more mathematical details. Each chapter ends with a number of exercises where those marked with '(*)' are more technical.

Multi-state survival data

Multi-state survival data (event history data) represent subjects followed over time for the occurrence of events of interest. The events occur in continuous time and an event is considered a transition between discrete states.

1.1 Examples of event history data

To support a more tangible discussion of the concepts highlighted above, we will in this section present a series of examples of event history data and, along the way, specify the scientific questions that the examples were meant to address, the events of interest (and, thereby, the states in the multi-state model), and the censoring events.

1.1.1 PBC3 trial in liver cirrhosis

The PBC3 trial was a multi-center randomized clinical trial conducted in six European hospitals (Lombard et al., 1993). Between January 1983 and January 1989, 349 patients with the liver disease primary biliary cirrhosis (PBC) were randomized to treatment with either Cyclosporin A (CyA, 176 patients) or placebo (173 patients). The purpose of the trial was to study the effect of treatment on the survival time, so, the event of interest is death of the patient. The censoring events include drop-out before the planned termination of the trial (end of December 1988) and being alive at the end of the trial. However, during the course

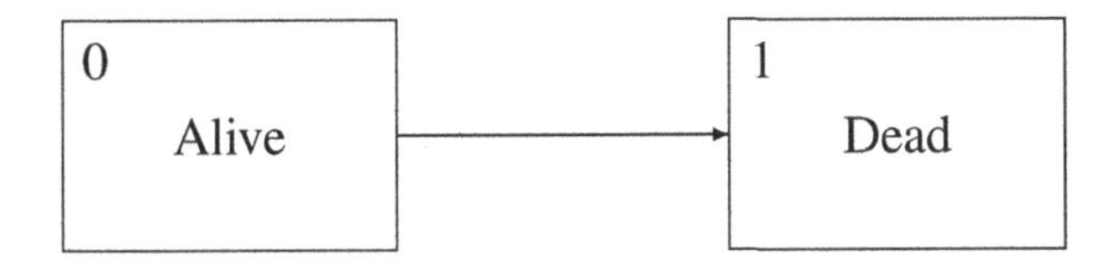

Figure 1.1 *The two-state model for survival data.*

of the trial, an increased use of liver transplantation as a possible treatment for patients with this disease forced the investigators to reconsider the trial design. Liver transplantation was primarily offered to severely ill patients and, therefore, censoring patients at the time of transplantation would likely leave the investigators with a sample of 'too well' patients that would no longer be representative of patients with PBC. This led them to redefine the main event of interest to be 'failure of medical treatment' defined as the *composite end-point* of either death or liver transplantation, whichever occurred first. This is because both death and the need of a liver transplantation signal that the medical treatment is no longer effective. Patients were followed from randomization until treatment failure, drop-out or January 1989; 61 patients died (CyA: 30, placebo: 31), another 29 were transplanted (CyA: 14, placebo: 15) and 4 patients were lost to follow-up before January 1989. For patients lost to follow-up and for those alive without having had a liver transplantation on January 1989, all that is known about time to failure was that it exceeds time from randomization to end of follow-up.

Figure 1.1 shows the general *two-state model for survival data* with states '0: Alive' and '1: Dead' and one possible transition from state 0 to state 1 representing the event 'death'. In the PBC3 trial, this model is applicable with the two states representing: (0) 'Alive without transplantation' and (1) 'Dead or transplantation' and the transition, $0 \rightarrow 1$, representing the event of interest – failure of medical treatment.

PBC3 was a randomized trial and, therefore, the explanatory variable of primary interest was the treatment indicator. However, in addition, a number of clinical, biochemical, and histological variables were recorded at entry into the study. Studying the distribution of such prognostic variables in the two treatment groups, it appeared that, in spite of the randomization, the CyA group tended to present with somewhat less favorable values of these variables than the placebo group. Therefore, evaluation of the treatment effect with or without adjustment for explanatory variables shows some differences to be discussed in later chapters.

Another option than defining the composite end-point 'failure of medical treatment' would be to study the two events 'death without transplantation' and 'liver transplantation' separately. This would enable a study of possibly different effects of treatment (and other covariates) on each of these separate events. This situation is depicted in Figure 1.2, showing the general *competing risks model.* Compared to Figure 1.1 it is seen that the initial state 'Alive' is the same whereas the final state 'Dead' is now split into a number, k separate states, transitions into which represent deaths from different causes. For the PBC3 trial,

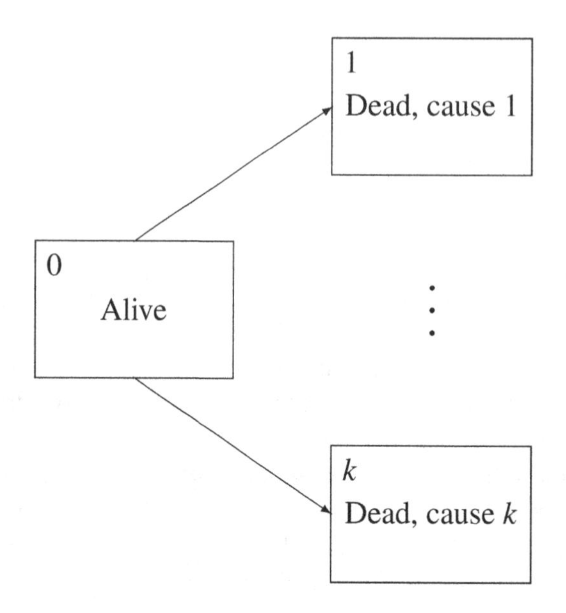

Figure 1.2 *The competing risks model with k causes of death.*

state 0 represents, as before, 'Alive without transplantation' and there are $k = 2$ final states representing, respectively, '1: Transplantation' and '2: Dead without transplantation'. The event 'liver transplantation' is a $0 \to 1$ transition and 'death without liver transplantation' a $0 \to 2$ transition. Some patients died after liver transplantation. However, the initial medical treatment (CyA or placebo) was no longer considered relevant after a transplantation, so, information on mortality after transplantation was not ascertained as a part of the trial and is not available.

1.1.2 Guinea-Bissau childhood vaccination study

The Guinea-Bissau study was a longitudinal study of women of fertile age and their prospectively registered children and was initiated in 1990 in five regions of Guinea-Bissau, West Africa. This observational study was set up to assess childhood and maternal mortality. In each region, 20 clusters of 100 women were selected and visited approximately every 6 months by a mobile team. The purpose of collecting these data, used by Kristensen et al. (2000), was to examine the association between routine childhood vaccinations and infant mortality. The main outcome was infant mortality over 6 months among 5,274 children between age 0-6 months at the initial visit (first visit by the mobile team). The recommended vaccination schedule for this age group in Guinea-Bissau at that time was BCG (Bacillus Calmette-Guérin) and polio at birth; DTP (diphtheria, tetanus, and pertussis) and oral polio at 6, 10, and 14 weeks. At the visits, vaccination status was ascertained by inspection of the immunization card. The authors analyzed mortality before next visit according to the vaccination status assessed at the initial visit.

Table 1.1 *Guinea-Bissau childhood vaccination study: Mortality during 6 months of follow up according to vaccination status (BCG or DTP) at initial visit among 5,274 children.*

	Died during follow-up		
Vaccinated	Yes	No	Total
No	95 (4.9%)	1847	1942
Yes	127 (3.8%)	3205	3332
Total	222 (4.2%)	5052	5274

As in the PBC3 example, there are two relevant states 'Alive' and 'Dead' as represented in Figure 1.1. The censoring events include out-migration between visits and being alive at the subsequent visit. Table 1.1 provides the basic mortality for vaccinated and non-vaccinated children.

This study was an *observational study*, as allocation to vaccination groups was not randomized. This means that any observed association between vaccination status and later mortality may be *confounded* because of uneven distributions of mortality risk factors among vaccinated and non-vaccinated children. Thus, there may be a need to adjust for covariates ascertained at the initial visit in the analysis of mortality and vaccinations.

In principle, information on vaccines received between visits was available for surviving children at the next visit. However, these extra data are discarded as, culturally, the belongings of deceased children, including immunization cards, are destroyed implying a differential information on vaccines given between visits and leading to *immortal time bias*.

1.1.3 Testis cancer incidence and maternal parity

In a registry-based study, Westergaard et al. (1998) extracted information from the Danish Civil Registration system on all women born in Denmark since 1935 (until 1978) who were alive when the system was established in April 1968. Based on this, a cohort of all (1,015,994) sons of those women who were alive in 1968 or born later was created, and this cohort was followed from April 1968 or date of birth (whichever came later) until a diagnosis of testicular cancer (ascertained in The Danish Cancer Registry, 626 cases), death (1.5%), emigration (1.3%), or end of study (end of 1992). The total follow-up time at risk was 15,981,967 years. The main purpose of the study was to address whether first-born sons have a higher incidence of testicular cancer than later born sons, and a secondary question was whether this potential association was present for two histological sub-types, seminomas (183 cases) and non-seminomas (443 cases). First-born sons provided 7,972,276 person-years at risk and 398 cases of testicular cancer, and the similar numbers for later born sons were 8,009,691 person-years and 228 cancer cases. A number of other potential risk factors for testis cancer, including age of the mother at time of birth and calendar time at birth of the son were also ascertained from the civil registration system.

The relevant multi-state model for this situation is the competing risks model (Figure 1.2) where the final states are 'Testis cancer' (possibly further split into seminomas and

Table 1.2 *PROVA trial in liver cirrhosis: Numbers of patients and numbers of events.*

Treatment group	Patients	Bleedings	Deaths without bleeding	Deaths after bleeding	Drop-out
Sclerotherapy only	73	13	13	5	5
Propranolol only	68	12	5	6	7
Both treatments	73	12	20	10	5
No treatment	72	13	8	8	3
Total	286	50	46	29	20

non-seminomas) and 'Death without testis cancer'. The censoring events are emigration and end-of-study. However, because of the rather large data set with more than a million cohort members, the raw data set with individual records was first tabulated according to the explanatory variables (including 'current age of the son') where, for each combination of these variables, the person-years at risk and numbers of seminomas and non-seminomas are given. In a similar fashion, the numbers of deaths could be tabulated; however, that information was not part of the available data and this has a number of consequences for the analyses that are possible to conduct for this study. This will be discussed later (Section 3.6).

1.1.4 PROVA trial in liver cirrhosis

The PROVA trial (PROVA Study Group, 1991) was a Danish-Norwegian multi-center, investigator-initiated clinical trial with the purpose of evaluating the prophylactic effect of propranolol (a beta-blocker) and/or sclerotherapy (a treatment where polidocanol is injected directly in the sub-mucosa next to the vein) on the occurrence of bleeding and death in patients with liver cirrhosis. Eligible patients, recruited from eleven hospitals, included those in whom cirrhosis was histologically verified, endoscopy had shown oesophageal varices, but a transfusion-requiring bleeding had not yet been observed. Between November 1985 and March 1989, 286 patients were randomized (1:1:1:1) as shown in Table 1.2 that also shows the numbers of events in each of the four treatment groups. Twenty patients dropped out of the trial without an event before the date of termination (end of 1989). At the end of follow-up, $286-(46+29+20)=191$ patients were still alive and in the trial (out of whom $50-29=21$ had experienced a bleeding).

Figure 1.3 shows the (irreversible or progressive) *illness-death model*. This multi-state model is also known as the *disability model*. Compared to Figure 1.1, it is now the initial state 0 that is split into separate states, and compared to Figure 1.2, a transition from state 1 to state 2 is now included. This model is applicable in the PROVA trial with state 0 representing 'Alive without bleeding', state 1 'Alive with bleeding' and state 2 'Dead'. The event 'bleeding' corresponds to a $0 \to 1$ transition, 'death without bleeding' to a $0 \to 2$ transition, and 'death after bleeding' to a $1 \to 2$ transition. Note that, in the trial, death after bleeding was not considered a primary end-point, but since these deaths were registered as secondary end-points (and since they are of clinical interest) they are included as events in the figure.

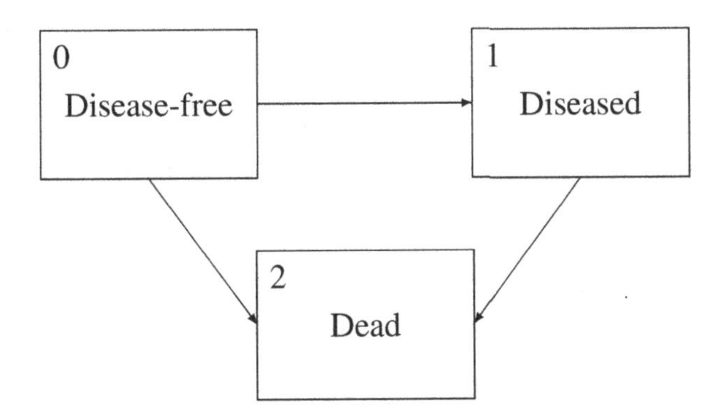

Figure 1.3 *The progressive illness-death model.*

As for the case with the PBC3 trial (Section 1.1.1), there were two censoring events: Drop-out and end-of-study. Furthermore, a number of potential explanatory variables were recorded at entry into the PROVA trial. These variables may be used when studying the prognosis of the patients.

1.1.5 Recurrent episodes in affective disorders

Psychiatric patients diagnosed with an affective disorder, i.e., unipolar disorder (depression) or bipolar disorder (manic-depression), often experience recurrent disease episodes after the initial diagnosis. Kessing et al. (2004) reported on follow-up of 186 unipolar and 220 bipolar patients who had been admitted to the Psychiatric Hospital, University of Zürich, Switzerland between 1959 and 1963. Here, we study the 119 of those patients who had their initial diagnosis in that period (98 unipolar and 21 bipolar patients). At follow-up times in 1963, 1965, 1970, 1975, 1980, and 1985, disease episodes were retrospectively ascertained via family doctors' reports, records from in- and out-patient services, and via patients or family members. Data on mortality and on dates of end of episodes were also collected. The purpose was to study the pattern of repeated disease episodes, in particular whether the disease course was deteriorating, and the event of primary interest was therefore the beginning of a new episode. Patients had on average 5.6 observed episodes (range from 1 to 26), 78 patients had died by 1985, 38 patients were still alive, and 3 were lost to follow-up before 1985. Figure 1.4 shows a multi-state model applicable in this situation, known as the *illness-death model with recovery*. Compared to Figure 1.3, a transition from state 1 to state 0, a 'recovery', is now included. If we think of an episode as an admission to hospital, then state 0 corresponds to 'Out of hospital' and state 1 to 'In hospital'. A transition from 0 to 1 is a hospital admission (initiation of a new episode – the event of primary interest) and a $1 \to 0$ transition is a discharge from hospital (end of an episode). From both states, a patient may die and, thereby, make a transition to state 2. Thus, the model depicts that there are periods where patients are not at risk for experiencing the event of primary interest.

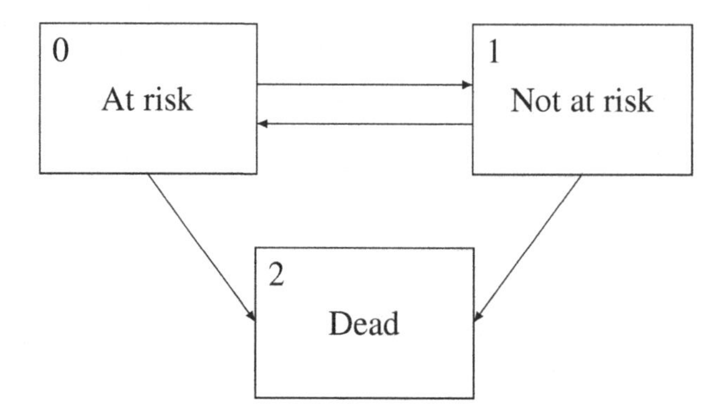

Figure 1.4 *The illness-death model with recovery, applicable for recurrent episodes with a terminal event, i.e., situations with a terminal event and with periods between times at which subjects are at risk for a new event.*

Sometimes, in spite of the fact that there are intervals between the 'at-risk periods', focus may be on times from the initiation of one episode to the initiation of the next rather on times between events. In such an approach, depicted in Figure 1.5, the interval not at risk is included in the time between events and, thereby by definition, there are no such intervals. Note that the terminal state has been re-labelled as 'D'. We will denote recurrent events where the events have a certain duration (as in this example) as *recurrent episodes*.

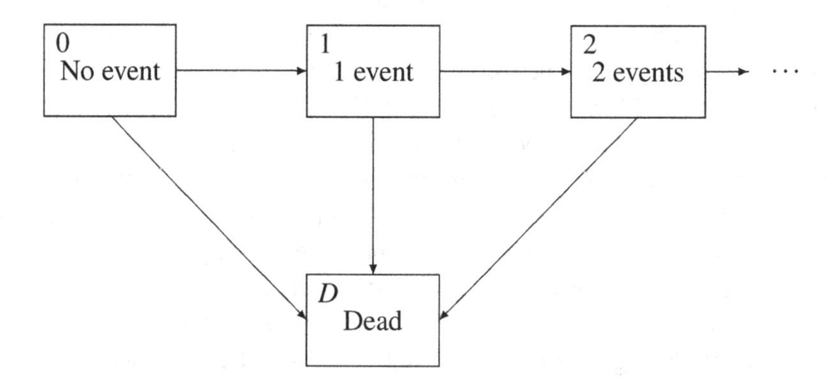

Figure 1.5 *A multi-state model for recurrent events with a terminal event and no intervals between at-risk periods.*

In these data, a number of explanatory variables that may affect the outcome and its association with the initial diagnosis (unipolar vs. bipolar) were recorded at the time when the initial diagnosis was given. These potential confounders include sex and age of the patient and calendar time at the initial diagnosis.

In situations without a terminal event, e.g., when mortality is negligible, the models in Figures 1.4 and 1.5 without the final 'Dead' state may be applicable.

1.1.6 LEADER cardiovascular trial in type 2 diabetes

The LEADER trial (Liraglutide Effect and Action in Diabetes: Evaluation of cardiovascular outcome Results) (Marso et al., 2016) was a company-initiated, double-blind randomized controlled multi-center trial investigating the cardiovascular effects of liraglutide, a glucagon like peptide-1 (GLP-1) receptor agonist approved for treatment of type 2 diabetes, versus placebo when added to standard of care in a population with type 2 diabetes and a high cardiovascular risk. A total of 9,340 subjects were randomized 1:1 to receive either liraglutide or placebo during the period from September 2010 through April 2012. Follow-up was terminated between August 2014 and December 2016, corresponding to a planned time on trial between 42 and 60 months. The median follow-up time was reported to be 3.8 years. The primary end-point was a three-component major adverse cardiovascular events (3-p MACE) composite end-point consisting of non-fatal stroke, non-fatal myocardial infarction (MI) or cardiovascular (CV) death. The primary analysis was a time-to-event analysis of time to first 3-p MACE. This end-point occurred in 608 out of 4,668 patients randomized to liraglutide and in 694 out of 4,672 placebo treated patients – a difference that was statistically significant when analyzed in a competing risks model (Figure 1.2), the competing event being death from non-CV causes.

Two of the components of 3-p MACE may occur repeatedly. Thus, recurrent MI and recurrent stroke (either fatal or non-fatal) are both possible, and a model like the one depicted in Figure 1.5 may be applicable. Here, one possibility would be to define the recurrent event as MI in which case the terminal event would be death from any cause (entry into state D). One could also be interested in 'recurrent 3-p MACE' where the state D would be non-CV death; however, one would then have to address the problem that the occurrence of one component of the recurrent end-point, namely CV death, would imply that no further events are possible. If an MI or stroke was fatal, this has been coded as an event (MI or 3-p MACE) occurring on a given calendar day, and then a cardiovascular death on the subsequent calendar day. Censoring was primarily caused by patients being alive at the end of follow-up. Thus, according to Marso et al. (2016), only 139 patients in the liraglutide group (3.0%) did not complete the study and the similar number in the placebo group was 159 (3.4%). Table 1.3 gives some key numbers of event counts observed in the trial.

1.1.7 Bone marrow transplantation in acute leukemia

For patients with certain hematological diseases such as leukemia, bone marrow (BM) transplantation (also called stem cell transplantation) is an often used treatment option. Briefly, the immune system of the patient is first affected by chemotherapy (the 'preconditioning' which removes disease symptoms) and, next, bone marrow from the donor is infused. Two serious and competing events are often studied in relation to the treatment: Relapse of the disease, i.e., return of the disease symptoms, and non-relapse mortality (death in remission), both of which signal that the treatment with BM is no longer effective. After a relapse, patients are given second line treatment to reduce the mortality. A complication

Table 1.3 *LEADER cardiovascular trial in type 2 diabetes: Observed myocardial infarctions (MI) and major adverse cardiovascular events (MACE).*

	Recurrent MI		Recurrent 3-p MACE	
	Liraglutide	Placebo	Liraglutide	Placebo
$\geq$ 1 event	292	339	608	694
(Total events	359	421	768	923)
Dead before 1st event	329	373	137	133
Censored before 1st event	4047	3960	3923	3845
Randomized	4668	4672	4668	4672

to the treatment is graft versus host disease (GvHD) where the infused donor cells react against the patient.

A data set compiled from the Center for International Blood and Marrow Transplant Research (CIBMTR) was analyzed by Andersen and Pohar Perme (2008) with the main purpose of studying how occurrence of the GvHD event affected relapse and death in remission. The CIBMTR is comprised of clinical and basic scientists who confidentially share data on their blood and bone marrow transplant patients with the CIBMTR Data Collection Center located at the Medical College of Wisconsin, Milwaukee, USA. The CIBMTR is a repository of information about results of transplants at more than 450 transplant centers worldwide. The present data set consists of 2,009 patients from 255 different centers who received an HLA-identical sibling transplant between 1995 and 2004 for acute myelogenous leukemia (AML) or acute lymphoblastic leukemia (ALL) and were transplanted in first complete remission, i.e., when the pre-conditioning has eliminated the leukemia symptoms. All patients received BM or peripheral blood (PB) stem cell transplantation. Table 1.4 gives an overview of the events observed during follow-up (until 2007). The $1,272$ $(= 2,009 - 737)$ patients who were still alive at that time are censored.

Figure 1.6 shows the states and events for this study. A number of potential prognostic variables for the events (GvHD, relapse and death) were ascertained at time of transplantation. These variables include disease type (ALL vs. AML), graft type (BM or BM/PB), and sex and age of the patient. Sometimes, GvHD is considered, not a state but rather a time-dependent covariate, in which case the diagram in Figure 1.3 would be applicable with states BMT, relapse and dead.

1.1.8 Copenhagen Holter study

In the Copenhagen Holter Study, men and women aged 55, 60, 65, 70, or 75 years and living in two postal regions in Copenhagen, Denmark were contacted during the period 1998–2000. These subjects received a questionnaire including items on cardiovascular risk factors and medical history. All respondents with more than 1 risk factor and a 60% sample of those with 0 or 1 risk factor were invited to a physical examination including a 48-hour continuous electrocardiogram recording ('Holter monitoring'). Larsen et al. (2015)

Table 1.4 *Bone marrow transplantation in acute leukemia: Events observed in 2,009 leukemia patients who underwent bone marrow transplantation (GvHD: Graft versus host disease).*

Event	No. of patients	Percentage
Relapse	259	12.9
Death	737	36.7
Relapse and death	232	89.6 of patients with relapse
GvHD	976	48.6
GvHD and relapse	91	9.3 of patients with GvHD
GvHD and death	389	39.9 of patients with GvHD

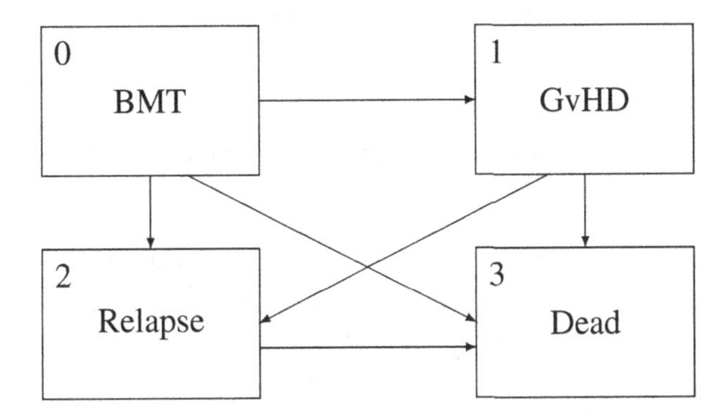

Figure 1.6 *Bone marrow transplantation (BMT) in acute leukemia: States and transitions (GvHD: Graft versus host disease).*

reported on a follow-up until 2013 of 678 participants with the purpose of studying the association between excessive supra-ventricular ectopic activity (ESVEA, a particular kind of irregular heart rhythm detected via the Holter monitoring) and later atrial fibrillation (AF, a serious heart arrhythmia affecting the blood circulation) and stroke. It was well known that ESVEA increases the incidence of AF, but one purpose of the study was to examine whether the incidence of stroke in patients with ESVEA was increased over and above what could be explained by an increase in the occurrence of AF. Events of AF, stroke and death during follow-up were ascertained via the Danish National Patient Registry. Figure 1.7 shows the possible states and transitions that can be studied based on these data. Note that, compared to Figure 1.6, this state diagram allows a $2 \rightarrow 1$ transition; however, such transitions do not impact the basic scientific question raised in the Copenhagen Holter Study. Table 1.5 shows the number of patients who were observed to follow the different possible paths through these states according to ESVEA at time of recruitment. From this table it appears that AF occurred in 18% of the patients with ESVEA and in 10% of those without, stroke occurred in 21% of patients with ESVEA and in 9% of those without. Among those who experienced AF without stroke, 25% of patients with ESVEA later had a stroke. The similar fraction for patients without ESVEA was 9%. An analysis of these interacting events must account for the fact that patients may also die without AF and/or stroke events.

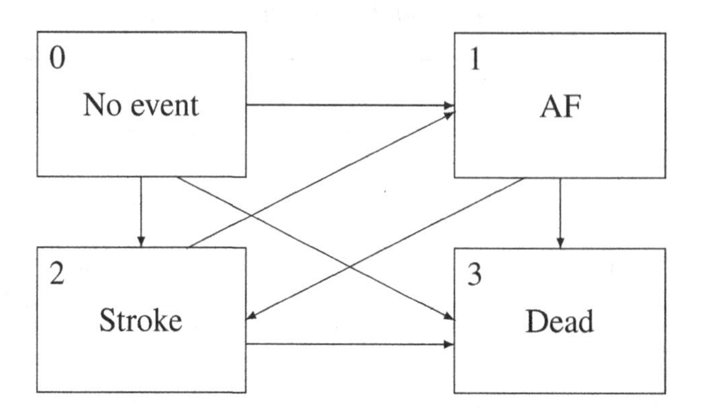

Figure 1.7 *Copenhagen Holter study: States and transitions (AF: Atrial fibrillation).*

Table 1.5 *Copenhagen Holter study: Number of patients following different paths (ESVEA: Excessive supra-ventricular ectopic activity; AF: Atrial fibrillation).*

	Number of patients		
Observed path	Without ESVEA	With ESVEA	Total
0	320	34	354
0 → AF	29	8	37
0 → Stroke	17	1	18
0 → AF → Stroke	3	1	4
0 → Stroke → AF	4	0	4
0 → Dead	158	32	190
0 → AF → Dead	20	4	24
0 → Stroke → Dead	25	14	39
0 → AF → Stroke → Dead	2	3	5
0 → Stroke → AF → Dead	1	2	3
Total	579	99	678

The Copenhagen Holter study was an observational study, so, adjustment for covariates (potential confounders) may be needed when examining the association between the 'exposure' ESVEA and later events like AF or stroke. A number of covariates were observed at the examination at the time of recruitment, including smoking status, age, sex, blood pressure, and body mass index. The follow-up in this study was, like in the testis cancer incidence study, Example 1.1.3, registry-based and in Denmark this means that, in principle, there is no loss to follow-up (except for the fact that patients may emigrate before the end of the study which, however, nobody did). As a consequence the only censoring event is end-of-study (alive in 2013). This data set will mainly be used for practical exercises throughout the book.

Examples of event history data

1. **PBC3 trial**: Randomized trial of effect of CyA vs. placebo on survival and liver transplantation in patients with Primary Biliary Cirrhosis ($n = 349$).
2. **Guinea-Bissau study**: Observational study of effect of childhood vaccinations on survival ($n = 5,274$).
3. **Testis cancer study**: Register study on the relationship between maternal parity and testicular cancer rates of their sons ($n = 1,015,994$).
4. **PROVA trial**: Randomized trial of effect of propranolol and/or sclerotherapy on the occurrence of bleeding and death in patients with liver cirrhosis ($n = 286$).
5. **Recurrent episodes in affective disorders**: Observational study of pattern of repeated disease episodes for patients with unipolar or bipolar disorder ($n = 119$).
6. **LEADER trial**: Randomized trial in type 2 diabetics with high cardiovascular risk – effect of liraglutide vs. placebo on cardiovascular events ($n = 9,340$).
7. **Bone marrow transplantation**: Observational study of effect of graft versus host disease (GvHD) on relapse and death in remission among bone marrow transplanted patients with leukemia ($n = 2,009$).
8. **Copenhagen Holter study**: Observational study of the association between excessive supra-ventricular ectopic activity (ESVEA) and later atrial fibrillation (AF) and stroke ($n = 678$).

1.2 Parameters in multi-state models

1.2.1 Choice of time-variable

In the examples in Section 1.1, we have seen how *multi-state models* may provide a suitable framework for describing event history data. Multi-state models are given by a number of states and possible transitions between these states that occur over time. For applications of multi-state models, for any set of data at hand, one must consider what is meant by *time*. In other words, a suitable *time origin*, a *time zero*, must be chosen.

In some cases, this choice is obvious. Thus, for randomized studies such as the PBC3, PROVA, and LEADER trials (Examples 1.1.1, 1.1.4, and 1.1.6), time zero is typically taken to be the time of randomization where participants fulfill the relevant inclusion criteria and where treatment is initiated. In clinical follow-up studies, there may also be an initiating event that defines the inclusion into the study and, therefore, serves as a suitable time origin. This includes the initial diagnosis of affective disorder in Example 1.1.5, the time of bone marrow transplantation in Example 1.1.7, and perhaps to a lesser extent, the time of initial clinical assessment in the Copenhagen Holter Study (Example 1.1.8).

However, in *observational studies* (Examples 1.1.2 and 1.1.3), the time of entry into the study is not necessarily a suitable time origin because that date may not be the time of any important event in the lifetime of the participants. In such cases, alternative time axes to be used for modeling include *age* or *calendar time*. Here, subjects are not always followed from the corresponding time origin (time of birth or some fixed calendar date), a situation known as *delayed entry* (or *left-truncation*), and subjects are only included into the study conditionally on being alive and event-free at the time of entry into the study. As an example

Table 1.6 *Small set of survival data.*

Subject number	Time from entry to exit	Status at time of exit 0 = censored, 1 = dead	Age at entry	Age at exit
1	5	1	12	17
2	6	0	0	6
3	7	1	0	7
4	8	1	10	18
5	9	0	6	15
6	12	0	6	18
7	13	1	9	22
8	15	1	3	18
9	16	1	8	24
10	20	0	0	20
11	22	0	0	22
12	23	1	2	25

we can consider the study of children in Guinea-Bissau (Example 1.1.2) where a choice of a primary time-variable for the survival analysis is needed. For the time-variable 'time since initial visit', say t, all children will be followed from the same time zero. This time-variable has the advantage of all risk factors being ascertained at the same time; however, the mortality rate for the children will likely not depend strongly on t. Thus, an alternative to using t as the time-variable would be to use (current) age of the children, a time-variable that will definitely affect the mortality rates. Some children were born into the study because the mother was followed during pregnancy and those children will be followed from age 0. Other children will only be included, i.e., being at risk of dying in the analysis, at a later age, namely the age at which the child was first observed to be in state 0 (initial visit). This is an example of delayed entry that will be further discussed in Section 2.3. Also for Example 1.1.3 (testis cancer incidence and maternal parity), there will be delayed entry when age is chosen as the primary time-variable because only boys born after 1968 are followed from birth.

For illustration, consider the small set of survival data provided in Table 1.6 and Figure 1.8. The subjects were followed from a time zero of entry (recruitment). Let t be the time since entry. Additionally, the age at time zero and at time of exit is given. Figure 1.8a depicts the survival data using time t (time since entry) as time-variable and Figure 1.8b the same survival data using age as time-variable, illustrating delayed entry. It is seen that the same follow-up intervals are represented in the two figures. These intervals, however, are re-allocated differently along the chosen time axis.

1.2.2 *Marginal parameters*

The small set of survival data in Table 1.6 is, in principle, a data set with a *quantitative* outcome variable, say T and one might, therefore, wonder whether quantities such as mean and standard deviation (SD) were applicable as descriptive parameters. However, the presence of censoring makes the calculation of a simple average as an estimator of the expected value $E(T)$ futile. Neither the average of all twelve observation times, nor those of the seven

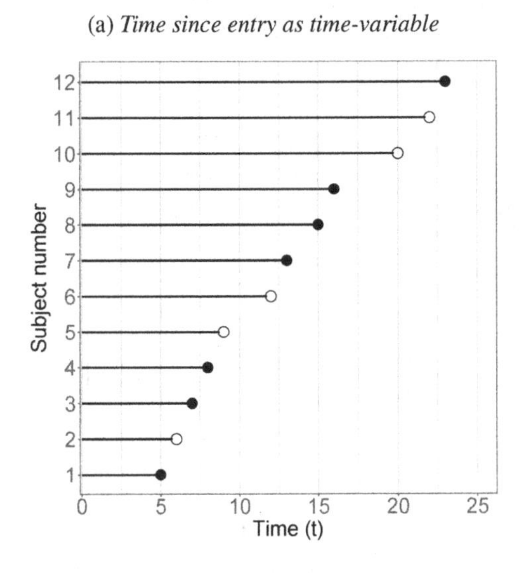

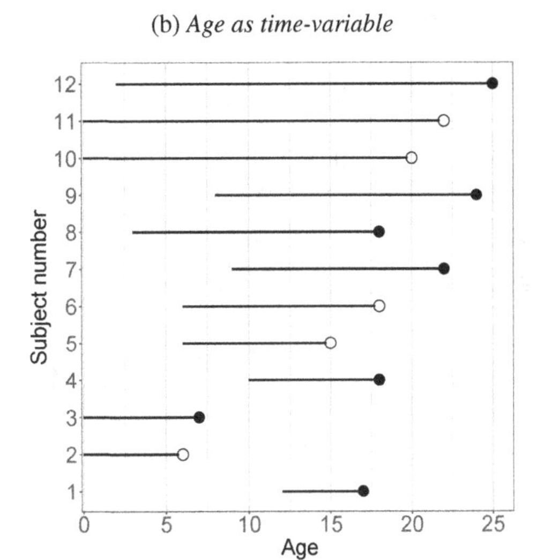

Figure 1.8 *Small set of survival data: Twelve subjects of whom seven died during the study (dots) and five were censored (circles).*

uncensored survival times are applicable as estimates of $E(T)$ (both will *under*-estimate the mean, see Exercise 1.1). Similarly, the relative frequency of observation times greater than, say $t = 10$, cannot be used as an estimator of the probability $P(T > t)$ of surviving past time t because of the censored observations $\leq t$ for which the corresponding true survival times T may or may not exceed t.

These considerations illustrate that other parameters and methods of estimation are required for multi-state survival data and, in the following, we will discuss such parameters. Having decided on a time zero, we let

$$(V(t), \quad t \geq 0)$$

be the *multi-state process* denoting, at time t, the state occupied at that time among a number of discrete states $h = 0, \ldots, k$. For the two-state model for survival data in Figure 1.1, the multi-state process at time t can take the values $V(t) = 0$ or $V(t) = 1$. One set of parameters of interest is the *state occupation* (or 'occupancy') *probabilities* at any time, t. Denote the probability (*risk*) of being in state h at time t as

$$Q_h(t) = P(V(t) = h);$$

then the sum of these over all possible states will be equal to 1

$$Q_0(t) + \cdots + Q_k(t) = \sum_{h=0}^{k} Q_h(t) = 1.$$

In the two-state model for survival data, Figure 1.1, with the random variable T being time to death, the state 0 occupation probability $Q_0(t)$ is the *survival function*, i.e.,

$$Q_0(t) = S(t) = P(T > t),$$

and $Q_1(t)$ is the *failure distribution function*, i.e.,

$$Q_1(t) = F(t) = P(T \leq t) = 1 - S(t).$$

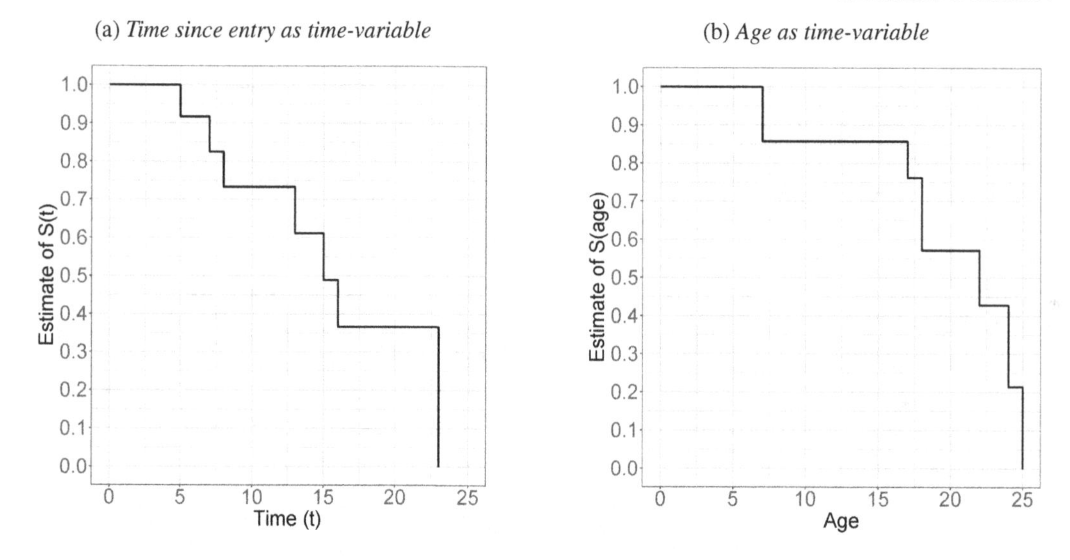

Figure 1.9 *Small set of survival data: Estimated survival functions.*

For the small set of survival data (Table 1.6), Figure 1.9 provides estimates of the survival function, $S(\cdot)$, with either (a) time since entry or (b) age as time-variable. The Kaplan-Meier estimator, which we will return to later in the book (Sections 4.1.1 and 5.1.1) was used for the estimation of $S(\cdot)$. Note that the shapes of the survival functions are somewhat different, illustrating the importance of choice of time zero. For any time point, the vertical distance from the curve up to 1 represents $F(t) = 1 - S(t)$.

The probabilities $Q_h(t)$ are examples of *marginal* parameters, i.e., at time t, their value is not conditional on the past history $(V(s), s < t)$ of the multi-state process (though they may involve *covariates* recorded at time zero). Other marginal parameters include the expected time, $\varepsilon_h(\cdot)$, spent in state h, either during all times, i.e., all the way up to infinity, $\varepsilon_h(\infty)$, or up to some threshold time $\tau < \infty$, $\varepsilon_h(\tau)$. The latter parameters have the property that they add up to τ, i.e.,

$$\varepsilon_0(\tau) + \cdots + \varepsilon_k(\tau) = \sum_{h=0}^{k} \varepsilon_h(\tau) = \tau,$$

because the time from 0 to τ has to be divided among the possible states. For the two-state model (Figure 1.1), $\varepsilon_0(\tau)$ is the *τ-restricted mean life time*, i.e., the expected time lived before time τ, and $\varepsilon_1(\tau) = \tau - \varepsilon_0(\tau)$ is the *expected time lost* before time τ. Figure 1.10 illustrates the estimated restricted mean life time for $\tau = 12$, the area under the survival curve, for the small set of survival data.

In cases where all subjects are in the same state '0' at time zero (which is the case for all the multi-state models depicted in Section 1.1), the distribution of the time (T_h) from time zero until (first) entry into another state h is another marginal parameter. Examples include the distribution of the survival time (time until entry into state 1 in Figure 1.1), time to event no. h in a recurrent events situation (e.g., Figure 1.5), or time to relapse or to GvHD in the model for the bone marrow transplantation data (Example 1.1.7, Figure 1.6). Note that, in the last two examples, not all subjects will eventually enter into these states and the entry

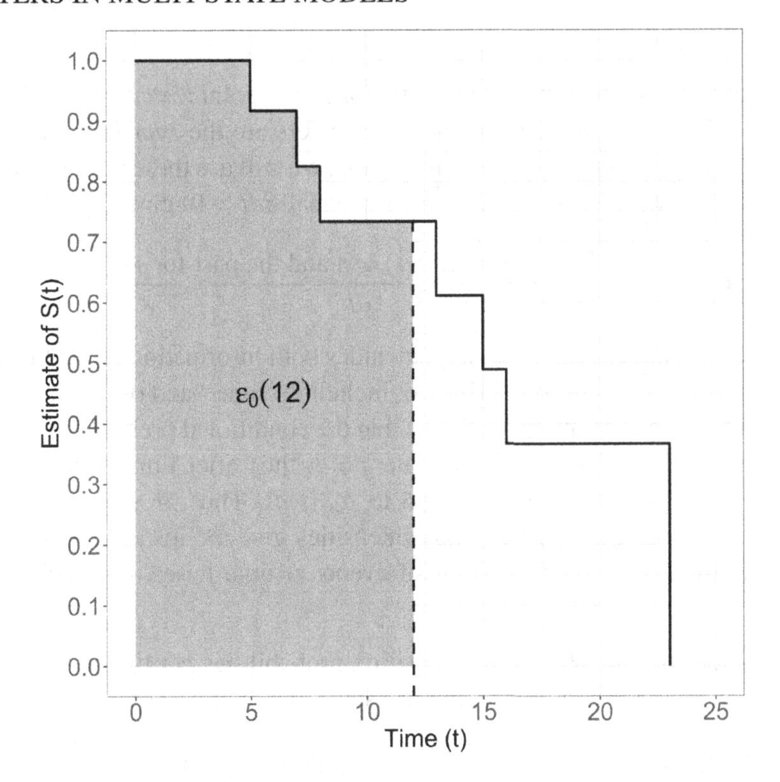

Figure 1.10 *Small set of survival data: Illustration of estimated restricted mean life time before time* $\tau = 12$.

times must be defined properly (formally, the value of these times may be infinite and T_h is denoted an *improper* variable).

For the recurrent events multi-state models (Figures 1.4 and 1.5), another marginal parameter of interest is the *expected number of events*, say $\mu(t) = E(N(t))$ before time t where $N(t)$ counts events before time t. In Figure 1.4, this is the *expected number of times* that state 1 is visited before time t.

Marginal parameters

Marginal parameters express the following type of quantities: You place yourself at time 0 and ask questions about aspects of the multi-state process $V(\cdot)$ at a later time point t (without consideration of what may happen between 0 and t). We will discuss the following marginal parameters:

- **State occupation probability**: $Q_h(t) = P(V(t) = h)$, the probability (risk) of being in state h at time t.
- **Restricted mean**: $\varepsilon_h(\tau)$, the expected time spent in state h before time τ.
- **Distribution of time** T_h to (first) entry into state h; relevant if everyone starts at time 0 in the same state (0).
- **Expected number of events** $\mu(t)$ before time t; relevant for a recurrent events process $N(\cdot)$, counting the number of events over time.

1.2.3 Conditional parameters

Another set of parameters in multi-state models for a time point t is *conditional* on the past up to that time. The most important of such parameters are the *transition intensities* (or *rates* or *hazards* – as emphasized in the Introduction, we will use these notions interchangeably). For two *different* states, h, j these are (for a 'small' $dt > 0$) given by

$$\alpha_{hj}(t) \approx \frac{P(V(t+dt) = j \mid V(t) = h \text{ and the past for } s < t)}{dt}, \tag{1.1}$$

where '|' is read 'given that'. Here, the *past* includes both information of the history on the multi-state process $(V(s), s < t)$ up to (but not including) time t and of possible *covariates* Z, recorded at time zero. The interpretation is that the conditional probability that a subject in state h at time t makes a transition to state $j \neq h$ 'just after time t' given the past is (approximately when $dt > 0$ is small) equal to $\alpha_{hj}(t)dt$. Thus, if subjects are followed over time then, at any time t, the transition intensities give the instantaneous conditional probabilities per time unit, given the past of events at time t, see, e.g., Andersen et al. (2021).

The transition intensities are short-term transition probabilities per time unit where, more generally, the *transition probability* for *any two* states (i.e., not necessarily distinct) is

$$P_{hj}(s,t) = P(V(t) = j \mid V(s) = h \text{ and the past before time } s),$$

i.e., the conditional probability of being in state j at time t given state h at the earlier time s and given the past at time s. In the common situation where all subjects are in an initial state (say, 0) at the time origin, $t = 0$ we have for any state h that

$$P_{0h}(0,t) = Q_h(t),$$

the state occupation probability at time t. However, more generally, transition probabilities are more complicated parameters than the state occupation probabilities because they may depend on the past at time $s > 0$. If $P_{hj}(s,t)$ only depends on the past via the state (h) occupied at time s then the multi-state process is said to be a *Markov process*. Note that the parameters $\alpha_{hj}(s)$ and $P_{hj}(s,t)$ involve conditioning on the past, but *never on the future* beyond time s. Indeed, conditioning on the future is a quite common mistake in multi-state survival analysis (e.g., Andersen and Keiding, 2012) – a mistake that we will sometimes refer to in later chapters.

The most simple example of a transition intensity is the hazard function, $\alpha_{01}(t) = \alpha(t)$, in the two-state model, in Figure 1.11. For some states, the transition intensities out of that state are all equal to 0, i.e., no transitions out of that state are possible. An example is the state 'Dead' in Figure 1.11 and such a state is said to be *absorbing*, whereas a state that is not absorbing is said to be *transient*, an example being the state 'Alive' in that figure. For the competing risks model (Figure 1.2), there is a transition intensity from the transient state 0 to each of the absorbing states, the *cause-specific hazards* for cause $h = 1, \dots, k$ having the interpretations $\alpha_{0h}(t)dt \approx$ the conditional probability of failure from cause h at time t given no failure before time t. In Figure 1.12, we have added the cause-specific hazards to the earlier Figure 1.2 in the same way as we did when going from Figure 1.1 to Figure 1.11.

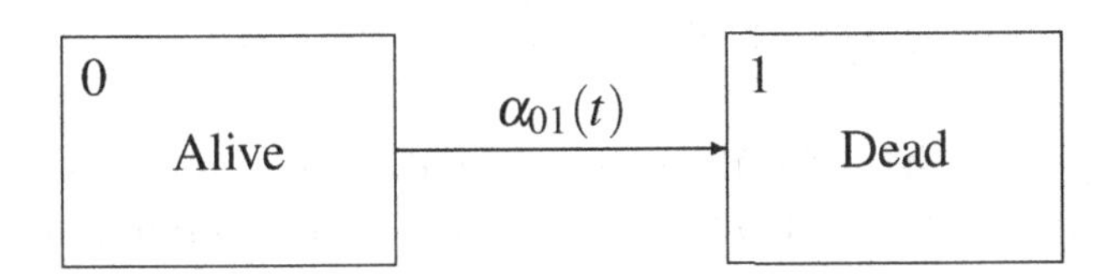

Figure 1.11 *The two-state model for survival data with hazard function for the* $0 \to 1$ *transition.*

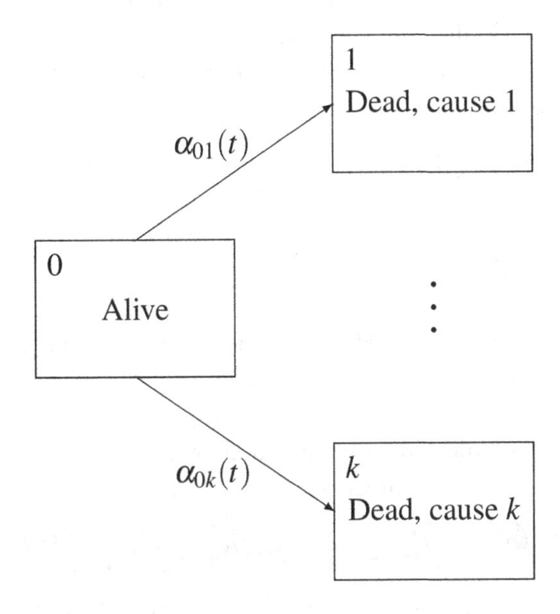

Figure 1.12 *The competing risks model with cause-specific hazard functions.*

In a similar way, transition intensities may be added to the other *box-and-arrow* diagrams in Section 1.1.

Intuitively, if one knows all transition intensities at all times. then both the marginal parameters and the transition probabilities may be calculated. This is because, by knowing the intensities, numerous paths for $V(t)$ may be generated by moving forward in time in small steps (of size dt), whereby $Q_h(t)$, $\varepsilon_h(\tau)$, and $P_{hj}(s,t)$ may be computed as simple averages over these numerous paths. This is, indeed, true and it is the idea behind *micro-simulation* that we will return to in Section 5.4. In some multi-state models, including the two-state model and the competing risks model (Figures 1.1, resp. 1.11 and 1.2, resp. 1.12), the marginal parameters may also be computed explicitly by certain mathematical expressions, e.g., the probability of staying in the initial state 0 in the two-state model (the *survival function)* is given by the formula

$$S(t) = Q_0(t) = \exp\left(-\int_0^t \alpha(u)du\right) \tag{1.2}$$

that expresses how to get the survival function from the hazard function. Likewise, for the competing risks model, the probability of being in the final, absorbing state $h = 1, \ldots, k$ at time t is given by

$$Q_h(t) = \int_0^t S(u)\alpha_{0h}(u)du, \tag{1.3}$$

where, in Equation (1.2), $\alpha = \alpha_{01} + \cdots + \alpha_{0k}$. This probability is frequently referred to as the *(cause h-) cumulative incidence function*, $F_h(t)$, a name that originates from epidemiology where that name means 'the cumulative risk of an event over time', see, e.g., Szklo and Nieto (2014, ch. 2). In Chapter 4, we will give intuitive arguments why Equations (1.2) and (1.3) look the way they do.

Models for multi-state survival data, e.g., *regression models* where adjustment for *covariates* is performed, may conveniently be specified via the transition intensities, the Cox (1972) regression model for survival data being one prominent such example. Intensity-based models are studied in Chapters 2 and 3. Having modeled all intensities, marginal parameters in simple multi-state models may be obtained by *plugging-in* the intensities into expressions like Equation (1.2) or (1.3). However, the marginal parameters may depend on the intensities in a non-simple fashion, and it is therefore of interest to aim at setting up direct regression models for the way in which, e.g., $\varepsilon_h(t)$, depends on covariates. Marginal models (both models based on plug-in and direct models) are the topic of Chapters 4 and 5 (see also Chapter 6 where such direct models are based on *pseudo-values*).

Conditional parameters

Conditional parameters for a multi-state process $V(\cdot)$ quantify, at time t, the future development of the process *conditionally* on the past of the process before t. We will discuss two types of conditional parameters:

1. **Transition intensities**: $\alpha_{hj}(t)$ gives the probability per time unit of moving to state j right after time t given that you are in state h at t and given the past up to t

$$\alpha_{hj}(t) \approx P(V(t+dt) = j \mid V(t) = h \text{ and the past for } s < t)/dt,$$

   (Equation 1.1). Transition intensities are only defined if j is different from h.

2. **Transition probabilities**: $P_{hj}(s,t)$ gives the probability of being in state j at time t, given that you were in state h at an earlier time point s and given the past history of $V(\cdot)$ up to that earlier time point. Transition probabilities are also defined if h and j are the same state.

1.2.4 Data representation

The multi-state survival data arising when observing a subject over time consist of a series of *times of transition* between states and the corresponding *types of transition*, i.e., from which state and to which state did the subject move at that time. For some subjects, this observed event history will end in an absorbing state from where no further transitions are possible, e.g., the subject is observed to die. However, there may be *right-censoring* in which case information on a time last seen alive, and the state occupied by the subject at that time will typically be available. Such data will have the format

$$((0,V(0)),(T_1,V(T_1)),(T_2,V(T_2)),\ldots,(X,V(X))),$$

with one record per subject where $T_1,T_2,\ldots,T_{N-1}$ are the observed times of transition and X is either the time T_N of the last transition into an absorbing state or the time, C of censoring and are said to be in *wide format* (or *marked point process format*). Such a format may typically be directly obtained from raw data consisting of *dates* where events happened (together with date of entry into the study and/or date of birth).

We will, in later chapters, typically assume that data for independent and identically distributed (*i.i.d.*) subjects $i=1,\ldots,n$ are observed and, in situations where data may be dependent, we will explicitly emphasize this.

Wide format is, however, less suitable as a basis for the analysis of the data, for which purpose data are transformed into *long format* (or *counting process format*) where each subject may be represented by several records. Here, each record typically corresponds to a given type of transition, say from state h to state j and includes time of entry into state h, time last seen in state h and information on whether the subject, at this latter time, made a transition to state j, the '(Start, Stop, Status)' triple.

As the name suggests, data in long format are closely linked to the mathematical representation of observations from multi-state processes via *counting processes*. Thus, for each possible transition, say from state h to another state j, data from a given subject, i may be represented as the counting process

$$N_{hji}(t) = \text{No. of direct } h \to j \text{ transitions observed for subject } i \text{ in the interval } [0,t],$$

together with the *indicator* of being *at risk* for that transition at time $t-$ (i.e., 'just before time t')

$$Y_{hi}(t) = I(\text{subject } i \text{ is in state } h \text{ at time } t-).$$

Here, the *indicator function* $I(\cdots)$ is 1 if $\cdots$ is true and 0 otherwise. Note that, for several multi-state models depicted in Section 1.1, the number of observed events of a given type is at most 1 for any given subject; an exception is the model for recurrent events in Figure 1.4 where each subject may experience several $0 \to 1$ and $1 \to 0$ transitions. Counting processes, $N(t) = N_{hj}(t)$ with more than one jump may also be constructed by adding up processes for individual subjects,

$$N(t) = \sum_i N_{hji}(t).$$

Likewise, the total *number at risk* at time t, $Y(t) = Y_h(t)$, is obtained by adding up individual at-risk processes,

$$Y(t) = \sum_i Y_{hi}(t).$$

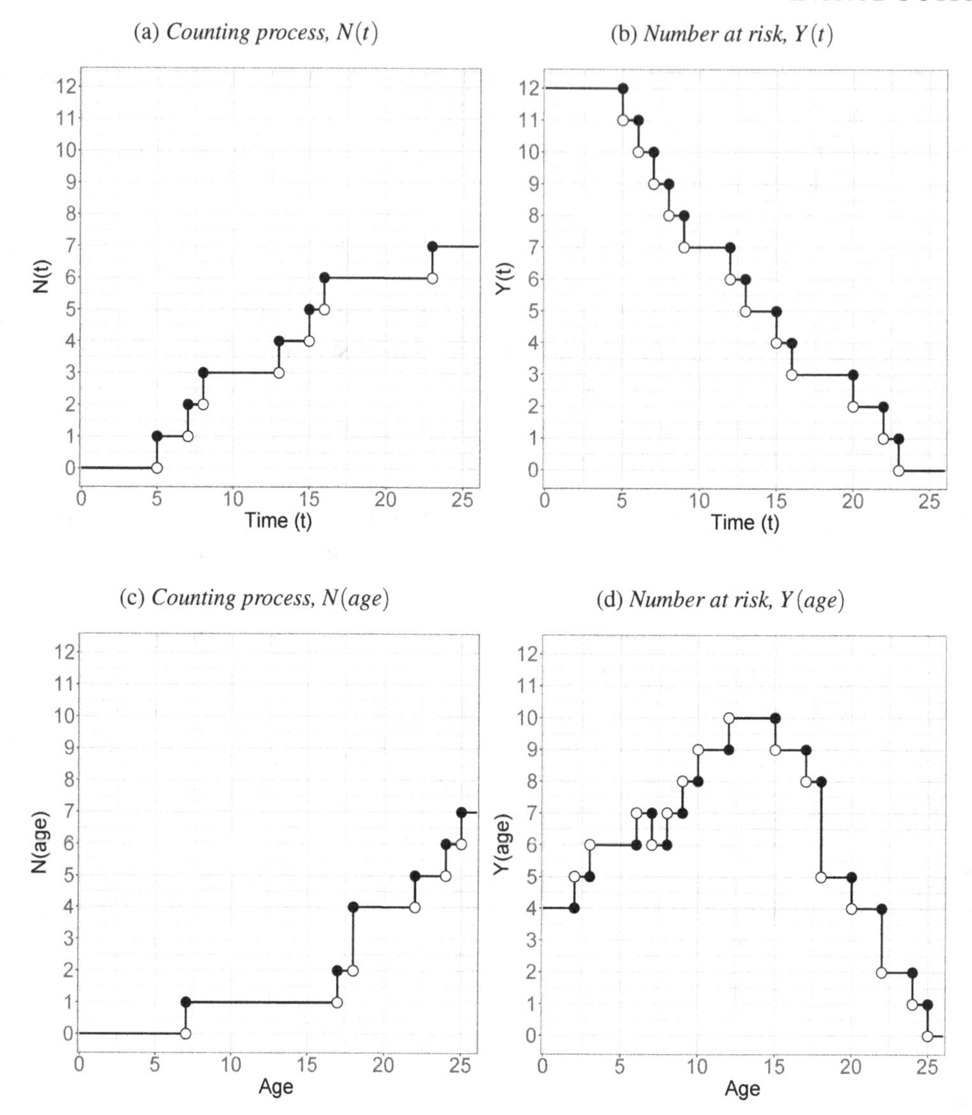

Figure 1.13 *Small set of survival data: Counting process and number at risk using t as time-variable (a-b) and age as time-variable (c-d).*

Figure 1.13 shows $N(t) = N_{01}(t)$ and $Y(t) = Y_0(t)$ for the small set of survival data from Table 1.6. When using t (time since entry) as time-variable the *at-risk function* $Y(\cdot)$ is monotonically decreasing, while using age as time-variable the at-risk function also increases due to the delayed entries.

For a small value of $dt > 0$ we define the *jump* at time T for the counting process $N(\cdot)$ to be $dN(T) = N(T + dt) - N(T-)$, i.e., $dN(t) = 1$ if $t = T$ is an observed time of transition and $dN(t) = 0$ otherwise. The representation using counting processes has turned out to be very useful when formulating estimators in multi-state models and their statistical properties.

We will next illustrate how typical raw data in wide format (with one record per subject) may be transformed into long format where each subject may be represented by several records. We will use the PROVA trial (Section 1.1.4) to illustrate the ideas and assume that the *dates* `doe`, `dob`, `dod`, and `dls` are defined as in Table 1.7.

Table 1.7 *PROVA trial in liver cirrhosis: Date variables available* (`NA`: *Not available*).

	Date		Description
1	`doe`		Date of entry into the study, i.e., date of randomization.
2	`dob`	($>$ `doe`)	Date of transfusion-requiring bleeding; if no bleeding was observed, then `dob=NA` . Time of bleeding is T_1 = `dob` $-$ `doe`.
3	`dod`	($>$ `doe`)	Date of death; if no death was observed, then `dod=NA`. If `dod` $\neq$ `NA`, then time of death is T_2 = `dod` $-$ `doe` and this is also the right-hand end-point, say X, for the interval of observation.
4	`dls`	($>$ `doe`)	Date last seen; if `dod=NA` then `dls` = date of censoring in which case the censoring time is C = `dls` $-$ `doe` and this equals the right-hand end-point, X for the interval of observation. If `dod` $\neq$ `NA`, then `dls` is equal to `dod`.

Note that the inequalities given in the table should all be checked with the data to make sure that the recorded dates are consistent with reality. Also, if both `dob` and `dod` are observed, then the inequality `dob` $\leq$ `dod` should hold. A data set with one line per subject containing these date variables is in wide format. From the basic date-variables, times and types of observed events may be defined as shown in Table 1.8, thereby taking the first step towards transforming the data into long format.

Table 1.8 *PROVA trial in liver cirrhosis: Observed transitions for different patterns of observed dates* (`NA`: *Not available*).

	Observed		Transition			Last seen	
	dob	dod	$0 \to 1$	$0 \to 2$	$1 \to 2$	State	Time
1	`NA`	`NA`	No	No	No	0	C
2	Yes	`NA`	At T_1	No	No	1	C
3	`NA`	Yes	No	At T_2	No	2	T_2
4	Yes	Yes	At T_1	No	At T_2	2	T_2

Here, times refer to time since entry, i.e., the time origin $t = 0$ is the time of randomization. The resulting counting processes and at-risk processes for the data in long format are shown in Table 1.9. Examples of how realizations of the multi-state process $V(t)$ would look like for specific values of T_1, T_2, C are shown in Figure 1.14.

Based on these observations, one may now construct three data sets – one for each of the possible transitions. Each record in the data set for the $h \to j$ transition has the structure

Table 1.9 *PROVA trial in liver cirrhosis: Counting processes and at-risk processes for different patterns of observed dates* (`NA`: *Not available*).

	dob	dod	$N_{01}(t)$	$N_{02}(t)$	$N_{12}(t)$	$Y_0(t)$	$Y_1(t)$
1	NA	NA	0	0	0	$I(C \geq t)$	0
2	Yes	NA	$I(T_1 \leq t)$	0	0	$I(T_1 \geq t)$	$I(T_1 < t \leq C)$
3	NA	Yes	0	$I(T_2 \leq t)$	0	$I(T_2 \geq t)$	0
4	Yes	Yes	$I(T_1 \leq t)$	0	$I(T_2 \leq t)$	$I(T_1 \geq t)$	$I(T_1 < t \leq T_2)$

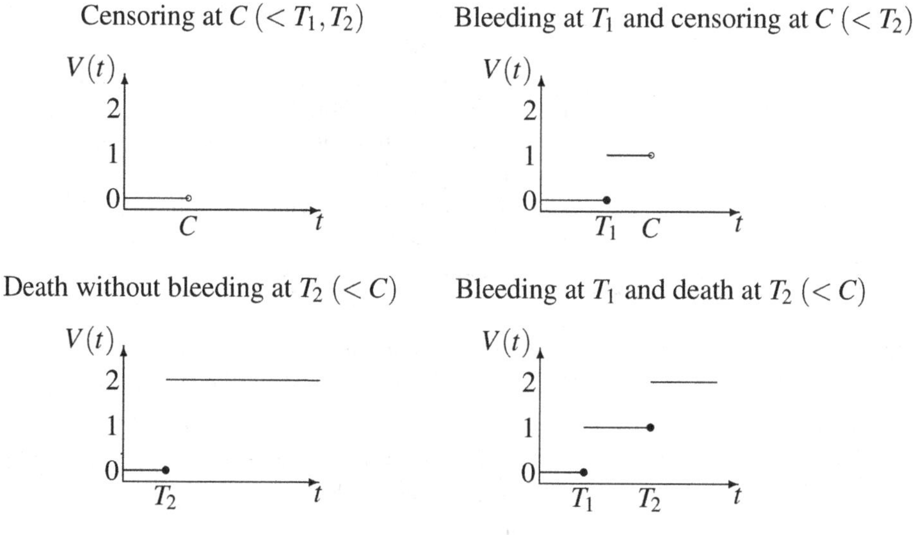

Figure 1.14 *PROVA trial in liver cirrhosis: The process* $V(t)$ *for different patterns of observed dates corresponding to the rows in Tables 1.8–1.10.*

(`Start, Stop, Status`) with

$$\begin{aligned}\texttt{Start} &= \text{Time of entry into } h,\\ \texttt{Stop} &= \text{Time last seen in } h,\\ \texttt{Status} &= \text{Transition to } j \text{ or not at time } \texttt{Stop}\end{aligned}$$

This is the data set in long format. In the PROVA example, each subject contributes at most with one record to each data set as shown in Table 1.10 where the `Status` variable is 1 if a $h \to j$ transition was observed at time `Stop` (i.e., `Status`=dN_{hj}(`Stop`)).

Note that an observed $0 \to 1$ transition also gives rise to a record in the data set for $0 \to 2$ transitions ending in no transition (and vice versa). Also note that, in the data set for the $1 \to 2$ transition, there is *delayed entry* meaning that subjects are not at risk of that transition from time zero but only from a later time point (`Start`, T_1) where the subject was first observed to be in state 1. Presence of delayed entry is closely connected to the choice of time-variable. Thus, had one chosen to consider the $1 \to 2$ transition intensity in the PROVA trial to depend primarily on time since bleeding (and not on time since

Table 1.10 *PROVA trial in liver cirrhosis: Records* (`Start, Stop, Status`) *in the three data sets for different patterns of observed dates* (`NA`: *Not available*).

	Observed		Data set		
	dob	dod	$0 \to 1$	$0 \to 2$	$1 \to 2$
1	NA	NA	$(0, C, 0)$	$(0, C, 0)$	NA
2	Yes	NA	$(0, T_1, 1)$	$(0, T_1, 0)$	$(T_1, C, 0)$
3	NA	Yes	$(0, T_2, 0)$	$(0, T_2, 1)$	NA
4	Yes	Yes	$(0, T_1, 1)$	$(0, T_1, 0)$	$(T_1, T_2, 1)$

randomization), then there would have been no delayed entry. The example also illustrates that even though a basic time origin is chosen for a given multi-state model (here time of randomization), there may be (later) transitions in the model for which another time origin may be more appropriate.

1.2.5 Target parameter

The scientific questions addressed through the examples in Section 1.1 most often involve a number of *covariates* and, as a result, both intensity models and direct models for marginal parameters will often be (multiple) *regression models*. In the regression models that we will be studying in later chapters, the covariates will always enter via a *linear predictor* combining (linearly) the *effects* of the individual covariates on some suitable function (the *link function*) of the multi-state model parameter via *regression coefficients*, see, e.g., Andersen and Skovgaard (2010, ch. 5).

As an example, we can consider a direct model for a state occupation probability (risk) $Q_h(t)$ at time t. Suppose that there are p covariates $Z_1, Z_2, \ldots, Z_p$ under consideration – then the model may be

$$\log(Q_h(t)) = \beta_0 + \beta_1 Z_1 + \beta_2 Z_2 + \cdots + \beta_p Z_p,$$

i.e., the covariate effects on $Q_h(t)$ are linear on the scale of the logarithm (the link function) of the risk. Note that we will often refer to a coefficient, β_j, as the 'effect' of the corresponding covariate Z_j – also in situations where a causal interpretation is not aimed at. The expression on the right-hand side of this equation is the linear predictor

$$\text{LP} = \beta_1 Z_1 + \beta_2 Z_2 + \cdots + \beta_p Z_p \tag{1.4}$$

and involves regression coefficients $\beta_1, \beta_2, \ldots, \beta_p$ (but not the intercept, β_0). The interpretation of a single β_j is, as follows. Consider two subjects differing 1 unit for covariate j and having identical values for the remaining covariates in the model. Then the difference between the log(risks) for those two subjects is

$$\begin{array}{rcllll} \beta_j & = & (\beta_0 + \beta_1 Z_1 + \cdots + \beta_{j-1} Z_{j-1} + & \beta_j (Z_j + 1) & + \cdots + \beta_p Z_p) \\ & - & (\beta_0 + \beta_1 Z_1 + \cdots + \beta_{j-1} Z_{j-1} + & \beta_j Z_j & + \cdots + \beta_p Z_p). \end{array}$$

Thus, $\exp(\beta_j)$ is the risk ratio for a 1 unit difference in Z_j *for given values of the remaining covariates in the model*. It is seen that not only does the interpretation of β_j depend on the chosen link function (here, the logarithm) but also on which other covariates

$(Z_1,\dots,Z_{j-1},Z_{j+1},\dots,Z_p)$ that are included in the model. Therefore, a regression coefficient, e.g., for a treatment variable, unadjusted for other covariates is likely to differ from one that is adjusted for sex and age. (One exception is when the model is linear, i.e., the link function is the *identity*, and treatment is randomized and, thereby, independent of sex and age and other covariates in which case the parameter is *collapsible*, see, e.g., Daniel et al., 2021). Nevertheless, for a number of reasons, regression models and their estimated coefficients are useful in connection with the analysis of multi-state survival data.

First of all, regression models *describe the association* between covariates and intensities or marginal parameters in multi-state models and insight may be gained from these associations when trying to understand the development of the process. In this connection, it is also of interest to compare estimates of β_j across models with different levels of adjustments, e.g., do we see similar associations with Z_j with or without adjustment for other covariates? Another major use of multi-state regression models is *prediction*, e.g., what is the estimated risk of certain events for a subject with given characteristics? These aspects will be further illustrated in the later chapters.

However, the proper answer to a scientific question posed need not be given by quoting a coefficient from a suitable regression model in which case other *target parameters* should be considered. We will see that regression models are still useful 'building blocks' when targeting alternative parameters. As an example of how a target parameter properly addressing the scientific question posed may be chosen, we can consider the PBC3 trial (Example 1.1.1). Here, the question of interest is whether treatment with CyA prolongs time to treatment failure and, since the study was randomized, this may be answered by estimating and comparing survival curves ($S(t)$ – see Section 1.2.2) for the CyA and placebo groups. However, as we shall see in Section 2.2.1, randomization was not perfect and levels of important prognostic variables (albumin and bilirubin) tended to be more beneficial in the placebo group than in the CyA group. For this reason (but also influenced by non-collapsibility), the estimated regression coefficients for treatment with or without adjustment for these two variables will differ. Also, estimated survival curves for treated and control patients will vary with their levels of albumin and bilirubin and it would be of interest to estimate one survival curve for each treatment group that properly accounts for the covariate imbalance between the groups. Such a parameter, the contrast (e.g., difference or ratio) between the survival functions in the two groups *had they had the same covariate distribution* may be obtained using the *g-formula* (e.g., Hernán and Robins, 2020, ch. 13) and works by averaging individually predicted curves over the observed distribution of albumin and bilirubin (Z_2, Z_3). Thus, two predictions are performed for each subject, i: One setting treatment (Z_1) to CyA and one setting treatment to placebo and in both predictions keeping the observed values (Z_{2i}, Z_{3i}) for albumin and bilirubin. The predictions for each value of treatment are then averaged over $i = 1,\dots,n$

$$\widehat{S}_j(t) = \frac{1}{n}\sum_{i=1}^{n} \widehat{S}(t \mid Z_1 = j, Z_2 = Z_{2i}, Z_3 = Z_{3i}), \quad j = \text{CyA, placebo.} \tag{1.5}$$

We will illustrate the use of the g-formula in later chapters. Here, a challenge will be to do inference for the treatment contrast (i.e., to assess the uncertainty in the form of a confidence interval) and, typically, a *bootstrap* procedure will be applied (e.g., Efron and Tibshirani, 1993, ch. 6).

In the final chapter of the book (Section 7.3), we will also discuss under what circumstances the resulting treatment contrast may be given a *causal* interpretation. There, we will also define what is meant by causal, discuss alternative approaches to *causal inference*, and under which assumptions (including that of *no unmeasured confounders*) a causal interpretation is possible.

1.3 Independent censoring and competing risks

In Section 1.1, the practical examples demonstrated that incomplete observation is inevitable when dealing with event history data, and in Section 1.2 we discussed the structure of the resulting data in the sample and the parameters in multi-state models that are typically targeted. In this section, we will be more specific about the target *population* for which inference is intended based on the observed, incomplete data and we will also discuss under which assumptions concerning the censoring mechanism that inference is valid.

We will first review some fundamental concepts from basic statistics. Statistics provides methods for doing inference on parameters attached to a population based on data from a *sample* from the population. Data are represented by random variables whose distribution is characterized by these parameters and, provided the sample is well-defined (e.g., random or representative), valid inference may be achieved. As a standard example, think of the distribution of blood pressure in women in a given age range with a given diagnosis, and where the parameters of interest are the mean and the standard deviation. If blood pressure measurements for a sample of such female patients are available, then inference for the parameters is often straightforward, and it is typically not hard to conceptualize the population from which the sample was drawn and that for which inference is intended.

Next, we turn to an example from event history analysis, namely a study of patients diagnosed with the chronic disease malignant melanoma (skin cancer), all given the standard treatment (radical surgery) and where interest focuses on the survival time from surgery, see Figure 1.1. By observing data from a sample of patients with malignant melanoma, one would aim at making inference on the survival time distribution in the population of such patients. However, because of the very nature of survival data, one has to face the additional complication (over and above randomness or representativeness of the sample) of incomplete observation of the random variables in question. At the time when data are to be analyzed, some patients may still be alive and all that is known about the survival time for such a patient is that it exceeds the time elapsed from time of surgery to time of analysis – the survival time is *right-censored.*

We saw in Section 1.2 that this incompleteness has consequences for the types of parameters on which one focuses (e.g., survival functions or hazards instead of means and standard deviations) and consequently for the types of statistical models that are typically used. However, it also has consequences for the conceptualization of the population from which the sample was drawn. For the malignant melanoma example, the population would be (all) patients with the disease who have undergone radical surgery and who are observed until failure. That is, in the *population* there is no incompleteness and the question naturally arises under what conditions the incomplete data in the *sample* allow valid inference for the population. The condition needed is known as *independent right-censoring*.

We discuss a definition of this concept further below, but let us first consider an extension of the example. Deaths among melanoma patients may either be categorized as death from the disease or as death from other causes, see Figure 1.2. Suppose that the scientific interest focuses on death from the disease. It is then a question of whether the competing risk of death from other causes can be considered as, possibly independent, right-censoring. For that to be the case, following the arguments just given, data should be considered as a sample from a population without censoring, i.e., a population where all melanoma patients are observed until death from the disease. We will therefore argue that the answer to the question is 'no' because the complete population without censoring (death from other causes) would be completely hypothetical: One can hardly imagine ever to obtain data from a population where melanoma patients cannot die from other causes.

The complete target population for which inference is intended is therefore one where *avoidable* causes of incompleteness (*right-censoring*) are absent. However, *non-avoidable* causes of incompleteness (*competing risks*) may, indeed, be present in the target population, and the corresponding events should therefore be included as possible transitions in the box-and-arrows diagram such as those shown in Section 1.1. Thus, if in the PBC3 trial (Example 1.1.1) one would study risk factors for transplantation, then the non-avoidable event of death without transplantation cannot be considered a, possibly independent, censoring mechanism because the population without this event would be completely hypothetical. A similar remark concerns an analysis of the event stroke in the Copenhagen Holter Study (Example 1.1.8) where considering death as censoring would be inappropriate. On the other hand, events like drop-out or end-of-study (e.g., in the PBC3 trial, Example 1.1.1, or the PROVA trial, Example 1.1.4) are examples of *avoidable* events (*censoring*).

An important question is whether such censoring events can be considered *independent censoring*. We will define independent censoring, as follows (see, e.g., Andersen et al., 1993, ch. III). Let, as in Section 1.2, $(V(t), t \geq 0)$ be the multi-state process. The transition intensities for this process are (approximately when $dt > 0$ is small)

$$\alpha_{hj}(t) \approx \frac{P(V(t+dt) = j \mid V(t) = h \text{ and the past for } s < t)}{dt}$$

for states $h \neq j$. Censoring at C is then independent if

$$\frac{P(V(t+dt) = j \mid V(t) = h, \text{ past for } s < t \text{ and } C > t)}{dt} \approx \alpha_{hj}(t), \tag{1.6}$$

i.e., if the additional information at time t that a subject is still uncensored does not alter the transition intensities. As a consequence of independent censoring, the subset of subjects still at risk (i.e., uncensored) at any time t represents the population at this time. Examples of independent censoring mechanisms include *random censoring* where C is independent of $V(t)$ and *type 2* censoring where remaining subjects are censored when a pre-specified number of events has been observed in the whole sample (Andersen et al., 1993, ch. III). Note that the definition of independent censoring involves the past and, thereby, the *covariates* that are included in the model for the transition intensities. This means that, roughly speaking, events and censoring should be conditionally independent given covariates.

The independent censoring condition can, unfortunately, typically *not be tested* based on the available censored data (except for the fact that it may be investigated if censoring depends on covariates - see Section 4.4.1). This is because the future events after censoring (that should be independent of the censoring time) are not observed. Investigation of independent censoring is therefore a matter of discussion in each single case. In the PBC3 and PROVA trials, two censoring mechanisms were operating: End-of-study and drop-out. The former (*administrative censoring*) can typically safely be taken to be independent (the fact that the study has reached its planned termination should have no consequence for future events for those still at risk). However, if there is some calendar time trend in patient recruitment (e.g., patients may tend to be less and less severely ill at time of recruitment as time passes), then the administrative censoring may only be independent conditionally on the time of recruitment, and calendar time should then be adjusted for in the analysis. Similarly, if censoring depends on other important prognostic variables, see Section 4.4.2 for further discussion. For censoring due to drop-out events, one must typically be more careful. In a trial, patients may drop out because of toxicity or because of lack of efficacy and, in both cases, knowing that a subject is still uncensored may carry information on the future event risk for that subject. If, in the PBC3 trial, one focuses on mortality, then censoring by liver transplantation cannot be considered independent because a liver transplantation was typically only offered to patients with a relatively poor prognosis, and this means that the information that a patient is still untransplanted tells that this patient is 'doing relatively well'.

A practical advice would be, first of all, to do the best to avoid drop-out and, if drop-outs do happen, then to record the reasons for drop-out in the patient file such that the problem of independent censoring can be discussed and such that covariates related to drop-out can be accounted for in the model for the event occurrence.

Multi-state model, competing risks, and censoring

A multi-state model is given by a number of different *states* that a subject can occupy and the possible *transitions* between the states. The transitions represent the *events* that may happen. Such a model can be depicted in a *box-and-arrow diagram* where the *transition intensities* may be indicated (e.g., Figures 1.11 and 1.12).

These diagrams show the possible states in a *completely observed population*, i.e., censoring is *not* a state in the model. If one particular transition is of interest, then other transitions in the multi-state model, possibly *competing* with that, are *non-avoidable* events that must be properly addressed in the analysis and should not be treated as a (potentially, avoidable) censoring.

1.4 Mathematical definition of parameters (*)

In the examples in Section 1.1, we have seen a number of *multi-state models* consisting of a finite number of *states* and possible *transitions* between some of these states. Section 1.2 gave an informal introduction to parameters associated with multi-state models and the present section discusses the mathematical definition of these parameters.

1.4.1 *Marginal parameters (*)*

We denote the multi-state process by $(V(t), t \geq 0)$, i.e., at time t, $V(t)$ is the state occupied by a generic subject. Usually, a small number, $k+1$ of states is considered and often labelled $0, 1, ..., k$. The *state space* is then the finite set

$$\mathcal{S} = \{0, 1, ..., k\}. \tag{1.7}$$

Corresponding to these states there are *state occupation* (or 'occupancy') *probabilities*

$$Q_h(t) = P(V(t) = h), \quad h \in \mathcal{S} \tag{1.8}$$

giving the marginal distribution over the states at time t, so we have for all t that $\sum_h Q_h(t) = 1$.

In the two-state model for survival data in Figure 1.1, $Q_0(t)$ is the probability of being still alive time t, the *survival function*, often denoted $S(t)$, and $Q_1(t) = 1 - Q_0(t)$ is the *failure distribution function*, $F(t) = 1 - S(t)$. In Figure 1.2, $Q_0(t)$ is also the survival function, $S(t)$, and $Q_h(t), h = 1, \ldots, k$ are the *cumulative incidence functions* for cause h, i.e., the probability $F_h(t)$ of failure from cause h before time t. The probability $F_h(t)$ is sometimes referred to as a *sub-distribution function* as $F_h(\infty) < 1$.

Another marginal parameter of interest, which may be obtained from the state occupation probabilities, is the *expected time spent in a given state* (expected length of stay). For state h, this is given by

$$\varepsilon_h(\infty) = E\left(\int_0^\infty I(V(t) = h)\mathrm{d}t\right) = \int_0^\infty Q_h(t)\mathrm{d}t. \tag{1.9}$$

Since we have to deal with right-censoring, whereby information about the process $V(t)$ for large values of time t is limited, *restricted means* are often studied, i.e.,

$$\varepsilon_h(\tau) = \int_0^\tau Q_h(t)\mathrm{d}t \tag{1.10}$$

for some suitable time threshold, $\tau < \infty$. This is the expected time spent in state h in the interval from 0 to τ. Since, for all t, $\sum_{h \in \mathcal{S}} Q_h(t) = 1$ it follows that $\sum_{h \in \mathcal{S}} \varepsilon_h(\tau) = \tau$.

For the two-state model for survival data (Figure 1.1), $\varepsilon_0(\infty)$ is the expected life time $E(T)$ and $\varepsilon_0(\tau)$ is the τ-restricted mean life time $E(T \wedge \tau)$, the expected time lived before time τ and, thus, $\varepsilon_1(\tau) = \tau - \varepsilon_0(\tau)$ is the *expected time lost* before time τ. For the competing risks model (Figure 1.2), $\varepsilon_0(\tau)$ is the τ-restricted mean life time and, for $h \neq 0$, $\varepsilon_h(\tau)$ is the expected time lost *due to cause h* before time τ (see Section 5.1.2). For the disability model (Figure 1.3), $\varepsilon_1(\tau)$ is the expected time lived with disability before time τ.

In the common situation where everyone is in the same state (0) at time $t = 0$ (i.e., $P(V(0) = 0) = 1$), the marginal distribution of the random variable

$$T_h = \inf_{t>0}\{V(t) = h\}, h \neq 0, \tag{1.11}$$

that is, the *time of first entry into state h*, $h \neq 0$, (which may be infinite) may also be of interest. For recurrent events, T_h is the time until the hth occurrence of the event, e.g., the time from diagnosis to episode no. h for the psychiatric patients discussed in Section 1.1.5 (Figure 1.5). However, the most important marginal parameter for a recurrent events process is the *expected number of events* in $[0,t]$

$$\mu(t) = E(N(t)), \tag{1.12}$$

where $N(t)$ is the number of recurrent events in $[0,t]$. For the model in Figure 1.4, this is the *expected number of visits* to state 1 in $[0,t]$.

The parameters defined in this section are called *marginal* since, at time t, they involve no conditioning on the past $(V(s), s < t)$ (though they may involve time-fixed covariates).

1.4.2 Conditional parameters (*)

To describe the *time-dynamics* of $V(t)$, one may use *conditional* parameters such as the *transition probabilities*

$$P_{hj}(s,t) = P(V(t) = j \mid V(s) = h), h, j \in \mathcal{S}, s < t. \tag{1.13}$$

Note that these probabilities do not necessarily correspond to direct transitions from h to j, thus, in Figure 1.3 there are two possible *paths* from state 0 to state 2: one going directly, and one going through state 1. A state h is said to be *absorbing* if no transitions out of the state are possible, i.e., for all $j \neq h$, the transition probabilities out of the state h are 0, $P_{hj}(s,t) = 0$ for all $s < t$. A state that is not absorbing is said to be *transient*. In Figure 1.1, state 1 is absorbing and state 0 is transient; in Figure 1.2, states 1 to k are absorbing and state 0 transient; while, in Figure 1.3, state 2 is absorbing and states 0 and 1 transient. In the situation where all subjects are in state 0 at time $t = 0$ $(P(V(0) = 0) = 1)$ we have that $P_{0h}(0,t) = Q_h(t)$.

The transition probabilities $P_{hj}(s,t)$ will, more generally, depend on the *past history* of the process $V(s)$ at time s and on *covariates* $\mathbf{Z}$. For the moment, we will restrict attention to *time-fixed* covariates $\mathbf{Z}$ recorded at time $t = 0$ and postpone the discussion of *time-dependent* covariates until Section 3.7. The past information available at time s for conditioning will be denoted $\mathcal{H}_s$. If all $P_{hj}(s,t)$ only depend on the past via the current state h (and, possibly, via time-fixed covariates), then the multi-state process is said to be a *Markov process*. As an example: If the probability, $P_{12}(s,t)$ of dying before time t for a patient in the bleeding state 1 at time s of the PROVA trial (Example 1.1.4) only depends on time since start of treatment (s), then the process is Markovian; if it depends on the time, $d = s - T_1$, elapsed in state 1 at time s, i.e., the time since onset of the bleeding episode (and possibly also on s), then the process is *semi-Markovian*.

For modeling purposes, as we shall see in later chapters, *transition intensities* $\alpha_{hj}(t)$ are convenient. These are given as the following limit

$$\alpha_{hj}(t) = \lim_{dt \to 0} \frac{1}{dt} P_{hj}(t, t+dt), \quad j \neq h, \tag{1.14}$$

which we assume to exist. That is, if $dt > 0$ is a 'small' time window, then

$$P_{hj}(t, t+dt) \approx \alpha_{hj}(t)dt, \quad j \neq h.$$

One reason why intensities are useful for modeling purposes is that they (in contrast to $P_{hj}(s,t)$ that must be between 0 and 1) can take on any non-negative value. For the two-state survival model, Figure 1.1, $\alpha_{01}(t)$ is the *hazard function*

$$\alpha(t) = \lim_{dt \to 0} P(T \leq t + dt \mid T > t)/dt \tag{1.15}$$

for the survival time $T = T_1$ (time until entry into state 1). For the competing risks model, Figure 1.2, $\alpha_{0h}(t)$ is *the cause-specific hazard*

$$\alpha_h(t) = \lim_{dt \to 0} P(T \leq t + dt, D = h \mid T > t)/dt, \tag{1.16}$$

where $D = V(\infty)$ is the cause of death and $T = \min_{h>0} T_h$ is the survival time (time of exit from state 0) with T_h defined in (1.11). Both of these multi-state processes are Markovian. The illness-death process of Figure 1.3 is non-Markovian if the intensity $\alpha_{12}(t)$ not only depends on t but also on the time, $d = t - T_1$, spent in state 1 at time t.

The transition intensities are the most basic parameters of a multi-state model in the sense that, in principle, if all transition intensities are specified, then all other parameters such as state occupation probabilities, transition probabilities, and expected times spent in various states may be derived. As we shall see in later chapters, the mapping from intensities to other parameters is sometimes given by explicit formulas, though this depends both on the structure of the states and possible transitions in the model and on the specific assumptions (e.g., Markov or non-Markov) made for the intensities. Examples of such explicit formulas include the survival function (1.2) for the two-state model for survival data and, more generally,

$$Q_0(t) = \exp\left(-\int_0^t \sum_h \alpha_h(u)du\right)$$

in the competing risks model (Figure 1.2). Also, the cause-*h cumulative incidence function* in the competing risks model, Equation (1.3), and the probability of being in the intermediate 'Diseased' state in both the Markov and semi-Markov illness-death model (Figure 1.3) (formulas to be given in Sections 5.1.3 and 5.2.4) are explicit functions of the intensities. A general way of going from intensities to marginal parameters (not building on a mathematical expression) is to use *micro-simulation* (Section 5.4).

1.4.3 Counting processes (*)

Closely connected to the transition intensities is the representation of observations from a multi-state process as *counting processes*. In later chapters, we will typically assume that observations from independent and identically distributed (*i.i.d.*) subjects $i = 1, \ldots, n$ are available and that the multi-state process $V_i(t)$ for subject i is observed on the time interval $[0, X_i]$. (In some cases, data may be dependent, and we will then make explicit remarks to this effect.) Here, $X_i = C_i \wedge T_i$, i.e., X_i is either equal to C_i, the observed *right-censoring time* for subject i, or X_i is the time (say, T_i) where subject i is observed to enter an absorbing state. In the latter case, the value of C_i may or may not be known depending on the situation, e.g., if for the two-state model for survival data, subject i is observed to die at time T_i, then it may not be known when (i.e., at time $C_i > T_i$) that subject would have been censored had it not died at T_i – that will depend on the actual right-censoring mechanism in the study. Thus, if in the PBC3 trial (Example 1.1.1), there had been no drop-out and, therefore, all censoring was caused by being alive at the end of 1988 (administrative censoring), then all potential censoring times would have been known.

We now consider a generic subject i and drop the index i for ease of notation. The multi-state process $(V(t), t \leq X)$ with state space $\mathscr{S}$ can then be represented by the counting processes

$$(N_{hj}(t), h, j \in \mathscr{S}, h \neq j, t \leq X) \tag{1.17}$$

where each $N_{hj}(t)$ counts the number of observed direct $h \to j$ transitions in the time interval $[0,t]$. If state h is absorbing, then $N_{hj}(t) = 0$ for all states $j \neq h$ and all values of time t. Note that, in our notation, we will not distinguish between the complete (uncensored) multi-state process and the censored process (observed until time X) and a similar remark goes for the counting processes derived from the multi-state process.

Let the past history of the multi-state process (including relevant covariates) at time t be the sigma-algebra $\mathscr{H}_t$. Then the (random) *intensity process*, $\lambda_{hj}(t)$ for $N_{hj}(t)$ (with respect to that history) under independent censoring is (approximately when $dt > 0$ is small)

$$E(dN_{hj}(t) \mid \mathscr{H}_{t-})/dt \approx \lambda_{hj}(t) = \alpha_{hj}(t)Y_h(t). \tag{1.18}$$

Here, the transition intensity $\alpha_{hj}(t)$ is some function of time t and the past history $(\mathscr{H}_{t-})$ for the interval $[0,t)$,

$$Y_h(t) = I(V(t-) = h) \tag{1.19}$$

is the indicator for the subject of being observed to be in state h just before time t, and

$$dN_{hj}(t) = N_{hj}(t) - N_{hj}(t-) \tag{1.20}$$

is the increment (0 or 1, the *jump*) for N_{hj} at time t. Since $\lambda_{hj}(t)$ is fixed given $(\mathscr{H}_{t-})$, Equation (1.18) implies that if we define

$$M_{hj}(t) = N_{hj}(t) - \int_0^t \lambda_{hj}(u)du \tag{1.21}$$

then

$$E(dM_{hj}(t) \mid \mathscr{H}_{t-}) = 0$$

from which it follows that the process $M_{hj}(t)$ in (1.21) is a *martingale*, i.e.,

$$E(M(t) \mid \mathscr{H}_s) = M(s), \quad s \leq t$$

see Exercise 1.4. The decomposition of the counting process in (1.21) into a martingale plus the integrated intensity process (the *compensator*) is known as the *Doob-Meyer decomposition* of $N_{hj}(\cdot)$. Since martingales possess a number of useful mathematical properties, including approximate large-sample normal distributions (e.g., Andersen et al., 1993, ch. II), this observation has the consequence that large-sample properties of estimators, estimating equations, and test statistics may be derived when formulated via counting processes. We will hint at this in later chapters.

1.5 Exercises

Exercise 1.1 Consider the small data set in Table 1.6 and argue why both the average of all (12) observation times and the average of the (7) uncensored times will likely underestimate the true mean survival time from entry into the study

Exercise 1.2

1. Consider the following records mimicking the Copenhagen Holter study (Example 1.1.8) in wide format and transform them into long format, i.e., create one data set for each of the possible transitions in Figure 1.7.

	Time of			
Subject	AF	stroke	death	last seen
1	NA	NA	NA	100
2	10	NA	NA	90
3	NA	20	NA	80
4	15	30	NA	85
5	NA	NA	70	70
6	30	NA	75	75
7	NA	35	95	95
8	25	50	65	65

2. Do the same for the entire data set.

Exercise 1.3 (*)

1. Derive Equations (1.2) and (1.3) for, respectively, the survival function in the two-state model (Figure 1.1) and the cumulative incidence function in the competing risks model (Figure 1.2).

2. Show, for the Markov illness-death model (Figure 1.3), that the state occupation probability for state 1 at time t, $Q_1(t)$, is

$$\int_0^t \exp\left(-\int_0^u (\alpha_{01}(x)+\alpha_{02}(x))dx\right)\alpha_{01}(u)\exp\left(-\int_u^t \alpha_{12}(x)dx\right)du.$$

Exercise 1.4 (*) Argue (intuitively) how the martingale property $E(M(t) \mid \mathscr{H}_s) = M(s)$ follows from $E(dM(t) \mid \mathscr{H}_{t-}) = 0$ (Section 1.4.3).

Chapter 2

Intuition for intensity models

In this chapter, we will give a non-technical introduction to models for intensities to be discussed in more mathematical details in Chapter 3. Along with the introduction of the models, examples will be given to illustrate how results from analysis of these models can be interpreted. More examples are given in Chapter 3. In Section 1.2, we argued that the *intensity* (or *rate* or *hazard*) is the basic parameter in multi-state models, so, estimation of the intensity is crucial. In the following, we will present intuitive arguments for the way in which models for the intensity may be analyzed for one transition at a time. An important point is that modeling of different intensities in a given multi-state model may be done *separately*, and in Section 3.1 we will present a mathematical argument (based on a likelihood factorization) why this is so. An intuitive argument is the following. The rate describes what happens *locally in time* among those who are at risk for a given type of event and because, in a small time interval from t to $t+dt$, only one event can happen, it is not needed to consider which other events may happen when focusing on a single event type. This means, e.g., that in a competing risks model, each single *cause-specific hazard* may be analyzed by, formally, censoring for the competing causes (and for avoidable censoring events, see Section 1.3). We will exemplify this in Section 2.4. This does not mean that 'censoring for competing causes' is considered an independent censoring mechanism, rather it should be thought of as a 'technical hack' in the estimation. However, for other parameters (such as the cumulative probability over time of dying from a specific cause, the cumulative incidence, cf. Equation (1.3)), different attention must be paid to deaths from other causes and to other kinds of loss to follow-up. This phenomenon has caused confusion in the literature on competing risks, and it will be an important topic for discussion in this book, see, e.g., Section 4.1.2.

2.1 Models for homogeneous groups

We will use the two-state model for the PBC3 trial (Section 1.1.1) and the composite endpoint 'failure of medical treatment', i.e., death or transplantation, as motivating example. Consider a patient (i) randomized to CyA treatment who is still event-free (alive without having had a liver transplantation) at time t since randomization. The intensity of the event 'failure of medical treatment' for that subject at time t, $\alpha_i(t)$ has the interpretation

$$\alpha_i(t)dt = P(i \text{ has an event before time } t+dt \mid i \text{ is event-free at } t).$$

2.1.1 Nelson-Aalen estimator

We will use the counting process notation (where $I(\cdots)$ is an *indicator function*, see Section 1.2)

$$dN_i(t) = I(i \text{ has an event in the interval } (t, t+dt))$$

and

$$Y_i(t) = I(i \text{ is event-free and uncensored just before time } t).$$

If we assume that censoring is *independent* (Equation (1.6)) and that all CyA treated patients have the same intensity $\alpha(t)$, then a natural estimator of the probability $\alpha(t)dt$ is the fraction

$$\frac{\text{No. of patients with an event in } (t, t+dt)}{\text{No. of patients at risk of an event just before time } t} = \frac{dN(t)}{Y(t)}.$$

Here, $dN(t) = \sum_i dN_i(t)$ and $Y(t) = \sum_i Y_i(t)$ are the total number of events at time t (typically 0 or 1) and the total number of subjects at risk at time t, respectively. This idea leads to the *Nelson-Aalen estimator* for the *cumulative hazard*

$$A(t) = \int_0^t \alpha(u)du,$$

as follows. Let $0 < X_1 \leq X_2 \leq \cdots \leq X_n$ be the ordered times of observation for the CyA treated patients, i.e., an X is either an observed time of failure or a time of censoring, whatever came first for a given subject. Then, for each such time, a term $dN(X)/Y(X)$ is added to the estimator, and the Nelson-Aalen estimator may then be written

$$\widehat{A}(t) = \frac{dN(X_1)}{Y(X_1)} + \frac{dN(X_2)}{Y(X_2)} + \cdots,$$

where contributions from all times of observation $\leq t$ are added up. Since only observed *failure* times ($dN(X) = 1$) effectively contribute to this sum (for a censoring at X, $dN(X) = 0$), the estimator may be re-written as

$$\widehat{A}(t) = \sum_{\text{event times, } X \leq t} \frac{1}{\text{No. at risk at } X}.$$

This estimates the *cumulative hazard* and on a plot of $\widehat{A}(t)$ against t, the 'approximate local slope' estimates the intensity at that point in time. To establish an interpretation of $A(t)$, we study the situation with survival data (Figure 1.1), where the following 'experiment' can be considered: Assume that one subject is observed from time $t = 0$ until failure (at time X_1, say). At that failure time, replace the first subject by another subject who is still alive and observe that second subject until failure (at time $X_2 > X_1$). At X_2, replace by a third subject who is still alive and observe until failure (at $X_3 > X_2$), and so on. In that experiment $A(t)$ is the *expected number of replacements* in $[0, t]$. In particular, we may note that $A(t)$ is not a probability and its value may exceed the value 1.

The *standard deviation* (or *standard error*) of the estimator $\widehat{A}(t)$ (which we will throughout abbreviate SD) can also be estimated, whereby confidence limits can be added to plots of $\widehat{A}(t)$, preferably by transforming symmetric limits for $\log(A(t))$. This amounts to a lower 95% confidence limit for $A(t)$ of $\widehat{A}(t)\exp(-1.96 \cdot \text{SD}/\widehat{A}(t))$ and an upper limit of

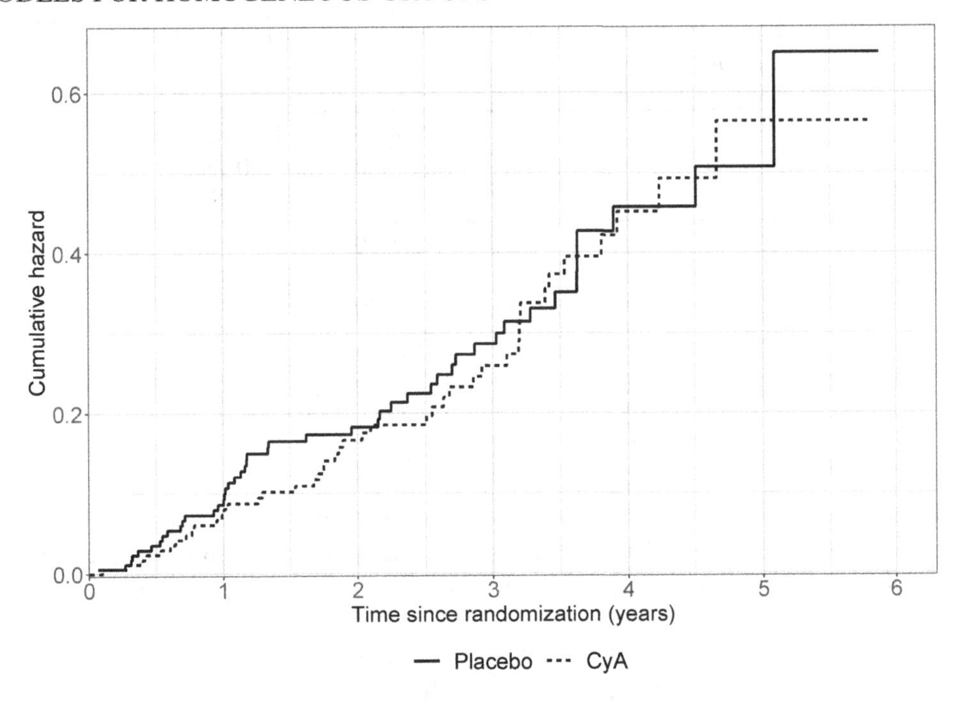

Figure 2.1 *PBC3 trial in liver cirrhosis: Nelson-Aalen estimates by treatment.*

$\widehat{A}(t)\exp(1.96 \cdot \text{SD}/\widehat{A}(t))$. The SD for the cumulative hazard estimate will increase most markedly during periods at which few subjects are at risk, i.e., $Y(t)$ is low.

Figure 2.1 shows the Nelson-Aalen estimates for the two treatment groups, CyA and placebo, from the PBC3 trial. It is seen that the curves for both treatment groups are roughly linear (suggesting that the hazards are roughly constant) and that they are quite equal (suggesting that CyA treatment does not affect the rate of failure from medical treatment in patients with PBC). This can be emphasized by estimating the SD of $\widehat{A}(t)$. Thus, at 2 years the estimates are, respectively, 0.183 (SD=0.035) in the placebo group and 0.167 (SD=0.034) in the CyA group, leading to the 95% confidence intervals $(0.126, 0.266)$, respectively $(0.112, 0.249)$. To simplify the figure, confidence limits have not been added to the curves.

2.1.2 Piece-wise constant hazards

Even though both hazards in the PBC3 trial seem rather constant, a model *assuming* this is typically considered to be too restrictive for practical purposes. However, a flexible extension of the constant hazard model exists, namely the *piece-wise constant hazards* (or *piece-wise exponential*) *model*. This builds on a division of time into 'suitable' intervals within which the hazard is assumed to be constant. To fit this model, a specification of a number, L of intervals in which $\alpha(t)$ is assumed to be constant is needed. This specification should, ideally, be pre-determined, i.e., without having looked at the data and the distribution of event times. This could, for example, be based on 'nice' values of time such as 5- or 10-year age intervals or yearly intervals of follow-up time. However, alternatively one sometimes attempts to choose intervals including roughly equal numbers of events. In each

Table 2.1 *PBC3 trial in liver cirrhosis: Events, risk time, and estimated hazards in a piece-wise exponential model by treatment group.*

Treatment	Interval	Events	Risk Time (in years)	Hazard (per 100 years)	
	ℓ (year)	D_ℓ	Y_ℓ	$\widehat{\alpha}_\ell$	SD
CyA	0-1	24	295.50	8.1	1.7
	2-3	18	137.67	13.1	3.1
	4-5	2	20.80	9.6	6.8
Placebo	0-1	27	287.08	9.4	1.8
	2-3	17	136.00	12.5	3.0
	4-5	2	23.66	8.5	6.0

such interval the value of the hazard may then be estimated by an *occurrence/exposure* rate obtained as the ratio between the *number of events* occurring in that interval and the *total ('exposure') time at risk* in the interval. Thus, if α_ℓ is the hazard in interval no. ℓ, then it is estimated by

$$\widehat{\alpha}_\ell = \frac{\text{No. of events in interval } \ell}{\text{Total time at risk in interval } \ell} = \frac{D_\ell}{Y_\ell}.$$

Note that the hazard has a *per time* dimension and that, therefore, whenever a numerical value of a hazard is quoted, the units in which time is measured should be given.

For the PBC3 data, working with two-year intervals of follow-up time, the resulting event and person-time counts together with the resulting estimated rates are shown in Table 2.1 together with their estimated standard deviation, $\text{SD} = \sqrt{D_\ell}/Y_\ell$ and depicted in Figure 2.2. It is seen that, judged from the SD values, the estimated hazards are, indeed, quite constant over time and between the treatment groups. The SD is smaller when the event count is large.

Figure 2.3 compares for the placebo group the estimated cumulative hazards using the two different models (a step function for the non-parametric model and a broken straight line for the piece-wise exponential model) and the two sets of estimates are seen to coincide well.

2.1.3 *Significance tests*

If a formal statistical comparison between the hazards in the two treatment groups is desired, either in the non-parametric case using the Nelson-Aalen estimators or in the case of the piece-wise constant hazard model, then this may be achieved via a suitable significance test. In the latter case, a standard *likelihood ratio test* (LRT) may be used which for the PBC3 data, splitting into two-year intervals of follow-up time, gives the value 0.31 yielding (using the χ^2_3-distribution, i.e., the chi-squared distribution with 3 DF (degrees of freedom)) a P-value of 0.96.

In the former case, the standard non-parametric test for comparing the hazards in the two treatment groups is the *logrank test.* This is obtained by, at each observation time, X (in

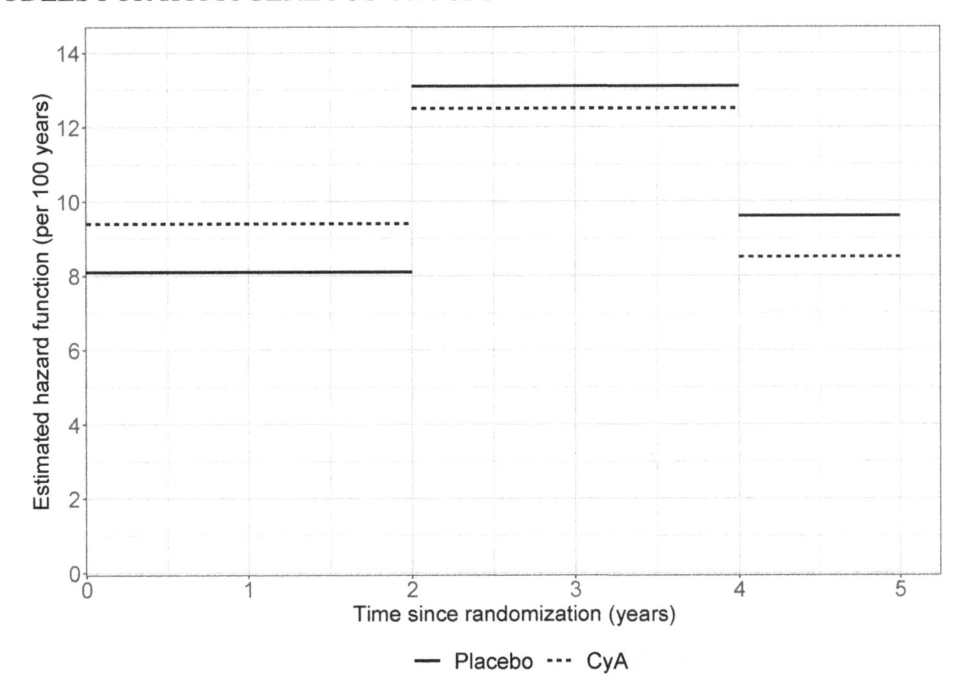

Figure 2.2 *PBC3 trial in liver cirrhosis: Estimated piece-wise exponential hazard functions by treatment group, see Table 2.1.*

either group, 0 or 1), setting up a two-by-two table summarizing the observations at that time, see Table 2.2.

Across the tables for different X, the *observed*, $dN_0(X)$, and *expected* (under the hypothesis of identical hazards in groups 0 and 1)

$$\frac{Y_0(X)}{Y_0(X)+Y_1(X)}(dN_0(X)+dN_1(X)),$$

numbers of failures from one group (here group 0) are added. Denote the resulting sums by O_0 and E_0, respectively. Also the variances

$$(dN_0(X)+dN_1(X))\frac{Y_0(X)Y_1(X)}{(Y_0(X)+Y_1(X))^2}$$

(if all failure times are distinct) are added across the tables to give v. Note that only observed *failure* times (in either group) effectively contribute to these sums. The two-sample logrank

Table 2.2 *Summary of observations in two groups at a time, X, of observation.*

Group	Died	Survived	Alive before
0	$dN_0(X)$	$Y_0(X)-dN_0(X)$	$Y_0(X)$
1	$dN_1(X)$	$Y_1(X)-dN_1(X)$	$Y_1(X)$
Total	$dN_0(X)+dN_1(X)$		$Y_0(X)+Y_1(X)$

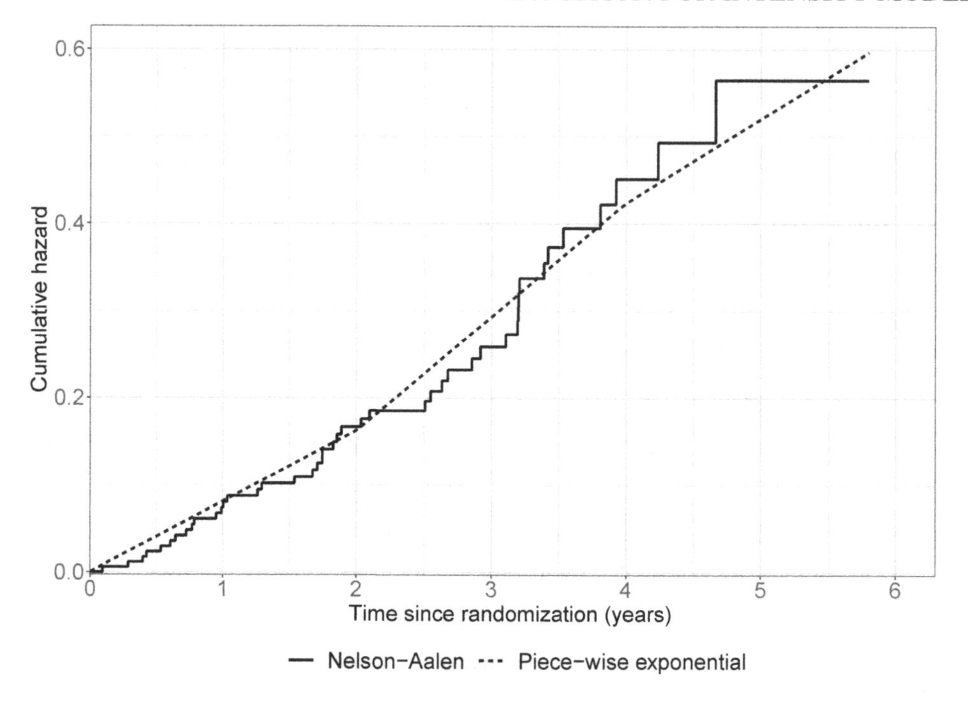

Figure 2.3 *PBC3 trial in liver cirrhosis: Estimated cumulative hazards for the placebo group.*

test statistic, to be evaluated in the χ^2_1-distribution, is then

$$\frac{(O_0 - E_0)^2}{v}.$$

For the PBC3 trial, the observed number of failures in the placebo group is $O_0 = 46$, the expected number is $E_0 = 44.68$, and the variance $v = 22.48$ leading to a logrank test statistic of 0.08 and the P-value 0.78. Note that the same results would be obtained by focusing, instead, on the CyA group (because $O_0 + O_1 = E_0 + E_1$ = total number of observed failures and, therefore, $(O_0 - E_0)^2 = (O_1 - E_1)^2$).

The logrank test can be extended to a comparison of more than two groups where, for comparison of k groups, the resulting test statistic is evaluated in the χ^2_{k-1}-distribution (e.g., Collett, 2015, ch. 2). Also a stratified version of the logrank test is available.

2.2 Regression models

In this section, we will study *regression models* describing the way in which hazard functions may depend on covariates. It will be an assumption throughout that censoring is independent *given* the covariates under study – see Section 1.3. As explained there, this assumption cannot be checked based on the censored event history data; however, if censoring depends on a covariate, then this should be included in the hazard model, see Section 4.4.2 for further discussion.

2.2.1 *Multiplicative regression models*

A more informative way of comparing the two groups than using a significance test is to *quantify* the discrepancies between the hazards in the two treatment groups via a *regression model.* Using a *hazard ratio* for this quantification leads to the *Cox proportional hazards regression model* (Cox model) when the starting point is the non-parametric model (Cox, 1972) and to the *'Poisson'* (or *piece-wise exponential*) *regression model* when the starting point is the model with a piece-wise constant hazard (e.g., Clayton and Hills, 1993, ch. 22-23).

Cox model

A Cox model for the PBC3 trial would assume that the hazard for the placebo group is $\alpha_0(t)$ and no assumptions concerning this hazard are imposed (it is a completely unspecified *non-parametric, baseline* hazard function). On the other hand, the hazard function (say, $\alpha_1(t)$) for the CyA group is assumed to be *proportional* to the baseline hazard, i.e., there exists a constant hazard ratio, say HR, such that, for all t,

$$\alpha_1(t) = \alpha_0(t)\,\mathrm{HR},$$

i.e.,

$$\frac{\alpha_1(t)}{\alpha_0(t)} = \mathrm{HR}.$$

This specifies a *regression model* because, for each patient (i) in the PBC3 trial, we can define an *explanatory variable* (or *covariate*) Z_i, as follows,

$$Z_i = \begin{cases} 0 & \text{if patient } i \text{ was in the placebo group} \\ 1 & \text{if patient } i \text{ was in the CyA group} \end{cases}$$

and then the Cox model for the hazard for patient i

$$\alpha_i(t) = \begin{cases} \alpha_0(t) & \text{if patient } i \text{ was in the placebo group} \\ \alpha_0(t)\,\mathrm{HR} & \text{if patient } i \text{ was in the CyA group} \end{cases}$$

can be written as the regression model

$$\alpha_i(t) = \alpha_0(t)\exp(\beta Z_i)$$

with $\mathrm{HR} = \exp(\beta)$. On the logarithmic scale, the model becomes

$$\log(\alpha_i(t)) = \log(\alpha_0(t)) + \beta Z_i.$$

Thus, the proportionality assumption is the same as a constant difference between the $\log(\text{hazards})$ at any time t. Figure 2.4 illustrates the proportional hazards assumption for a binary covariate Z, both on the hazard scale (a) and the $\log(\text{hazard})$ scale (b).

Because the Cox model combines a non-parametric baseline hazard with a parametric specification of the covariate effect, it is often called *semi-parametric.*

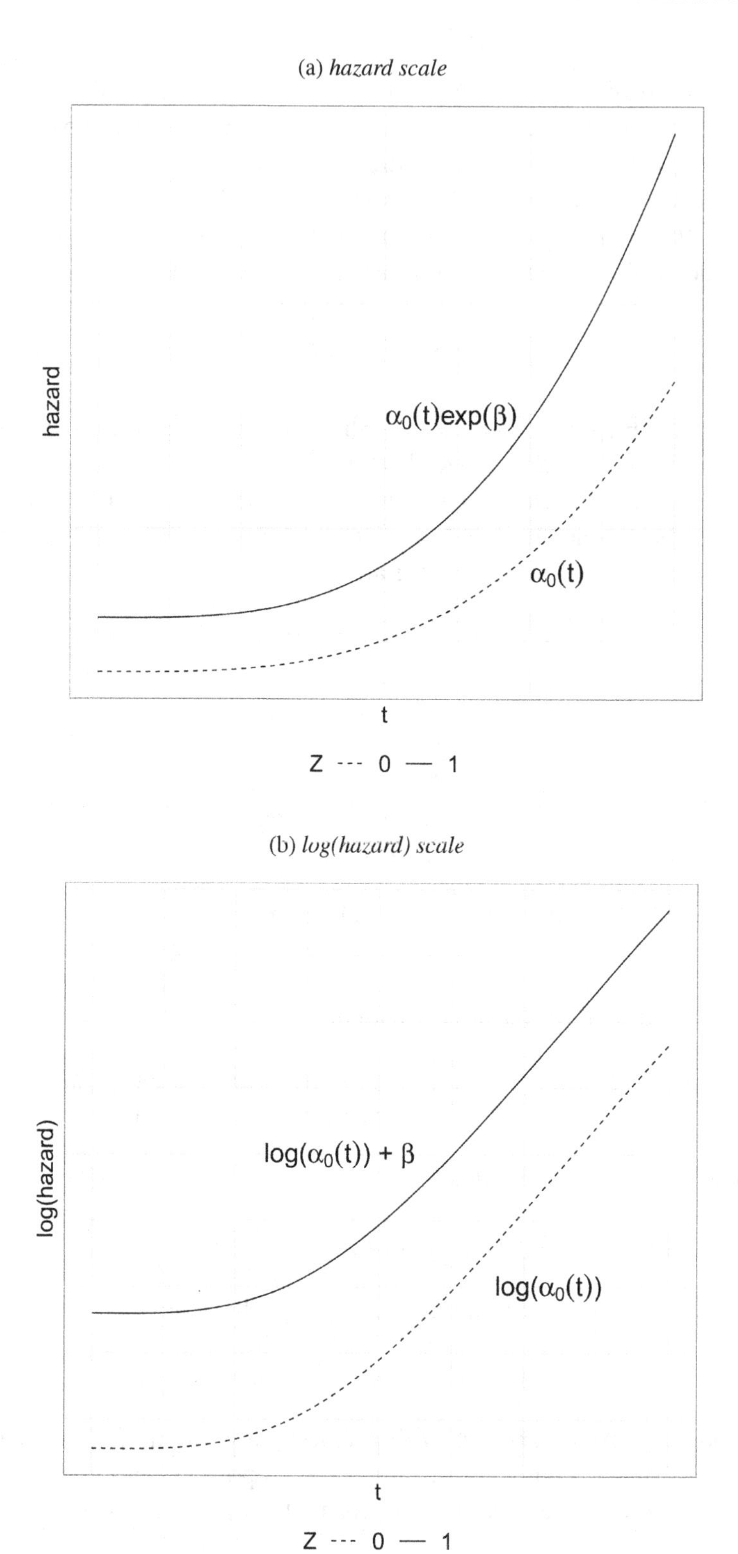

Figure 2.4 *Illustrations of the assumptions for a Cox model for a binary covariate Z.*

To estimate the hazard ratio, HR (or the *regression coefficient* $\beta = \log(\text{HR})$), the *Cox log-partial likelihood function* $l(\beta)$ is maximized. This is

$$l(\beta) = \sum_{\text{event times, } X} \log\left(\frac{\exp(\beta Z_{\text{event}})}{\sum_{j \text{ at risk at time } X} \exp(\beta Z_j)}\right) \tag{2.1}$$

and the intuition behind this is, as follows. At any event time, X, the covariate value, say Z_{event}, for the individual with an event at that time 'is compared' to that of the subjects (j) who were at risk for an event at that time (i.e., still event-free and uncensored, including the failing subject). Thus, if 'surprisingly often', the individual having an event is placebo-treated compared to the distribution of the treatment variable Z_j among those at risk, then this signals that placebo treatment is a risk factor. The set of subjects who are at risk of the event at a given time t is denoted the *risk set*, $R(t)$.

For the PBC3 data, a Cox model including the treatment indicator, Z yields an estimated regression coefficient of $\widehat{\beta} = -0.059$ with an estimated standard deviation of 0.211, leading to an estimated hazard ratio of $\exp(-0.059) = 0.94$ with 95% confidence limits from $0.62(= \exp(-0.059 - 1.96 \cdot 0.211))$ to $1.43(= \exp(-0.059 + 1.96 \cdot 0.211))$. This contains the null value HR = 1, in accordance with the logrank test statistic. The estimated SD is known as a *model-based* standard deviation since it follows from the *likelihood function* $l(\beta)$. In the Cox model, the cumulative baseline hazard may be estimated using a 'Nelson-Aalen-like' estimator, known as the *Breslow estimator*:

$$\widehat{A}_0(t) = \sum_{\text{event times, } X \leq t} \frac{1}{\sum_{j \text{ at risk at time } X} \exp(\widehat{\beta} Z_j)}. \tag{2.2}$$

For the PBC3 data, $A_0(t)$ is the cumulative hazard in the placebo group, and the estimate is shown in Figure 2.5. Note that, compared to Figure 2.1, there are many more steps in the Breslow estimate. This is because *all* event times, i.e., in either treatment group, give rise to a jump in the baseline hazard estimator. The intuition is that, due to the proportional hazards assumption, an event for a CyA treated patient also contains information about the hazard in the placebo group.

Multiple Cox regression

The PBC3 trial was randomized and a regression analysis including only the treatment variable may be reasonable. In observational studies focusing on an exposure covariate, there may be confounders for which adjustment is needed when estimating the association with exposure. This leads to the need of performing *multiple (Cox) regression* analyses. Because the randomization in the PBC3 trial was not completely successful (it turned out that CyA treated patients, in spite of the randomization, tended to have slightly less favorable values of important prognostic variables like serum bilirubin and serum albumin, see Table 2.3), we will illustrate multiple Cox regression using this example. The joint effect of p covariates, as explained in Section 1.2.5, is summarized in a *linear predictor*

$$\text{LP} = \beta_1 Z_1 + \beta_2 Z_2 + \cdots + \beta_p Z_p$$

(see Equation (1.4)) in which each covariate Z_j enters via a regression coefficient β_j. The resulting model for a subject with covariates $Z_1, Z_2, \ldots, Z_p$ is then given by the hazard

$$\alpha_0(t) \exp(\text{LP})$$

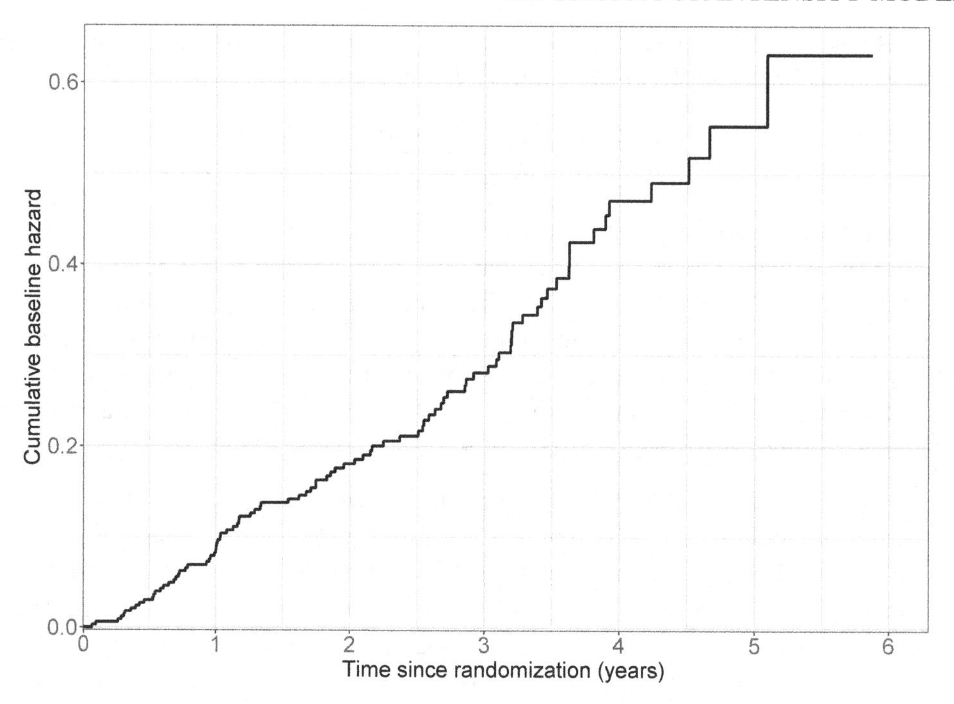

Figure 2.5 *PBC3 trial in liver cirrhosis: Breslow estimate for the cumulative baseline hazard in a Cox model including only treatment.*

where the *regression coefficients* have the following interpretation (see Section 1.2.5). Consider two subjects differing 1 unit for covariate j and having identical values for the remaining covariates in the model. Then the ratio between their hazards at any time t is

$$\frac{\alpha_0(t)\exp(\beta_1 Z_1 + \cdots + \beta_{j-1}Z_{j-1} + \beta_j(Z_j+1) + \cdots + \beta_p Z_p)}{\alpha_0(t)\exp(\beta_1 Z_1 + \cdots + \beta_{j-1}Z_{j-1} + \beta_j(Z_j) \quad + \cdots + \beta_p Z_p)} = \exp(\beta_j).$$

Thus, $\exp(\beta_j)$ is the hazard ratio for a 1 unit difference in Z_j at any time t and for given values of the remaining covariates in the model. Furthermore, the *baseline hazard* $\alpha_0(t)$ is the hazard function for subjects where the linear predictor equals 0.

Table 2.3 *PBC3 trial in liver cirrhosis: Average covariate values by treatment group.*

Treatment (n)	Albumin (g/L)	Bilirubin (μmol/L)
CyA (176)	37.51	48.56
Placebo (173)	39.26	42.34

If, for the PBC3 data we add the covariates Z_2 = albumin and Z_3 = bilirubin to the model including only the treatment indicator, then the results in Table 2.4 are obtained.

It is seen that, for given value of the variables albumin and bilirubin, the log(hazard ratio) comparing CyA with placebo is now numerically much larger and with a 95% confidence interval for $\exp(\beta_1)$ which is $(0.391, 0.947)$ and, thus, excludes the null value. The interpretation of the coefficient for albumin is that the hazard ratio when comparing two subjects differing 1 g/L is $\exp(-0.116) = 0.89$ for given values of treatment and bilirubin.

Table 2.4 *PBC3 trial in liver cirrhosis: Estimated coefficients (and SD) from a Cox model.*

Covariate		$\widehat{\beta}$	SD
Treatment	CyA vs. placebo	-0.496	0.226
Albumin	per 1 g/L	-0.116	0.021
Bilirubin	per 1 μmol/L	0.00895	0.00098

Table 2.5 *PBC3 trial in liver cirrhosis: Estimated coefficients (and SD) from a Poisson regression model.*

Covariate		$\widehat{\beta}$	SD
Treatment	CyA vs. placebo	-0.475	0.224
Albumin	per 1 g/L	-0.112	0.021
Bilirubin	per 1 μmol/L	0.00846	0.00094

Poisson model

In a similar way, a multiplicative regression model can be obtained with the starting point being the piece-wise constant hazards model. For the PBC3 data, the model including only treatment is

$$\alpha_i(t) = \alpha_0(t)\exp(\beta Z_i),$$

but now the baseline hazard, instead of being completely unspecified as in the Cox model, is assumed to be constant in, e.g., 2-year intervals of follow-up time

$$\alpha_0(t) = \begin{cases} \alpha_1 & \text{if } t < 2, \\ \alpha_2 & \text{if } 2 \leq t < 4, \\ \alpha_3 & \text{if } 4 \leq t. \end{cases}$$

The resulting regression model is known as *Poisson* or *piece-wise exponential* regression. The reason behind the name 'Poisson' is a bit technical and will be given in Section 3.3. Estimates of the parameters $\beta, \alpha_1, \alpha_2, \alpha_3$ are obtained by referring to the *maximum likelihood principle* (which also motivates the parameter estimates in the Cox model). For a simple Poisson model, including only treatment, the estimates are (with 95% confidence limits) $\exp(\widehat{\beta}) = 0.942\,(0.623, 1.424)$, $\widehat{\alpha}_1 = 0.088\,(0.067, 0.115)$, $\widehat{\alpha}_2 = 0.128\,(0.092, 0.178)$, $\widehat{\alpha}_3 = 0.090\,(0.034, 0.239)$ (the latter three expressed in the unit 'per 1 year') close to what was seen in Table 2.1. In this model, the time intervals appear as a categorical covariate.

Multiple Poisson regression is now straightforward and is given by the following hazard for a subject with covariates $Z_1, Z_2, \ldots, Z_p$

$$\alpha_0(t)\exp(\text{LP})$$

where the linear predictor LP is given by Equation (1.4) and the baseline hazard $\alpha_0(t)$ (the hazard function when $\text{LP} = 0$) is assumed piece-wise constant. For the PBC3 data, adding albumin and bilirubin to the model yields the estimates shown in Table 2.5 which are seen to be quite close to those from the similar Cox model (Table 2.4).

2.2.2 Modeling assumptions

Whenever the effect on some outcome of several explanatory variables is obtained by combining the variables into a linear predictor, some assumptions are imposed:

- The effect of a quantitative covariate on the linear predictor is *linear*.
- For each covariate, its effect on the linear predictor is independent of other variables' effects, i.e., there are *no interactions* between the covariates.

Since these assumptions are standard in models with a linear predictor, there are standard ways of checking them. Thus, as discussed, e.g., by Andersen and Skovgaard (2006, ch. 4-5), to check linearity, extended models including non-linear effects, such as splines or polynomials, may be fitted and compared statistically to the simple model with a linear effect. To examine interactions, interaction terms may be added to the linear predictor and the resulting models may be compared statistically to the simple additive model.

We will exemplify goodness-of-fit investigations using the data from the PBC3 trial.

Checking linearity

We will illustrate how to examine linearity of a quantitative covariate, Z, using either a *quadratic effect* or *linear splines*. Both in the Cox model and in the Poisson model, either the covariate Z^2 or covariates of the form

$$Z_j = (Z - a_j) \cdot I(Z > a_j), \quad j = 1, \dots, s$$

may be added to the linear predictor. Here, the covariate values $a_1 < \cdots < a_s$ are *knots* to be selected. If no particular clinically relevant cut-points are available, then one would typically use certain percentiles as knots. The spline covariate Z_j gives, for subjects who have a value of Z that exceeds the knot a_j, how much the value exceeds a_j. For subjects with $Z \leq a_j$, the spline covariate is $Z_j = 0$ and the linear predictor now depends on Z as a *broken straight line*. Here, the interpretation of coefficient no. j is the change in slope at the knot a_j, and the coefficient for Z is the slope below the first knot, a_1. Linearity, therefore, corresponds to the hypothesis that all coefficients for the added spline covariates are equal to zero. In a model with a quadratic effect, i.e., including both Z and Z^2, the corresponding coefficients (say, β_1 and β_2) do not, themselves, have particularly simple interpretations. However, the fact that a positive β_2 suggests that the best fitting quadratic curve for the covariate is a convex ('happy') parabola, while a negative β_2 suggests that the best fitting parabola for the covariate is concave ('bad tempered') does give some insight into the dose-response relationship between the linear predictor and Z. In both cases, the extreme point for the parabola (a minimum if $\beta_2 > 0$ and a maximum if $\beta_2 < 0$) corresponds to $Z = -\beta_1/(2\beta_2)$, a fact that may give further insight.

For albumin, there is a normal range from $25g/L$ and up, and we choose $s = 1$ knot placed at $a_1 = 25$. For bilirubin, the normal range is 0 to 17.1μmol/L and we let $s = 3$ and $a_1 = 17.1$, $a_2 = 2 \times 17.1$, and $a_3 = 3 \times 17.1$. Table 2.6 shows the results for both the Cox model and the Poisson model. It is seen that, for albumin, there is no evidence against linearity in either model which is also illustrated in Figure 2.6 where the estimated linear predictors under linearity and under the spline model are shown for the Poisson case.

Table 2.6 *PBC3 trial in liver cirrhosis: Estimated coefficients (and SD) from Cox and Poisson models modeling the effects of albumin, bilirubin, and* $\log_2$*(bilirubin) using linear splines (S) or as quadratic (Q); LRT denotes the appropriate likelihood ratio test for linearity. All models included albumin and bilirubin (modeled as described) and treatment.*

		Cox model		Poisson model	
	Covariate	$\widehat{\beta}$	SD	$\widehat{\beta}$	SD
S	Albumin	-0.0854	0.045	-0.0864	0.045
	> 35 g/L	-0.0557	0.073	-0.0474	0.072
	LRT	0.60 (1 DF), $P = 0.44$		0.44 (1 DF), $P = 0.51$	
Q	Albumin	-0.1295	0.213	-0.1388	0.210
	Albumin^2	0.000195	0.00298	0.000371	0.00293
	LRT	0.0042 (1 DF), $P = 0.95$		0.02 (1 DF), $P = 0.90$	
S	Bilirubin	0.0624	0.062	0.0617	0.062
	$> 17.1 \mu$mol/L	-0.0146	0.085	-0.0168	0.085
	$> 2 \times 17.1 \mu$mol/L	-0.0026	0.053	0.0027	0.053
	$> 3 \times 17.1 \mu$mol/L	-0.0400	0.026	-0.0428	0.026
	LRT	24.40 (3 DF), $P < 0.001$		24.54 (3 DF), $P < 0.001$	
Q	Bilirubin	0.0200	0.0033	0.0200	0.0033
	Bilirubin^2	-0.000031	0.0000091	-0.000032	0.0000091
	LRT	12.35 (1 DF), $P < 0.001$		13.34 (1 DF), $P < 0.001$	
S	$\log_2$(bilirubin)	0.201	0.465	0.198	0.466
	$> 17.1 \mu$mol/L	0.935	0.915	0.882	0.912
	$> 2 \times 17.1 \mu$mol/L	-0.386	1.293	-0.234	1.278
	$> 3 \times 17.1 \mu$mol/L	-0.181	0.988	-0.314	0.971
	LRT	1.61 (3 DF), $P = 0.66$		1.71 (3 DF), $P = 0.63$	
Q	$\log_2$(bilirubin)	0.582	0.500	0.628	0.498
	$(\log_2(\text{bilirubin}))^2$	0.0072	0.043	0.0016	0.042
	LRT	0.03 (1 DF), $P = 0.87$		0.00 (1 DF), $P = 0.97$	

For bilirubin, however, linearity describes the relationship quite poorly as illustrated both by the likelihood ratio tests and Figure 2.7 (showing the linear predictors for the Poisson model under linearity and with linear splines). Both the negative coefficients from the models with quadratic effects and this figure suggest that the effect of bilirubin should rather be modeled as some concave function. The maximum point for the parabola corresponds to a bilirubin value of $0.0200/(2 \cdot 0.000031) = 322.6$ which is compatible with the figure. The concave curve could be approximated by a logarithmic curve and Table 2.6 (and Figure 2.8) show the results after a $\log_2$-transformation and using the same knots. It should be noticed that any logarithmic transformation would have the same impact on the results, and we chose the $\log_2$-transformation because it enhances the interpretation, as will be explained in what follows. Since the linear spline has no systematic deviations from a straight line, linearity after log-transformation is no longer contraindicated, and Table 2.7 shows the

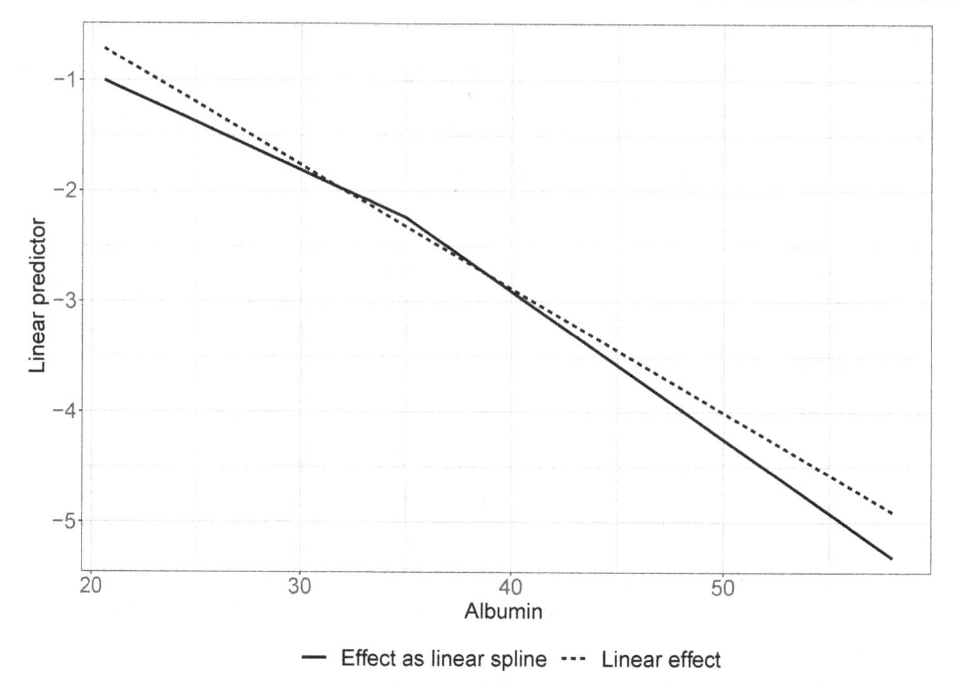

Figure 2.6 *PBC3 trial in liver cirrhosis: Linear predictor as a function of albumin in two Poisson models. Both models also included treatment and bilirubin.*

Table 2.7 *PBC3 trial in liver cirrhosis: Estimated coefficients (and SD) from Cox and Poisson models with linear effects of albumin and* $\log_2$*(bilirubin).*

		Cox model		Poisson model	
Covariate		$\widehat{\beta}$	SD	$\widehat{\beta}$	SD
Treatment	CyA vs. placebo	-0.574	0.224	-0.546	0.223
Albumin	per 1 g/L	-0.091	0.022	-0.087	0.022
$\log_2$(bilirubin)	per doubling	0.665	0.074	0.647	0.073

estimates from Cox and Poisson models including treatment, albumin and $\log_2$(bilirubin). The interpretation of the Cox-coefficient for the latter covariate is that the hazard increases by a factor of about $\exp(0.665) = 1.94$ when comparing two subjects where one has twice the value of bilirubin compared to the other (and similarly for the coefficient from the Poisson model).

Checking interactions

In the models from Table 2.7, we will now study potential treatment-covariate interactions. In Table 2.8, interactions between treatment and, in turn, albumin and $\log_2$(bilirubin) have been introduced, as follows. The covariate Z (albumin or $\log_2$(bilirubin)) is replaced by two covariates:

$$Z_{(0)} = \begin{cases} Z & \text{if treatment is placebo,} \\ 0 & \text{if treatment is CyA,} \end{cases}$$

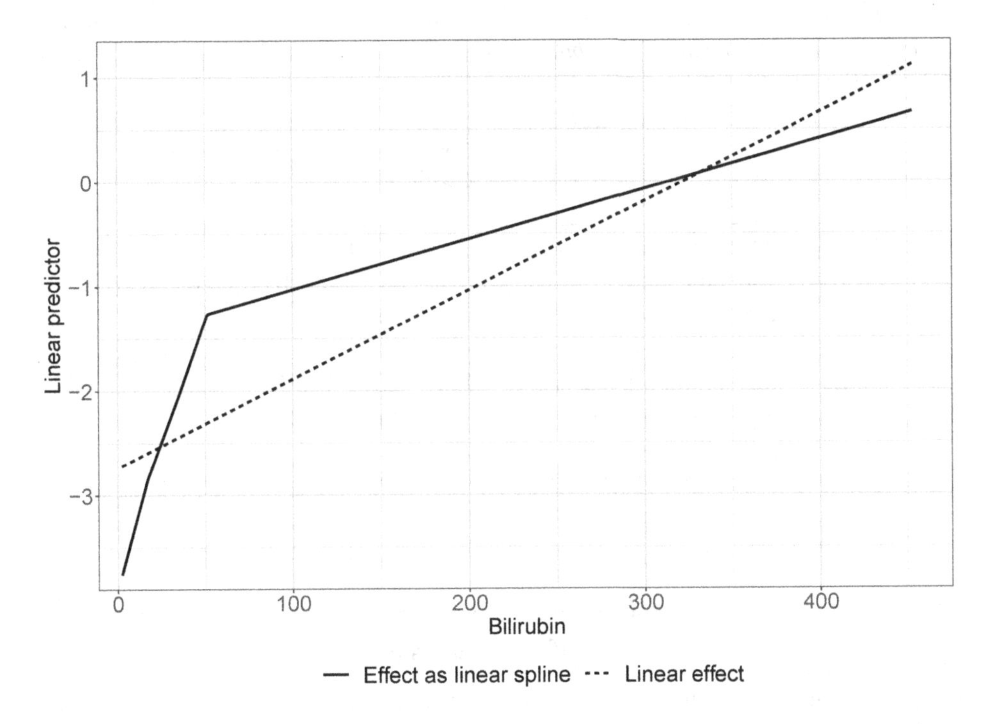

Figure 2.7 *PBC3 trial in liver cirrhosis: Linear predictor as a function of bilirubin in two Poisson models. Both models also included treatment and albumin.*

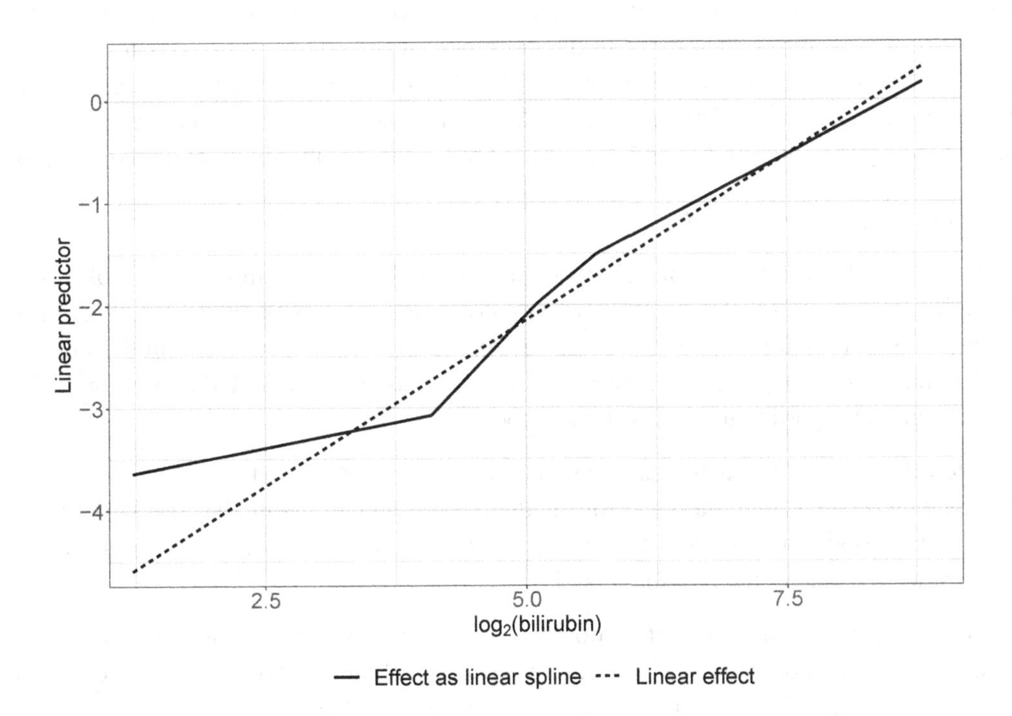

Figure 2.8 *PBC3 trial in liver cirrhosis: Linear predictor as a function of* $\log_2$*(bilirubin) in two Poisson models. Both models also included treatment and albumin.*

Table 2.8 *PBC3 trial in liver cirrhosis: Cox and Poisson models with examination of interaction between treatment and albumin or* $\log_2$*(bilirubin); LRT denotes the likelihood ratio test for the hypothesis of no interaction.*

		Cox model		Poisson model	
	Covariate	$\widehat{\beta}$	SD	$\widehat{\beta}$	SD
$Z_{(0)}$	Albumin, placebo	-0.081	0.034	-0.076	0.034
$Z_{(1)}$	Albumin, CyA	-0.097	0.027	-0.094	0.028
	LRT	0.13, $P = 0.71$		0.17, $P = 0.68$	
$Z_{(0)}$	$\log_2$(bilirubin), placebo	0.726	0.099	0.704	0.097
$Z_{(1)}$	$\log_2$(bilirubin), CyA	0.593	0.106	0.580	0.105
	LRT	0.86, $P = 0.35$		0.78, $P = 0.38$	

and

$$Z_{(1)} = \begin{cases} Z & \text{if treatment is CyA,} \\ 0 & \text{if treatment is placebo.} \end{cases}$$

The model, additionally, includes a main effect of treatment. However, since the interpretation of this is the hazard ratio for treatment when $Z = 0$, its parameter estimate was not included in the table. From Table 2.8 it is seen that, for both models, the interactions are quite small both judged from the separate coefficients in the two treatment groups and the corresponding likelihood ratio tests. If a more satisfactory interpretation of the main effect of treatment is required, then the Z in the definition of the interaction covariates $Z_{(0)}$ and $Z_{(1)}$ can be replaced by a *centered* covariate, $Z - \bar{Z}$, where $\bar{Z}$ is, e.g., an average Z-value. In that case, the main effect of treatment is the hazard ratio at $Z = \bar{Z}$. Since centering does not change the coefficients for $Z_{(0)}$ and $Z_{(1)}$, we did not make this modification in the analysis.

Checking proportional hazards

Both Cox and Poisson regression models impose the additional assumption of *proportional hazards*, i.e., the multiplicative effect of any covariate is constant over time. Since this corresponds to no interaction between covariate and time and since 'time' in the Poisson model enters as a categorical explanatory variable, tests for no interaction are applicable for examining proportional hazards in that model.

For the PBC3 trial, introducing such interactions leads to likelihood ratio tests (all with 2 DF since there are 3 time intervals) for the three covariates in the model, see Table 2.9. It is seen that proportional hazards give a reasonable description of how all covariates affect the hazard over time in the Poisson model.

In the Cox model, the time effect is modeled via the non-parametric baseline hazard, and examination of proportional hazards requires special techniques. We will return to a more detailed discussion of such methods later in the book (e.g., Sections 3.7 and 5.7) and here just mention a graphical technique (based on a stratified Cox model).

Table 2.9 *PBC3 trial in liver cirrhosis: Examination of proportional hazards in a Poisson model including treatment, albumin, and* $\log_2$*(bilirubin); LRT denotes the likelihood ratio test.*

	Treatment		Albumin		$\log_2$(bilirubin)	
Interval (year)	$\widehat{\beta}$	SD	$\widehat{\beta}$	SD	$\widehat{\beta}$	SD
0-1	-0.562	0.291	-0.110	0.028	0.710	0.093
2-3	-0.462	0.345	-0.052	0.035	0.558	0.121
4-5	-1.266	1.230	-0.065	0.153	0.305	0.482
LRT	0.43, $P = 0.81$		1.80, $P = 0.41$		1.58, $P = 0.45$	

For the Cox model, an alternative to assuming proportional hazards for treatment is to stratify by treatment, leading to the *stratified Cox model*

$$\alpha_i(t) = \alpha_{j0}(t)\exp(\text{LP}), \text{ when } i \text{ is in stratum } j. \tag{2.3}$$

Here, the linear predictor no longer includes treatment and the stratum is $j = 0$ for placebo treated patients and $j = 1$ for patients from the CyA group. The effect of treatment is via the two separate baseline hazards $\alpha_{00}(t)$ for placebo and $\alpha_{10}(t)$ for CyA, and these two baseline hazards are not assumed to be proportional. Rather, like the baseline hazard in the unstratified Cox model, they are completely unspecified. Figure 2.9 illustrates the assumptions behind the stratified Cox model for two strata and one binary covariate Z both on the hazard and the log(hazard) scale.

By estimating the cumulative baseline hazards $A_{00}(t)$ and $A_{10}(t)$ separately, the proportional hazards assumption may be investigated. This may be done graphically by plotting $\widehat{A}_{10}(t)$ against $\widehat{A}_{00}(t)$ where, under proportional hazards, the resulting curve should be close to a straight line through the point $(0,0)$ with a slope equal to the hazard ratio for treatment. Note that, in Equation (2.3), the effect of the linear predictor (i.e., of the variables albumin and $\log_2$(bilirubin)) is the same in both treatment groups. Inference for this model builds on a stratified Cox log partial likelihood where there are separate risk sets for the two treatment groups (details to be given in Section 3.3).

We fitted the stratified Cox model to the PBC3 data which resulted in coefficients (SD) -0.090 (0.022) for albumin and 0.663 (0.075) for $\log_2$(bilirubin) close to what we have seen before. Figure 2.10 shows the *goodness-of-fit plot* and suggests that proportional hazards for treatment fits the PBC3 data well. The slope of the straight line in the plot is the estimated hazard ratio for treatment $\exp(-0.574)$ found in the unstratified model. Similar investigations could be done for albumin and $\log_2$(bilirubin); however, for these quantitative covariates, one would need a categorization in order to create the strata. For such covariates, other ways of examining proportional hazards are better suited and will be discussed in Sections 3.7 and 5.7.

2.2.3 *Cox versus Poisson models*

In the PBC3 trial, almost identical results were found for the Cox models and the corresponding Poisson models. This is not surprising because, in this example, the hazard seemed rather time-constant. However, the similarity between the two types of model tends

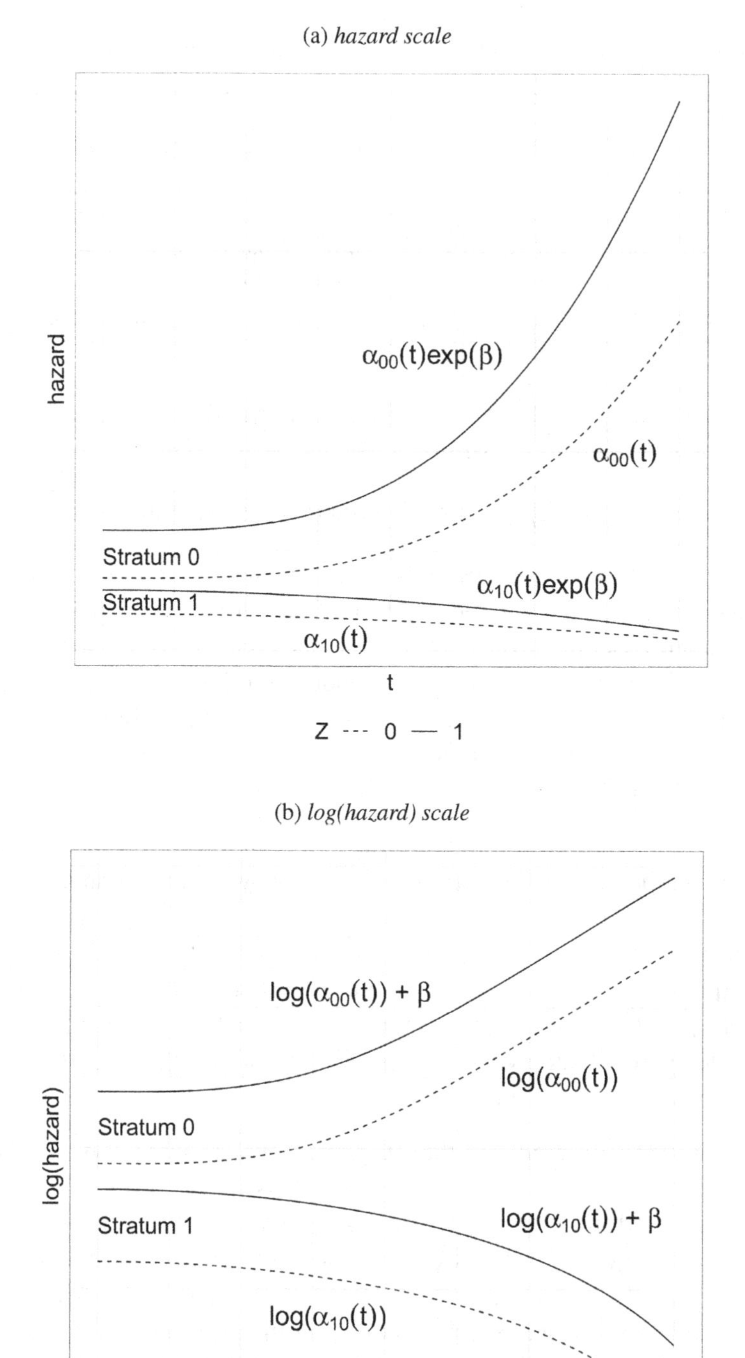

Figure 2.9 *Illustrations of the assumptions for a stratified Cox model for two strata and one binary covariate Z.*

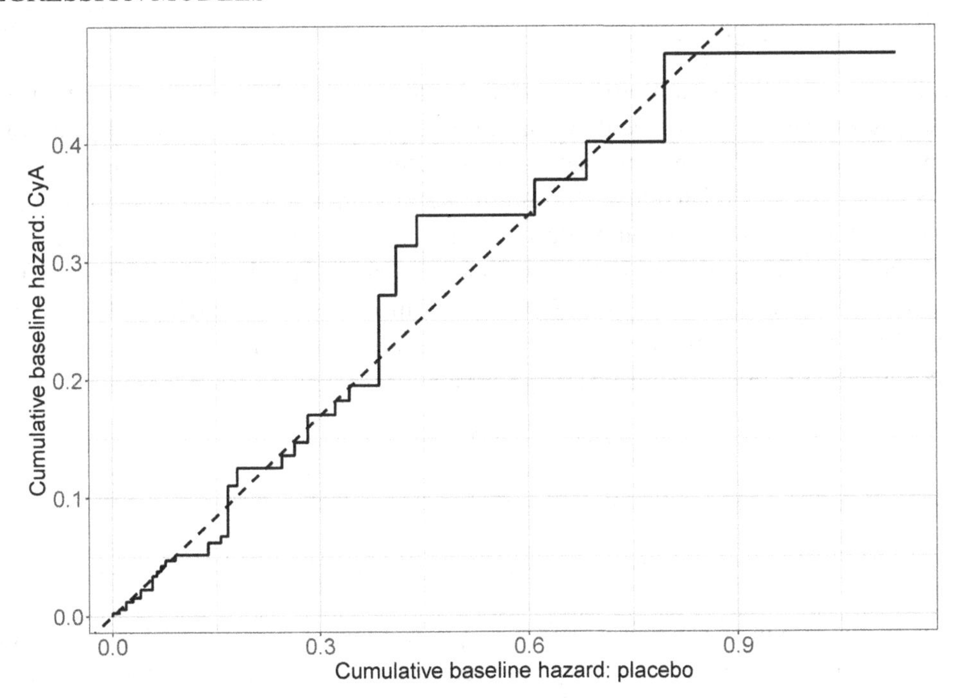

Figure 2.10 *PBC3 trial in liver cirrhosis: Cumulative baseline hazard for CyA plotted against cumulative baseline hazard for placebo in a stratified Cox model. The straight line has slope* $\exp(-0.574) = 0.563$.

to hold quite generally (depending, though, to some extent on how the time-intervals for the Poisson model are chosen). This is because any (hazard) function may be approximated by a piece-wise constant function. Given this fact, the choice between the two types of model is rather a matter of convenience. Some pros and cons may be given:

- In the Cox model a choice of time-intervals is not needed.
- For the Poisson model, estimates of the time-variable are given together with covariate estimates. In the Cox model, the (cumulative) baseline hazard needs special consideration.
- In the Poisson model, examination of proportional hazards is an integrated part of the analysis requiring no special techniques.
- Some problems may involve *several time-variables* (e.g., Example 1.1.4). For the Cox model, one of these must be selected as the 'baseline' time-variable, and the others can then be included as time-dependent covariates (Section 3.7). For the Poisson model, several (categorized) time-variables may be accounted for simultaneously.
- The Poisson model with categorical covariates may be fitted to a tabulated (and typically much smaller) data set (Sections 3.4 and 3.6.4).

2.2.4 *Additive regression models*

Both the Cox model and the Poisson model resulted in hazard ratios as measures of the association between a covariate and the hazard function. Furthermore, we saw in the PBC3 example (and tried to argue beyond that study in Section 2.2.3) that these two multiplicative models were so closely related that the resulting hazard ratios from either would be similar. However, other hazard regression models exist and may sometimes provide a better fit to a given data set and/or provide estimates with a more useful and direct interpretation. One such class of models is the class of *additive hazard* models among which the *Aalen model* (Aalen, 1989) is the most frequently used. In this model, the hazard function for a subject with covariates $(Z_1, Z_2, \dots, Z_p)$ is given by the sum

$$\alpha(t) = \alpha_0(t) + \text{LP}(t),$$

where the (now time-dependent) linear predictor is

$$\text{LP}(t) = \beta_1(t)Z_1 + \cdots + \beta_p(t)Z_p.$$

Here, both the baseline hazard $\alpha_0(t)$ and the *regression functions* $\beta_1(t), \dots, \beta_p(t)$ are unspecified functions of time, t. The interpretation of the baseline hazard, like the baseline hazard $\alpha_0(t)$ in the Cox model, is the hazard function for a subject with $\text{LP}(t) = 0$, while the value, $\beta_j(t)$, of the jth regression function is the *hazard difference* at time t for two subjects who differ by 1 unit in their values for Z_j and have identical values for the remaining covariates:

$$\begin{aligned} \beta_j(t) &= (\alpha_0(t) + \beta_1(t)Z_1 + \cdots + \beta_j(t)(Z_j + 1) + \cdots + \beta_p(t)Z_p) \\ &- (\alpha_0(t) + \beta_1(t)Z_1 + \cdots + \beta_j(t)(Z_j) \quad + \cdots + \beta_p(t)Z_p) \end{aligned}$$

(see also Section 1.2.5). In the Cox model, the cumulative baseline hazard could be estimated using the Breslow estimator. Likewise, in the Aalen model the cumulative baseline hazard and the cumulative regression functions

$$A_0(t) = \int_0^t \alpha_0(u)du,\ B_1(t) = \int_0^t \beta_1(u)du, \dots, B_p(t) = \int_0^t \beta_p(u)du$$

can be estimated. More specifically, the *change in* $(A_0(t), B_1(t), \dots, B_p(t))$ at an observed event time, X, is estimated by multiple linear regression. The subjects who enter this linear regression are those who are at risk at time X, the outcome is 1 for the subject with an event and 0 for the others, and this outcome is regressed linearly on the covariates for those subjects. The resulting estimators $\widehat{A}_0(t), \widehat{B}_1(t), \dots, \widehat{B}_p(t)$ at time t are obtained by adding up the estimated changes for event times $X \leq t$. Since the estimated change at an event time, X need not be positive, plots of the estimates $\widehat{B}_j(t)$ against t need not be increasing.

Figure 2.11 shows both the estimated cumulative baseline hazard and the estimated cumulative treatment effect in an Aalen model for the PBC3 data including treatment as the only covariate. The estimated treatment effect is equipped with 95% point-wise confidence limits. It is seen that the cumulative baseline hazard is roughly linear suggesting (as we have seen in previous analyses) that the baseline hazard is roughly constant. In this model including only one binary explanatory variable, the estimated cumulative baseline hazard

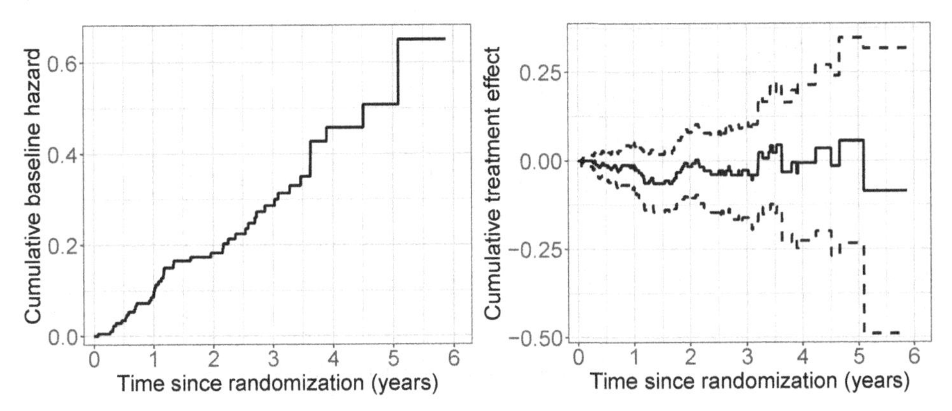

Figure 2.11 *PBC3 trial in liver cirrhosis: Estimated cumulative baseline hazard and cumulative regression function for treatment (with 95% point-wise confidence limits) in an additive hazard model.*

is the Nelson-Aalen estimate in the placebo group, cf. Figure 2.1. The cumulative treatment effect, $\widehat{B}_1(t)$ (judged from the confidence limits) is close to 0, still in accordance with previous analyses. The estimator in this model is the difference between the Nelson-Aalen estimators for the CyA group and the placebo group, see Figure 2.1. A significance test derived from the model confirms that there is no evidence against a null effect of treatment in this model ($P = 0.75$).

It is seen that the Aalen model is very flexible including completely unspecified (non-parametric) baseline hazard and covariate effects and that the estimates from the model are entire curves that may be hard to communicate (though $\exp(-B_1(t))$ is the ratio between survival functions, see Equation (1.2)). It is, thus, of interest to simplify the model, e.g., by restricting the regression functions to be time-constant (e.g., Martinussen and Scheike, 2007, ch. 5). The hypothesis of a time-constant hazard difference $\beta_1(t) = \beta_1$ may also be tested within the model and results in a P-value of 0.62 and an estimate $\widehat{\beta}_1 = -0.0059$ $(\text{SD} = 0.021)$ per year corresponding to $P = 0.78$. Note that this coefficient has a 'per time' dimension relating to the units in which the time-variable was recorded. Thus, if somehow 10,000 person-years were collected for both the treated group and for the control group then, according to this estimate, 59 fewer treatment failures are expected in the treated group.

The simple additive model with a time-varying treatment effect may be extended with more covariates like albumin and bilirubin. This leads to the estimated cumulative regression functions shown in Figure 2.12. To interpret such a curve one should (as it was the case for the Nelson-Aalen estimator) focus on its *local slope* which at time t is the approximate hazard difference at that time when comparing subjects differing by one unit of the covariate. It is seen that these slopes are generally negative for treatment and albumin and positive for bilirubin in accordance with earlier results. To enhance readability of the figure, confidence limits have been excluded. However, significance tests for the three regression functions (Table 2.10) show that both of the biochemical variables are quite significant, but treatment is not. Inspection of Figure 2.12 suggests that the regression functions are roughly constant

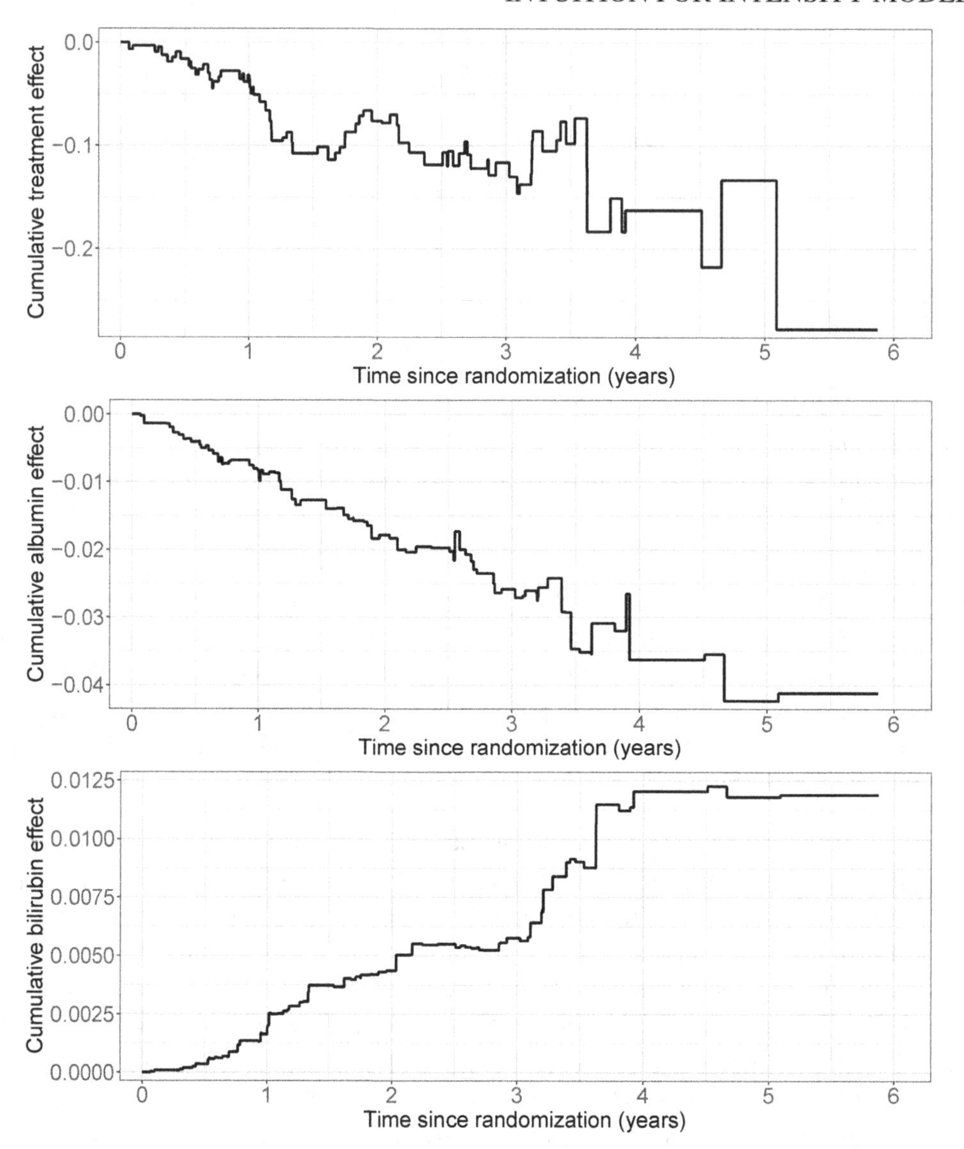

Figure 2.12 *PBC3 trial in liver cirrhosis: Estimated cumulative regression functions for treatment, albumin, and bilirubin in an additive hazard model.*

(roughly linear estimated cumulative effects) and this is also what formal significance tests indicate (Table 2.10).

Even though there is not evidence against a constant effect for any of the three covariates, we first consider a model with a constant effect of treatment (Z) and time-varying effects of albumin (Z_2) and bilirubin (Z_3)

$$\alpha(t) = \alpha_0(t) + \beta_1 Z + \beta_2(t) Z_2 + \beta_3(t) Z_3.$$

Here, the estimated hazard difference for treatment is $\widehat{\beta}_1 = -0.040$ per year (0.020) $P = 0.05$. This model imposes the assumptions for the linear predictor of linear effects of albumin and bilirubin and no interactions between the included covariates – now

Table 2.10 *PBC3 trial in liver cirrhosis: P-values and estimated coefficients (and SD) from additive hazard models including treatment and linear effects of albumin and bilirubin.*

	P-value for		Estimated constant effect per year	
Covariate	Covariate effect	Constant effect	$\widehat{\beta}$	SD
Treatment	0.112	0.69	-0.041	0.020
Albumin	0.006	0.96	-0.0084	0.0022
Bilirubin	<0.001	0.16	0.0023	0.00038

assumptions referring to the additive hazard scale. These assumptions may be tested using standard methods. Adding quadratic effects of, respectively, albumin and bilirubin to the model with a constant effect of treatment results in P-values for linearity of 0.065 for albumin and 0.05 for bilirubin. There seems to be no strong evidence against linearity. In the models including quadratic effects, the estimated (constant) treatment effects are, respectively, $\widehat{\beta}_1 = -0.042\,(0.020)$, and $\widehat{\beta}_1 = -0.040\,(0.021)$ (per year). This means that more flexible models for the biochemical variables do not substantially change the estimated treatment effect. A test for no interaction was performed in the model where all three covariate effects are constant (Table 2.10, last column). This was done along the same lines as for the Cox model and identified no important interactions between treatment and albumin ($P = 0.76$) and between treatment and bilirubin ($P = 0.08$).

Like the Cox model, an additive hazard regression model having some or all regression functions constant ($\beta_j(t) = \beta_j$) and an unspecified baseline hazard $\alpha_0(t)$ is *semi-parametric*.

2.2.5 *Additive versus multiplicative models*

From the analyses of the PBC3 data, we have seen that both a Cox model with a non-parametric baseline and time-constant hazard *ratios* and an additive model with a non-parametric baseline and time-constant hazard *differences* provide reasonable fits to the data. This is in spite of the fact that the two models are mathematically incompatible (unless the effects of the quantitative covariates are null and the baseline hazard is constant) and should be interpreted to the effect that the methods used for investigating lack-of-fit may not be sufficiently effective to detect minor model departures. A potential advantage of using additive hazards models is that the coefficients relate more directly to *absolute differences* in event occurrence than coefficients from multiplicative models for which the impact on absolute event occurrence depends strongly on the magnitude of the baseline hazard.

The multiplicative models assuming either a non-parametric baseline (Cox) or a piece-wise constant baseline (Poisson) were quite similar in terms of their estimates, and one may wonder if an additive hazard model with a piece-wise constant baseline could also be studied? The answer is 'yes', however, the algorithms for fitting such a model may be quite sensitive to details concerning starting values to avoid negative predicted hazards during the iterations. A simple model, including only treatment (Z) is

$$\alpha(t) = \alpha_0(t) + \beta_1 Z$$

Table 2.11 *PBC3 trial in liver cirrhosis: Estimated time-constant coefficients and SD (per year) from an additive hazard model with piece-wise constant baseline hazard including treatment and linear effects of albumin and bilirubin.*

Covariate	$\widehat{\beta}$	SD
Treatment	-0.050	0.062
Albumin	-0.0083	0.0048
Bilirubin	0.0020	0.00064

where

$$\alpha_0(t) = \begin{cases} \alpha_1 & \text{if } t < 2, \\ \alpha_2 & \text{if } 2 \leq t < 4, \\ \alpha_3 & \text{if } 4 \leq t. \end{cases}$$

The estimates in this model are: $\widehat{\alpha}_1 = 0.091\,(0.016), \widehat{\alpha}_2 = 0.132\,(0.024), \widehat{\alpha}_3 = 0.094\,(0.046)$ (all rates *per year*, quite close to the similar estimates from the multiplicative Poisson model) and $\widehat{\beta}_1 = -0.0073\,(0.021)$ (per year, quite close to the insignificant treatment effect in the Aalen model with a time-constant hazard difference). When including the two biochemical variables, however, the convergence of the algorithm is questionable but nevertheless leads to estimates quite similar to those from the time-constant Aalen model but with larger estimated standard deviations, see Table 2.11.

2.3 Delayed entry

So far, in this chapter, we have shown how models for a single intensity may be set up, analyzed, and checked. As illustrating example, the PBC3 trial was used with focus on the composite end-point 'failure of medical treatment'. The PBC3 trial was a randomized trial and, therefore, the time origin for the intensity models was naturally chosen to be the time of randomization and all subjects were observed and followed from this time origin. Recall Example 1.1.2 in which children in Guinea-Bissau were visited by a mobile team and followed for about six months at which time deaths and migrations in the intermediate period were ascertained at a second visit by the team. To relate the mortality rate to vaccinations at baseline, a Cox model may be used with time at first visit as the time origin and censoring for emigrations and end of follow-up (alive at time of second visit). Table 2.12 shows estimated regression coefficients in a model including an indicator for being BCG vaccinated at baseline and age at baseline as a quantitative variable. It is seen that BCG vaccination is associated with a reduced mortality rate while age at baseline is formally statistically insignificant. However, because of its confounding effect, adjustment for age at baseline may still be reasonable when estimating the BCG effect: Unadjusted for age, the latter is $\widehat{\beta} = -0.282\,(0.135)$, somewhat different from the adjusted estimate.

Figure 2.13 shows the estimated cumulative hazard for an unvaccinated child aged 3 months at baseline. It is seen that the curve is roughly linear and, therefore, the estimated baseline hazard is a roughly constant function of follow-up time. This illustrates a general feature of observational studies like the Bissau study, namely that, as discussed in Section 1.2.1, recruitment into such a study is not an important event in the life of the participants. Therefore, time since recruitment (follow-up time in the Bissau study) may be unlikely to affect

Table 2.12 *Guinea-Bissau childhood vaccination study: Estimated coefficients (and SD) from Cox models using either follow-up time or age as the time-variable.*

		Follow-up time		Age	
Covariate		$\widehat{\beta}$	SD	$\widehat{\beta}$	SD
BCG	yes vs. no	-0.353	0.144	-0.356	0.141
Age	months	0.055	0.038		

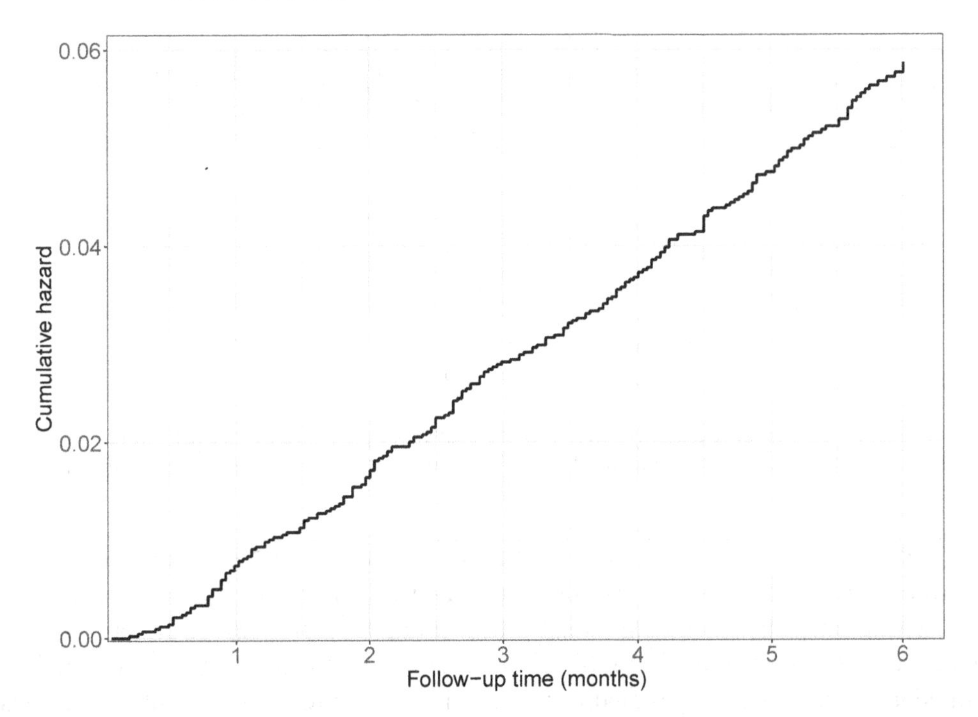

Figure 2.13 *Guinea-Bissau childhood vaccination study: Estimated cumulative hazard in a Cox model for a BCG-unvaccinated child aged 3 months at baseline. Time origin is time of first visit.*

the intensity of future events and to use the non-parametric baseline hazard in a Cox model to model its effect may not be an optimal use of that model's possibilities. In the Bissau study and in many other observational studies, a general alternative to using time since recruitment as the time-variable for intensity models is to use (current) age. That is, the time origin is now time of birth, time of entry is age at entry and time of event or censoring is the corresponding age. Inference in this case has to be performed taking *delayed entry* into account.

It turns out that intensity modeling as discussed earlier in this chapter, carries through with simple modifications. Thus, the following types of model may be studied for the Bissau data. If, for the ith child, a_i is the age at entry then the mortality rate at age $a, a > a_i$ could be $\alpha_1(a)$ if the child was BCG vaccinated before age a_i and $\alpha_0(a)$ otherwise. If these rates are not further specified then their cumulatives can be estimated using the Nelson-Aalen estimator and they may be non-parametrically compared using the logrank test (with delayed entry). In both cases, all that is needed is a modification of the risk set at any age a

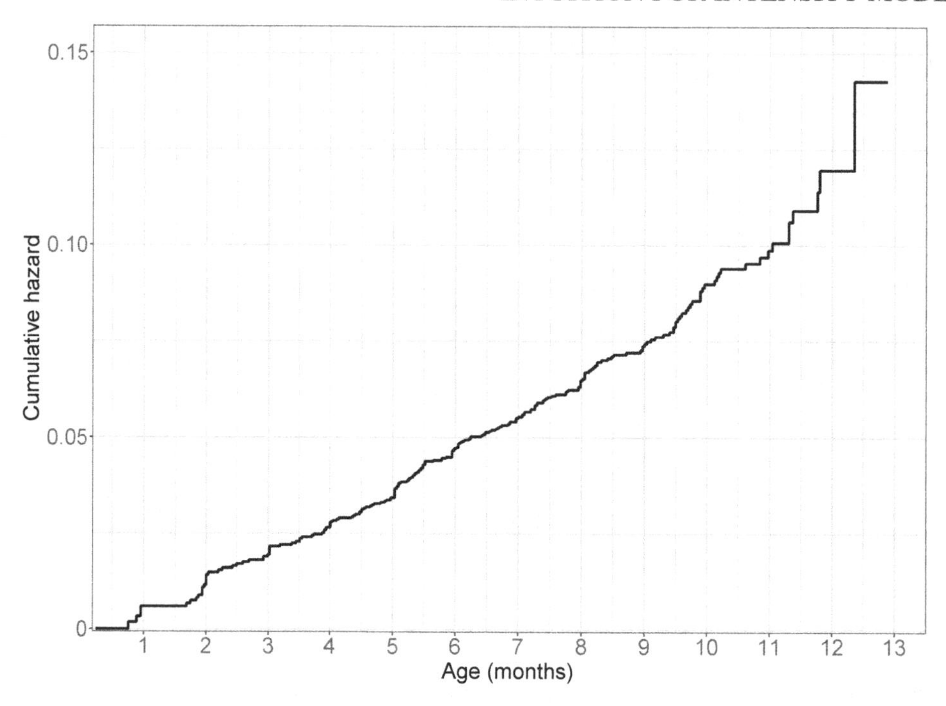

Figure 2.14 *Guinea-Bissau childhood vaccination study: Breslow estimate of the cumulative baseline hazard in a Cox model with current age as time-variable, i.e., time origin is time of birth.*

which should now be those children, i for whom $a_i < a$ and $a_i + X_i \geq a$ where, as previously, X_i is time (since entry) of event or censoring. That is, the risk set at age a includes those subjects who have an age at entry smaller than a and for whom event or censoring has not yet occurred (see Figure 1.8b). Similarly, a Cox model $\alpha_i(a) = \alpha_0(a)\exp(\beta Z_i)$ may be fitted using the same risk set modification, and inference for a model with a piece-wise constant intensity should be based on numbers of events and person-years at risk in suitably chosen age intervals. Also, additive hazard models may be adapted to handle delayed entry using similar risk set modifications.

Table 2.12 also shows the estimated BCG coefficient in a Cox model with age as the time-variable and it is seen to be very similar to that from the model using follow-up time. Figure 2.14 shows the Breslow estimate of the cumulative baseline hazard (i.e., for an unvaccinated child) in this model. The curve appears slightly convex (bending upwards) in accordance with the positive coefficient for age at entry in the Cox model with follow-up time as time-variable (Table 2.12). This figure also illustrates another feature of intensity models for data with delayed entry, namely that, for small values of time (here, ages less that about 1 month) there may be few subjects at risk and, thereby, few or no events. As a consequence, cumulative hazard plots may appear flat (as here) or have big steps. In such situations, one may choose to present plots where the time axis begins in a suitable value $a_0 > 0$ of age where the risk set is sufficiently large – a choice that has no impact of the shape of the curve (i.e., the estimated hazard itself) for values of $a > a_0$.

In Section 1.3, the concept of *independent censoring* was discussed. Recall that the meaning of that concept is that, at any time t, subjects who were censored at that time should have the same failure rate as those who survived beyond t. Handling delayed entry as just described relies on a similar assumption of *independent delayed entry*. The assumption is here that, at any time t, subjects who enter at that time should have the same failure rate as those who were observed to be at risk prior to t. The consequence of independent delayed entry is that, at any time t, the observed risk set at that time is representative for all subjects in the population who are still alive at t.

Time zero

When analyzing multi-state survival data, a time zero must be chosen. For randomized studies, the time of randomization where participants fulfill the relevant inclusion criteria and where treatment is initiated is the standard choice. In clinical follow-up studies there may also be an initiating event that defines the inclusion into the study and may, therefore, serve as time zero. In observational studies, the time of entry into the study is not necessarily a suitable time zero because that date may not be the time of any important event in the life time of the participating subjects. In such cases, an alternative time axis to be used is (current) *age*. For this, subjects are not always followed from the corresponding time zero (age at birth), and data may be observed with *delayed entry*. This means that subjects are only included into the study conditionally on being alive and event-free at the time of entry into the study. Analysis of intensities in the presence of delayed entry requires a modification of the risk sets, and care must be taken when the risk set, as a consequence, becomes small for small values of age.

2.4 Competing risks

We have seen how the intensity of a single event may be analyzed with main emphasis on the composite end-point 'failure of medical treatment' in the PBC3 trial. However, completely similar inference can be made for the two cause-specific hazards in that study, i.e., for the events 'transplantation' and 'death without transplantation'. As explained in the introduction to this chapter, each of these cause-specific hazards may be analyzed by, formally, censoring for the competing cause (and for the avoidable censoring events drop-out and end of study). We will now study Cox models for the PBC3 trial, separately for these two end-points which, together, constitute the composite end-point. Table 2.13 shows the results together with a re-analysis of the composite end-point, now also including the covariates sex and age. These covariates did not have different distributions in the two treatment groups and were not included in previous analyses. It is seen that treatment, sex, albumin, and bilirubin have quite similar effects on the two competing end-points, and these effects are then emphasized when studying the composite end-point. This illustrates what we discussed in Section 1.3, namely that transplantation was primarily offered to patients with a poor prognosis (e.g., low albumin and high bilirubin). However, looking at the age effects we see that the situation is different: High age means a higher hazard of death, but high age is also associated with a lower rate of transplantation. This means that, even though transplantation was primarily offered to patients with a poor prognosis, it was less of an option if the patient was old. Overall, the composite end-point is more common for older

Table 2.13 *PBC3 trial in liver cirrhosis: Estimated coefficients (and SD) from Cox models for death without transplantation, transplantation, and failure of medical treatment, respectively.*

Event type	Covariate		$\widehat{\beta}$	SD
Death without transplantation	Treatment	CyA vs. placebo	-0.420	0.268
	Albumin	per 1 g/L	-0.070	0.029
	$\log_2$(Bilirubin)	per doubling	0.692	0.093
	Sex	male vs. female	-0.486	0.319
	Age	per year	0.073	0.016
Transplantation	Treatment	CyA vs. placebo	-0.673	0.413
	Albumin	per 1 g/L	-0.094	0.039
	$\log_2$(Bilirubin)	per doubling	0.832	0.147
	Sex	male vs. female	-0.204	0.563
	Age	per year	-0.048	0.021
Failure of medical treatment	Treatment	CyA vs. placebo	-0.510	0.223
	Albumin	per 1 g/L	-0.071	0.023
	$\log_2$(Bilirubin)	per doubling	0.738	0.078
	Sex	male vs. female	-0.585	0.267
	Age	per year	0.031	0.012

patients because there were more deaths than transplantations. Such patterns can of course only be observed when actually separating the components of the composite end-point.

Note that, because the hazard for the composite end-point, $\alpha(t)$ is the sum $\alpha_1(t)+\alpha_2(t)$ of the cause-specific hazards, a Cox model for $\alpha(t)$ may be mathematically incompatible with Cox models for the cause-specific hazards. However, such a potential lack of fit may not be sufficiently serious to invalidate conclusions.

Readers familiar with the Fine-Gray regression model may wonder why that model is not discussed under this section's heading of competing risks. The explanation is that the Fine-Gray model is not a hazard-based model, and we will, therefore, postpone the discussion of this model to Chapters 4 and 5 where the focus is on models for marginal parameters, such as the cumulative incidence function.

2.5 Recurrent events

Similar remarks as those in the previous section go for the other examples introduced in Section 1.1 where each transition intensity may be analyzed separately. This also includes the recurrent events in Examples 1.1.5 and 1.1.6, see Figures 1.4 and 1.5. However, in these examples several transitions correspond to occurrence of the same event – an affective episode (a psychiatric admission) in Example 1.1.5 and a major cardiovascular event or a myocardial infarction in Example 1.1.6. It may seem more natural to model a general admission intensity (from state 0 to state 1 in Figure 1.4) than setting up separate models for the first, second, etc. occurrence of the event (e.g., $0 \to 1, 1 \to 2, \ldots$ transitions in Figure 1.5). We will illustrate these points using the data from both of these examples.

2.5.1 Recurrent episodes in affective disorders

Table 2.14 (left column with heading 'Time since diagnosis') shows estimated regression coefficients for bipolar versus unipolar disease based on separate Cox models for first, second, third, and fourth re-admission. The time-variable was time since initial diagnosis, so, models are fitted with delayed entry since a patient is only at risk for admission no. $h = 1,2,3,4$ after discharge from admission no. $h-1$. It is seen that bipolar patients generally have increased re-admission intensities; however, there is a large variability among the estimates for different values of the number of episodes. As an alternative to using time since initial diagnosis as the time-variable, one may use time since latest discharge from psychiatric hospital (known as *gap time models*) in which case there is no delayed entry. Table 2.14 (right column) also shows estimated regression coefficients for bipolar versus unipolar disease based on separate Cox models for first, second, third, and fourth re-admission using the gap time-variable. Results are quite similar to those obtained using time since initial diagnosis. For both sets of models, we see both a considerable variation among different values of h and an increasing SD with increasing h. The latter is due to the diminishing number of patients at risk for later admissions.

It is, therefore, of interest to try to summarize the effects across repeated events. One way of doing this is via the model

$$\alpha_{01}(t \mid Z) = \alpha_{01,0}(t)\exp(\beta Z)$$

with a single baseline $0 \to 1$ transition intensity and a common effect of the covariate Z ($Z = 1$ for bipolar patients and $Z = 0$ for unipolar patients). Table 2.14 (left column) also shows the resulting $\widehat{\beta}$ summarizing the h-specific estimates into a single effect, and leading to a markedly reduced SD (all re-admissions are studied here, not just the first four episodes). This model is a simple version of what is sometimes referred to as the *AG model*, so named after Andersen and Gill (1982), see, e.g., Amorim and Cai (2015), where re-admissions for any patient are independent of previous episodes. More involved models for these data will be studied in later sections (e.g., Section 3.7.5). A way of accounting for previous episodes is to *stratify*, leading to the model

$$\alpha_{01h}(t \mid Z) = \alpha_{01,h0}(t)\exp(\beta Z)$$

for episode no. $h = 1,2,3,\cdots$ where there are separate baseline hazards ($\alpha_{01,h0}(t)$ for episode no. h) but only a single regression coefficient for Z. This model (known as the *PWP model* after Prentice, Williams and Peterson, 1981, see Amorim and Cai, 2015) is seen to provide a smaller coefficient for Z than the AG model (Table 2.14, bottom line). This is because, by taking the number of previous episodes into account (here, via stratification), some of the discrepancy between bipolar and unipolar patients disappears since the occurrence of repeated episodes is itself affected by the initial diagnosis. Similar models (AG and PWP) may be set up for gap times and the results are also shown in the right column of Table 2.14. These models rely on the assumption that gap times are independent, not only among patients – a standard assumption – but also within patients. The latter assumption may not be reasonably fulfilled and we will in later sections (e.g., Sections 3.7.5 and 3.9) discuss models where this assumption is relaxed. A discussion of the use of gap time models for recurrent events was given by Hougaard (2022), recommending (at least

Table 2.14 *Recurrent episodes of affective disorder: Estimated coefficients (and SD) from Cox models per episode, AG model, and PWP model for bipolar vs. unipolar disease.*

		Time since diagnosis		Gap time model	
Model	Episode	$\widehat{\beta}$	SD	$\widehat{\beta}$	SD
Cox model	1	0.356	0.250	0.399	0.249
	2	0.189	0.260	0.217	0.258
	3	-0.117	0.301	-0.111	0.287
	4	1.150	0.354	0.596	0.318
AG model		0.366	0.094	0.126	0.094
PWP model		0.242	0.112	0.028	0.100

for data from randomized trials) not to use gap time models. A further complication arising when studying recurrent events is that censoring may depend on the number of previous events (see, e.g., Cook and Lawless, 2007; sect. 2.6).

2.5.2 *LEADER cardiovascular trial in type 2 diabetes*

Analyses similar to those discussed in the previous section were presented for recurrent myocardial infarctions (MI) in the LEADER trial (Example 1.1.6) by Furberg et al. (2022). Figure 2.15 shows the Nelson-Aalen estimates for the cumulative hazards, and Table 2.15

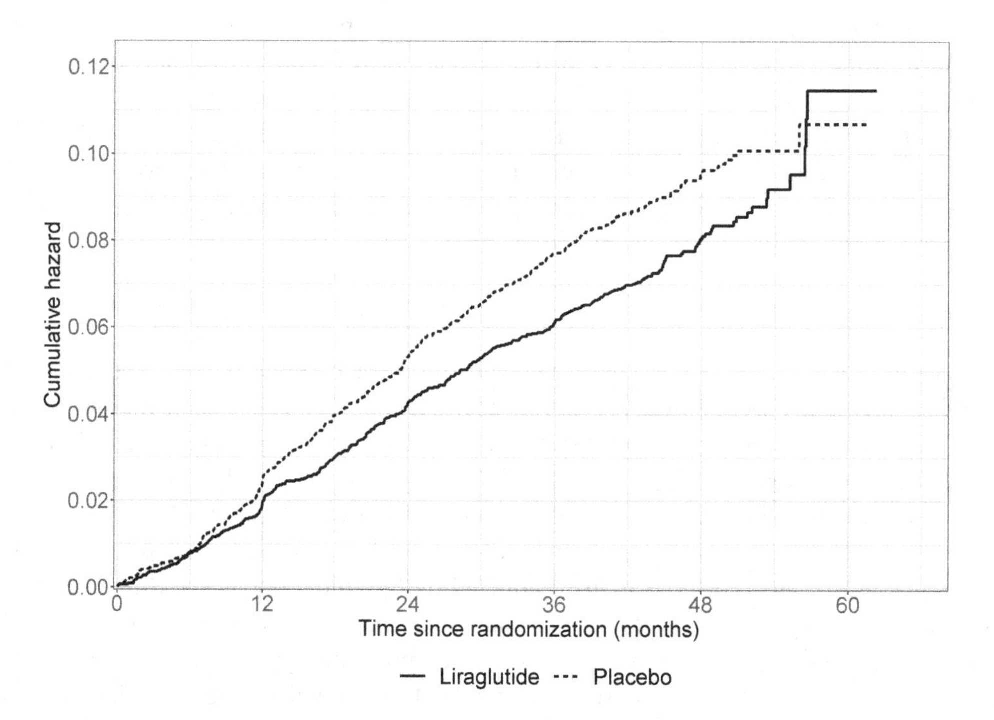

Figure 2.15 *LEADER cardiovascular trial in type 2 diabetes: Estimated cumulative hazards (Nelson-Aalen estimates) of recurrent myocardial infarctions by treatment group.*

Table 2.15 *LEADER cardiovascular trial in type 2 diabetes: Estimated coefficients (and SD) from models for the hazard of recurrent myocardial infarctions for liraglutide vs. placebo.*

Model		$\widehat{\beta}$	SD
Cox model	1st event	-0.159	0.080
AG model	Cox type	-0.164	0.072
	Piece-wise constant	-0.164	0.072
PWP model	2nd event	-0.047	0.197
	3rd event	-0.023	0.400
	4th event	0.629	0.737
	5th event	-0.429	1.230
	All events	-0.130	0.072

summarizes results from regression models. The coefficients for liraglutide versus placebo from a Cox model for time to first event and from an AG model are quite similar; however, with the latter having a smaller SD. Notice that a Cox type model and a model with a piece-wise constant intensity give virtually identical results. As in the previous example, we can see that estimates from PWP-type models with event-dependent coefficients become highly variable, and that the coefficient from a PWP model with a common effect among different event numbers gets numerically smaller compared to that from an AG model. One may argue that the PWP models with event-dependent coefficients are not relevant for estimating the treatment effect in a randomized trial because, for later event numbers, patients are no longer directly comparable between the two treatment groups due to different selection into the groups when the treatment, indeed, has an effect.

Intensity, hazard, rate

The basic parameter in multi-state models is the intensity (also called hazard or rate). The intensity describes what happens locally in time and conditionally on the past among those who are at risk for a given type of event at that time. Specification of all intensities allows simulation of realizations of the multi-state process. Intensities may typically be analyzed one transition at a time using methods as described in this chapter. These include:

- Non-parametric estimation using the Nelson-Aalen estimator.
- Parametric estimation using a model with a piece-wise constant hazard.
- Multiplicative regression models:
  - Cox model with a non-parametric baseline hazard (including the AG model for recurrent events).
  - Poisson model with a piece-wise constant baseline hazard.
- Aalen (additive) regression model.

2.6 Exercises

Exercise 2.1 Consider the data from the Copenhagen Holter study (Example 1.1.8).

1. Estimate non-parametrically the cumulative hazards of death for subjects with or without ESVEA.
2. Make a non-parametric test for comparison of the two.
3. Make a similar analysis based on a model where the hazard is assumed constant within 5-year intervals.

Exercise 2.2 Consider the data from the Copenhagen Holter study.

1. Make a version of the data set enabling an analysis of the composite end-point of stroke or death without stroke ('stroke-free survival', i.e., define the relevant `Time` and `Status` variables), see Section 1.2.4.
2. Estimate non-parametrically the cumulative hazards of stroke-free survival for subjects with or without ESVEA.
3. Make a non-parametric test for comparison of the two.
4. Make a similar analysis based on a model where the hazard is assumed constant within 5-year intervals.

Exercise 2.3 Consider the data from the Copenhagen Holter study and the composite end-point stroke-free survival.

1. Fit a Cox model and estimate the hazard ratio between subjects with or without ESVEA.
2. Fit a Poisson model where the hazard is assumed constant within 5-year intervals and estimate the hazard ratio between subjects with or without ESVEA.
3. Compare the results from the two models.

Exercise 2.4 Consider the data from the Copenhagen Holter study and the composite end-point stroke-free survival.

1. Fit a Cox model and estimate the hazard ratio between subjects with or without ESVEA, now also adjusting for sex, age, and systolic blood pressure (sysBP).
2. Fit a Poisson model where the hazard is assumed constant within 5-year intervals and estimate the hazard ratio between subjects with or without ESVEA, now also adjusting for sex, age, and sysBP.
3. Compare the results from the two models.

Exercise 2.5

1. Check the Cox model from the previous exercise by examining proportional hazards between subjects with or without ESVEA and between men and women.
2. Check for linearity on the log(hazard)-scale for age and sysBP.
3. Do the same for the Poisson model.

Exercise 2.6 Consider the data from the Copenhagen Holter study and focus now on the mortality rate after stroke.

1. Estimate non-parametrically the cumulative hazards for subjects with or without ESVEA using the time-variable 'time since recruitment'.
2. Assume proportional hazards and estimate the hazard ratio between subjects with or without ESVEA.
3. Repeat these two questions using now the time-variable 'time since stroke' and compare the results.

Exercise 2.7

1. Consider the data from the Copenhagen Holter study and fit Cox models for the cause-specific hazards for the outcomes stroke and death without stroke including ESVEA, sex, age, and sysBP.
2. Compare with the results from Exercise 2.4 (first question).

Exercise 2.8 Consider the data on repeated episodes in affective disorder, Example 1.1.5.

1. Estimate non-parametrically the cumulative event intensities for unipolar and bipolar patients.
2. Fit an AG-type model and estimate, thereby, the ratio between intensities for unipolar and bipolar patients, adjusting for year of diagnosis.
3. Fit a PWP model and estimate, thereby, the ratio between intensities for unipolar and bipolar patients, adjusting for year of diagnosis.
4. Compare the results from the two models.

Chapter 3

Intensity models

This chapter gives some of the mathematical details behind the intensity models introduced in Chapter 2. The corresponding sections are marked with '(*)'. The chapter also contains a less technical section dealing with examples (Section 3.6) and, finally, time-dependent covariates (Section 3.7) and models with shared parameters (Section 3.8) are introduced, as well as random effects (*frailty*) models for situations where an assumption of independence among observations may not be justified (Section 3.9).

3.1 Likelihood function (*)

In Section 1.4, we described the way in which data obtained by observing a multi-state process $V(t)$ could be represented using *counting processes*. Thus, for each pair of states $h, j, h \neq j$, the process $N_{hj}(t)$ counts the number of observed direct $h \to j$ transitions in $[0,t]$. Further, the distribution of each $N_{hj}(t)$ is dynamically described by the intensity process

$$\lambda_{hj}(t) \approx E(dN_{hj}(t) \mid \mathscr{H}_{t-})/dt = P(dN_{hj}(t) = 1 \mid \mathscr{H}_{t-})/dt$$

which, under independent censoring, equals

$$\lambda_{hj}(t) = \alpha_{hj}(t)Y_h(t).$$

Here, $Y_h(t) = I(V(t-) = h)$ is the indicator for being in state h just before time t and the *transition intensity* $\alpha_{hj}(t)$ is some non-negative function of the past ($\mathscr{H}_{t-}$) and of time t containing parameters (say, $\boldsymbol{\theta}$) to be estimated. Estimation is based on observing $(N_{hji}(t), Y_{hi}(t))$ over some time interval for n independent subjects $i = 1, \ldots, n$ possibly, together with covariates $\mathbf{Z}_i$ for those subjects. As explained in Section 1.4, the right-hand end-point of the time interval of observation for subject i is either an observed time, C_i of right-censoring or it is a point in time (say, T_i) where the multi-state process $V_i(t)$ reaches an absorbing state, e.g., when subject i dies. We will denote the time interval of observation for subject i by $[0, X_i]$ with $X_i = T_i \wedge C_i$. There may also be *delayed entry* (Sections 1.2 and 2.3), i.e., the observation of subject i begins at a later time point $B_i > 0$ and i is only observed conditionally on not having reached an absorbing state by time B_i. For the moment, we will discuss the case of no delayed entry and return to the general case below. We assume throughout that both censoring and delayed entry are *independent* (Sections 1.3 and 2.3), possibly given relevant covariates.

For a given state space $\mathscr{S} = \{0, 1, \ldots, k\}$ and a given transition structure as indicated by the box-and-arrow diagrams of Chapter 1, there is a certain number of possible transitions and we index these by $v = 1, \ldots, K$. Splitting the time interval of observation, $t \leq X_i$, into (small) sub-intervals, each of length $dt > 0$ we have, for each subject, a (long!) sequence of multinomial experiments

$$\big(1; dN_{1i}(t), \ldots, dN_{Ki}(t), 1 - \sum_v dN_{vi}(t)\big).$$

These correspond to events of type $v = 1, \ldots, K$ or to no event between time t and $t + dt$ and have probability parameters conditionally on the past $\mathscr{H}_{t-}$ given by the intensity processes

$$\big(\lambda_{1i}(t)dt, \ldots, \lambda_{Ki}(t)dt, 1 - \sum_v \lambda_{vi}(t)dt\big).$$

The index parameter of the multinomial distribution equals 1 because, in continuous time, at most 1 event can happen at time t. The probability, conditionally on the past $\mathscr{H}_{t-}$, of observing a given configuration $(dN_1(t), \ldots, dN_K(t))$ of events at time t, where either exactly one of the $dN_v(t)$ equals 1 or they are all equal to 0, is then

$$L_i(t) = \big(\lambda_{1i}(t)dt\big)^{dN_{1i}(t)} \cdots \big(\lambda_{Ki}(t)dt\big)^{dN_{Ki}(t)} \times \big(1 - \sum_v \lambda_{vi}(t)dt\big)^{1 - \sum_v dN_{vi}(t)}$$

and, therefore, the contribution to the *likelihood function* from subject i is the product

$$L_i = \prod_{t \leq X_i} L_i(t)$$

over all such intervals for $t \leq X_i$. There will only be a finite number of intervals with an event and letting $dt \to 0$, the last factor

$$\prod_{t \leq X_i} (1 - \sum_v \lambda_{vi}(t)dt)^{1 - \sum_v dN_{vi}(t)}$$

reduces to the *product-integral* $\exp(-\sum_v \int_0^{X_i} \lambda_{vi}(u)du)$ (Andersen et al., 1993, ch. II). We will discuss the product-integral in more detail in Section 5.1. The likelihood contribution from subject i then becomes

$$L_i = \prod_{t \leq X_i} \big((\lambda_{1i}(t)dt)^{dN_{1i}(t)} \cdots (\lambda_{Ki}(t)dt)^{dN_{Ki}(t)}\big) \exp(-\sum_v \int_0^{X_i} \lambda_{vi}(u)du). \tag{3.1}$$

Equation (3.1) is the *Jacod formula* for the likelihood based on observing a multivariate counting process for subject i, $(N_{vi}(t), t \leq X_i; v = 1, \ldots, K)$ (Andersen et al., 1993, ch. II). For *independent* subjects, the overall likelihood is

$$L = \prod_i L_i.$$

Some brief remarks are in order here; for more details, the reader is referred to Andersen et al. (1993; ch. III).

1. In the case of delayed entry at time $B_i > 0$, the νth observed counting process for subject i is

$$\int_{B_i}^{t} dN_\nu(u) = N_\nu(t) - N_\nu(B_i)$$

and, in order to compute the associated likelihood contribution L_i in Equation (3.1), it must be assumed that the intensity $\lambda_{\nu i}(t)$ for $t > B_i$ does not depend on the unobserved past from the time interval $[0, B_i)$. Such an assumption must be discussed on a case-by-case basis.

2. There may be *covariates*, $\mathbf{Z}_i$ to be included in the model. These could be *time-fixed* covariates, observed at time of entry into the study, and in that case the intensities in the likelihood are conditional on these covariates. Here, it is assumed that the distribution of the covariates carries no information on the parameters of interest (i.e., the $\boldsymbol{\theta}$'s implicitly appearing in the $\alpha_\nu(t)$ functions). There could also be *time-dependent* covariates, $\mathbf{Z}_i(t)$ that are functions of the past history $\mathscr{H}_{t-}$ including covariates observed at entry – known as *adapted* time-dependent covariates (see Section 3.7.1). In some cases there are (non-adapted) time-dependent covariates that depend on other information on each subject in which case the likelihood is more complicated. This and other aspects of time-dependent covariates will be discussed in Section 3.7.

3. We assume throughout that censoring is *independent*, i.e., (Section 1.3) the multi-state process is conditionally independent of censoring times for given covariates. Under this assumption, observation of censoring times gives rise to factors in the likelihood, and the factors arising from observation of the (censored) multi-state process (the Jacod formula) will depend correctly on the parameter $\boldsymbol{\theta}$ from the complete population.

4. It has been assumed in (3.1) that the factor in the likelihood arising from observation of the time of censoring, C_i for subject i does not depend on the parameters, $\boldsymbol{\theta}$. Following Kalbfleisch and Prentice (1980, ch. 5), we will denote this assumption *non-informative censoring*. Note the difference between this concept and that of independent censoring discussed in Section 1.3. What was there defined as independent censoring is in some texts, somewhat confusingly, denoted non-informative censoring, but we have, like Andersen et al. (1993, ch. III), followed the notation introduced by Kalbfleisch and Prentice. Similar remarks go for the observation of a potential time, B_i of delayed entry.

5. The likelihood (3.1) *factorizes* into a product over transition types ν

$$\begin{aligned} L_i &= \prod_{t \le X_i} \left((\lambda_{1i}(t)dt)^{dN_{1i}(t)} \cdots (\lambda_{Ki}(t)dt)^{dN_{Ki}(t)}\right) \times \exp(-\sum_\nu \int_0^{X_i} \lambda_{\nu i}(u)du) \\ &= \prod_\nu \left(\exp(-\int_0^{X_i} \lambda_{\nu i}(u)du) \prod_{t \le X_i} (\lambda_{\nu i}(t)dt)^{dN_{\nu i}(t)}\right). \end{aligned} \quad (3.2)$$

This has the consequence that data may be analyzed by considering *one transition at a time* (unless some parameters of interest are *shared* between several transition types – a situation that will be discussed in Section 3.8).

6. For the special case of survival data (Figure 1.1) there is only one type of event and the index ν may be omitted. There is also at most one event for subject i (at time X_i) and in

this case the likelihood contribution from subject i reduces to

$$L_i = \exp\big(-\int_0^{X_i} \lambda_i(u)du\big)\big(\lambda_i(X_i)\big)^{dN_i(X_i)}.$$

With $D_i = dN_i(X_i)$, and f_i and S_i, respectively, the density and survival function for the survival time T_i, this becomes

$$L_i = f_i(X_i)^{D_i} S_i(X_i)^{(1-D_i)}.$$

It is seen that, for each v, the factor (3.2) in the general multi-state likelihood has the same structure as the likelihood contribution from a subject, i in the two-state model (however, possibly with more than one jump for each subject).

Likelihood factorization

Intensities are modeled in continuous time and at most one event can happen at any given time. This has the consequence that the likelihood function (the Jacod formula) *factorizes* and intensities may be modeled one at a time.

Under the assumption of *independent censoring*, the multi-state process is conditionally independent of the censoring times given covariates. This has the consequence that, as long as the covariates that create conditional independence are accounted for, the censoring times will also give rise to separate factors in the likelihood, and intensities in the multi-state process may be analyzed without specifying a model for censoring.

As a consequence of the likelihood factorization, models for the $\alpha_v(t)$ can be set up for one type of transition v ($\sim h \to j$) at a time. Possible such models (to be discussed in detail in the following sections) include (3.3)-(3.6) below, where for ease of notation we have dropped the index v. First,

$$\alpha(\cdot) \text{ is completely unspecified} \tag{3.3}$$

(Section 3.2). Here, the time-variable needs specification, the major choices being the baseline time-variable t, or current duration in the state h, i.e., $d = t - T_h$ where T_h (as in Section 1.4) denotes the time of entry into state h. In both cases, $\alpha(\cdot)$ may depend on categorical covariates by stratification, and we shall see in Section 3.2 how non-parametric estimation of the cumulative hazard $A(t) = \int_0^t \alpha(u)du$ (using the *Nelson-Aalen estimator*) is performed.

$$\alpha(\cdot) \text{ is piece-wise constant} \tag{3.4}$$

(Section 3.4). The time-variable needs specification and, in contrast to (3.3), it is now assumed that time is split into a number, L, of intervals given by $0 = s_0 < s_1 < \cdots < s_L = \infty$ and in each interval $\alpha(\cdot)$ is constant

$$\alpha(u) = \alpha_\ell \text{ for } s_{\ell-1} \le u < s_\ell.$$

The likelihood, as we shall see in Section 3.4, now leads to estimating α_ℓ by *occurrence/exposure rates*. As it was the case for (3.3), the hazard may depend on a categorical covariate using stratification.

$$\alpha(\cdot \mid \mathbf{Z}) = \alpha_0(\cdot)\exp(\boldsymbol{\beta}^\mathsf{T}\mathbf{Z}) \text{ Cox regression model} \tag{3.5}$$

(Section 3.3). In (3.5), the hazard is allowed to depend on time-fixed *covariates*, $\mathbf{Z}$ assuming that all hazards are *proportional* and the *baseline hazard*, $\alpha_0(\cdot)$ is completely unspecified (as in (3.3)). The resulting *Cox regression model* is said to be *semi-parametric* because it contains both the finite parameter vector $\boldsymbol{\beta} = (\beta_1, \ldots, \beta_p)^\mathsf{T}$ and the non-parametric component $\alpha_0(\cdot)$. This model may be generalized by allowing the covariates to depend on time. Thus, in a recurrent events process (Example 1.1.5) the intensity of events can be modeled as depending on the number of previous events, see Section 3.7. Though the Cox model (3.5) is the most frequently used regression model, alternatives do exist. Thus (3.4) and (3.5) may be combined into a multiplicative hazard regression model with a piece-wise constant baseline hazard, known as *piece-wise exponential* or *Poisson* regression (Section 3.4). Finally, as an alternative to the multiplicative model (3.5), an *additive* hazard model

$$\alpha(\cdot \mid \mathbf{Z}) = \alpha_0(\cdot) + \boldsymbol{\beta}(\cdot)^\mathsf{T}\mathbf{Z} \tag{3.6}$$

may be studied (Section 3.5). Here, all *regression functions* $\beta_j(\cdot)$ may be time-dependent (leading to the *Aalen model*) or some or all of the $\beta_j(\cdot)$ may be time-constant. The baseline hazard $\alpha_0(\cdot)$ is typically taken to be unspecified like in the Cox model, an alternative being to assume it to be piece-wise constant.

3.2 Non-parametric models (*)

3.2.1 Nelson-Aalen estimator (*)

In this section, we consider a single type of event and the model for the intensity process for the associated counting process $N_i(t)$ is

$$\lambda_i(t) = \alpha(t)Y_i(t)$$

with $Y_i(t) = I(B_i < t \leq X_i)$ for some entry time $B_i \geq 0$. That is, there is a common hazard, $\alpha(t)$ for all subjects and this hazard is not further specified. In other words, we have a *non-parametric model* for the intensity. The likelihood based on observing independent subjects $i = 1, \ldots, n$ is

$$L = \prod_i \prod_t (\alpha(t)Y_i(t))^{dN_i(t)} \exp(-\int_0^\infty \alpha(u)Y_i(u)du),$$

leading to the log-likelihood

$$\log(L) = \sum_i \Big(\int_0^\infty \log\big(\alpha(t)Y_i(t)\big)dN_i(t) - \int_0^\infty \alpha(t)Y_i(t)dt\Big). \tag{3.7}$$

Here, integrals are written with a lower limit of 0 and an upper limit equal to ∞ because the indicators $Y_i(t)$ take care of the proper range of integration, $(B_i, X_i]$ for subject i. Formally, the derivative with respect to a single $\alpha(t)$ is

$$\sum_i (\frac{dN_i(t)}{\alpha(t)} - Y_i(t)dt) \tag{3.8}$$

leading to the *score equation*

$$\sum_i \big(dN_i(t) - Y_i(t)\alpha(t)dt\big) = 0 \tag{3.9}$$

with the solution

$$\widehat{\alpha(t)dt} = \frac{\sum_i dN_i(t)}{\sum_i Y_i(t)}.$$

Thus, the *Nelson-Aalen estimator* for the *cumulative hazard* $A(t) = \int_0^t \alpha(u)du$ is

$$\widehat{A}(t) = \int_0^t \frac{\sum_i dN_i(u)}{\sum_i Y_i(u)} \tag{3.10}$$

and it has a *maximum likelihood interpretation.* Note that $\sum_i Y_i(t)$ is the number of subjects observed to be at risk at time t and if X is an observed event time for $(N_1(t), \ldots, N_n(t))$ then, for each such time, a term $1/(\sum_i Y_i(X))$ is added to the estimator. On a plot of $\widehat{A}(t)$ against t, the approximate local slope estimates the hazard at that point in time, see Figure 2.1 for an example. Note that this slope may be estimated directly by *smoothing* the Nelson-Aalen estimator (e.g., Andersen et al., 1993; ch. IV).

3.2.2 *Inference (*)*

If we let $N = \sum_i N_i$, $Y = \sum_i Y_i$ then, by the Doob-Meyer decomposition (1.21) we have that

$$\begin{aligned}
&\widehat{A}(t) - \int_0^t I(Y(u) > 0)\alpha(u)du \\
&= \int_0^t I(Y(u) > 0)\frac{\alpha(u)Y(u)du + dM(u)}{Y(u)} - \int_0^t I(Y(u) > 0)\alpha(u)du \\
&= \int_0^t I(Y(u) > 0)\frac{dM(u)}{Y(u)}
\end{aligned}$$

showing that the Nelson-Aalen estimator minus its target parameter (slightly modified because of the possibility that the risk set may become empty, i.e., $Y(u) = 0$) is a martingale integral and, thereby, itself a martingale (Andersen et al., 1993, ch. IV). From this, point-wise confidence limits for $A(t)$ may be based on asymptotic normality for martingales and on the following model-based variance estimator. By formally taking the derivative of (3.8) with respect to $\alpha(s)$, we get 0 for $s \neq t$ and $-dN(t)/\alpha(t)^2$ for $s = t$. Plugging in the Nelson-Aalen increments $dN(t)/Y(t)$, this leads to the following estimator for $\text{var}(\widehat{A}(t))$

$$\widehat{\sigma}^2(t) = \int_0^t \frac{dN(u)}{(Y(u))^2}. \tag{3.11}$$

Point-wise confidence limits based on (3.11) typically build on symmetric confidence limits for $\log(A(t))$, i.e., a 95% confidence interval for $A(t)$ is

$$\text{from } \widehat{A}(t)\exp(-1.96\widehat{\sigma}(t)/\widehat{A}(t)) \text{ to } \widehat{A}(t)\exp(1.96\widehat{\sigma}(t)/\widehat{A}(t)).$$

Simultaneous *confidence bands* can also be constructed (e.g., Andersen et al., 1993; ch. IV).

As mentioned in Section 3.1, stratification according to a categorical covariate, Z is possible and separate Nelson-Aalen estimators can be calculated for each category. Comparison among two or more cumulative hazards may be performed using the *logrank test* or other non-parametric tests, as follows. For the two-sample test, let $\widehat{A}_j(t) = \int_0^t dN_j(u)/Y_j(u)$ be the Nelson-Aalen estimator in group $j = 0, 1$, where $N_j = \sum_{i \text{ in group } j} N_i$ counts failures in group j and $Y_j = \sum_{i \text{ in group } j} Y_i$ is the number at risk in that group. A general class of non-parametric test statistics for the null hypothesis $H_0 : A_0(u) = A_1(u), u \leq t$ can then be based on the process

$$\int_0^t K(u)\left(d\widehat{A}_1(u) - d\widehat{A}_0(u)\right), \tag{3.12}$$

where $K(\cdot)$ is some weight function of the observations in the interval $[0, t)$ which is 0 whenever $Y_1(u) = 0$ or $Y_0(u) = 0$. Under H_0 and using (1.21), the process (3.12) reduces to the martingale

$$\int_0^t K(u)\left(\frac{dM_1(u)}{Y_1(u)} - \frac{dM_0(u)}{Y_0(u)}\right),$$

whereby conditions for asymptotic normality under H_0 of the test statistic may be found (Andersen et al., 1993; ch. V), see Exercise 3.1.

The most common choice of weight function is

$$K(t) = \frac{Y_0(t)Y_1(t)}{Y_0(t) + Y_1(t)}$$

leading to the logrank test

$$\text{LR}(t) = N_1(t) - \int_0^t \frac{Y_1(u)}{Y_0(u) + Y_1(u)}(dN_0(u) + dN_1(u)).$$

Evaluated at $t = \infty$, this has the interpretation as

$$\text{LR}(\infty) = \text{'Observed' - 'Expected' (in group 1)},$$

as explained in Section 2.1.3.

The statistic $\text{LR}(\infty)$ is normalized using the (hypergeometric) variances v_i

$$(dN_0(X_i) + dN_1(X_i))\frac{Y_0(X_i)Y_1(X_i)}{(Y_0(X_i) + Y_1(X_i))^2}\frac{Y_0(X_i) + Y_1(X_i) - dN_0(X_i) - dN_1(X_i)}{Y_0(X_i) + Y_1(X_i) - 1}$$

added across all observation times (X_i) to give v. Note that the last factor in v_i equals 1 when exactly 1 failure is observed at X_i, i.e., when $dN_0(X_i) + dN_1(X_i) = 1$. The resulting two-sample logrank test statistic to be evaluated in the χ_1^2-distribution is

$$\frac{\text{LR}(\infty)^2}{v}.$$

The logrank test is most powerful against *proportional hazards* alternatives, i.e., when $\alpha_1(t) = \text{HR}\alpha_0(t)$ for some constant *hazard ratio* HR, but other choices of weight function $K(t)$ provide non-parametric tests with other power properties.

Along the same lines, non-parametric tests for comparison of intensities $\alpha_j(t)$ among $m > 2$ groups may be constructed, as well as a stratified tests (Andersen et al., 1993, ch. V).

3.3 Cox regression model (*)

Often, analyses of multi-state models involve several covariates, in which case stratification as discussed in Section 3.2, is no longer feasible and some model specification of how the covariates affect the hazard is needed. In the *Cox regression model* (Cox, 1972) this specification is done using *hazard ratios* in a multiplicative model while, at the same time, keeping the way in which time affects the hazard unspecified. Thus, in the Cox model the hazard function for subject i with covariates $\mathbf{Z}_i = (Z_{i1}, \ldots, Z_{ip})^\mathsf{T}$ is

$$\alpha(t \mid \mathbf{Z}_i) = \alpha_0(t)\exp(\boldsymbol{\beta}^\mathsf{T}\mathbf{Z}_i), \tag{3.13}$$

where the *baseline* hazard $\alpha_0(t)$ is not further specified and the effect of the covariates is via the *linear predictor* $\text{LP}_i = \boldsymbol{\beta}^\mathsf{T}\mathbf{Z}_i$ involving p *regression coefficients* $\boldsymbol{\beta} = (\beta_1, \ldots, \beta_p)^\mathsf{T}$ (Section 1.2.5). Inference for the baseline hazard and regression coefficients builds on the Jacod formula

$$L_i = \exp\left(-\int_0^\infty Y_i(t)\alpha_0(t)\exp(\boldsymbol{\beta}^\mathsf{T}\mathbf{Z}_i)dt\right)\prod_t \left(Y_i(t)\alpha_0(t)\exp(\boldsymbol{\beta}^\mathsf{T}\mathbf{Z}_i)\right)^{dN_i(t)}$$

for the likelihood contribution from subject i (Section 3.1). This leads to the total log-likelihood

$$\sum_i \left(\int_0^\infty \log\left(Y_i(t)\alpha_0(t)\exp(\boldsymbol{\beta}^\mathsf{T}\mathbf{Z}_i)\right)dN_i(t) - \int_0^\infty Y_i(t)\alpha_0(t)\exp(\boldsymbol{\beta}^\mathsf{T}\mathbf{Z}_i dt\right)$$

and differentiation with respect to a single $\alpha_0(t)$ (along the same lines as were used when deriving the Nelson-Aalen estimator in Section 3.2.1) leads to the *score equation*

$$\sum_i \left(dN_i(t) - Y_i(t)\alpha_0(t)\exp(\boldsymbol{\beta}^\mathsf{T}\mathbf{Z}_i)dt\right) = 0. \tag{3.14}$$

For fixed $\boldsymbol{\beta}$, this has the solution

$$\widehat{\alpha_0(t)}dt = \frac{\sum_i dN_i(t)}{\sum_i Y_i(t)\exp(\boldsymbol{\beta}^\mathsf{T}\mathbf{Z}_i)} \tag{3.15}$$

which is identical to the Nelson-Aalen increments in the case of no covariates ($\boldsymbol{\beta} = \mathbf{0}$). Inserting (3.15) into the likelihood $\prod_i L_i$ yields the *profile likelihood*

$$\text{PL}(\boldsymbol{\beta}) \times \exp\left(-\int_0^\infty \sum_i dN_i(t)\right)\prod_t \left(\sum_i dN_i(t)\right)^{\sum_i dN_i(t)}$$

where the first factor

$$\text{PL}(\boldsymbol{\beta}) = \prod_i \prod_t \left(\frac{Y_i(t)\exp(\boldsymbol{\beta}^\mathsf{T}\mathbf{Z}_i)}{\sum_j Y_j(t)\exp(\boldsymbol{\beta}^\mathsf{T}\mathbf{Z}_j)}\right)^{dN_i(t)} \tag{3.16}$$

is the *Cox partial likelihood* (Cox, 1975) and the second factor does not depend on the $\boldsymbol{\beta}$-parameters. To estimate $\boldsymbol{\beta}$, $\text{PL}(\boldsymbol{\beta})$ is maximized by computing the *Cox score*

$$\frac{\partial}{\partial \boldsymbol{\beta}} \log(\text{PL}(\boldsymbol{\beta})) = \sum_i \int_0^\infty \left(\mathbf{Z}_i - \frac{\sum_j Y_j(t)\mathbf{Z}_j\exp(\boldsymbol{\beta}^\mathsf{T}\mathbf{Z}_j)}{\sum_j Y_j(t)\exp(\boldsymbol{\beta}^\mathsf{T}\mathbf{Z}_j)}\right)dN_i(t) \tag{3.17}$$

and solving the resulting score equation. This leads to the Cox maximum partial likelihood estimator $\widehat{\boldsymbol{\beta}}$ and inserting this into (3.15) yields the *Breslow estimator* of the cumulative baseline hazard $A_0(t) = \int_0^t \alpha_0(u)du$

$$\widehat{A}_0(t) = \int_0^t \frac{\sum_i dN_i(u)}{\sum_i Y_i(u)\exp(\widehat{\boldsymbol{\beta}}^\mathsf{T} \mathbf{Z}_i)} \tag{3.18}$$

(Breslow, 1974). Note that the sums '$\sum Y_j(t)$...' in (3.15)-(3.18) are effectively sums over the *risk set*

$$R(t) = \{j : Y_j(t) = 1\} \tag{3.19}$$

at time t.

Large-sample inference for $\widehat{\boldsymbol{\beta}}$ may be based on standard likelihood results for PL($\boldsymbol{\beta}$). A crucial step is to note that, evaluated at the true regression parameter and considered as a process in t when based on the data from $[0,t]$, (3.17) is a *martingale*. (Andersen and Gill, 1982; Andersen et al., 1993, ch. VII), see Exercise 3.2. Thus, model-based standard deviations of $\widehat{\boldsymbol{\beta}}$ may be obtained from the second derivative of $\log(PL(\boldsymbol{\beta}))$ and the resulting Wald tests (as well as score- and likelihood ratio tests) are also valid. Also, joint large-sample inference for $\widehat{\boldsymbol{\beta}}$ and $\widehat{A}_0(t)$ is available (Andersen and Gill, 1982; Andersen et al., 1993, ch. VII). For a simple model including only a binary covariate, the score test reduces to the logrank test discussed in Section 3.2.2 (see Exercise 3.3). If the simple model is a stratified Cox model (see (3.20) below), the score test reduces to a stratified logrank test.

Since the Cox model (3.13) includes a linear predictor $\text{LP}_i = \boldsymbol{\beta}^\mathsf{T}\mathbf{Z}_i$, there are general methods available for checking some of the assumptions imposed in the model, such as linear effects (on the log(hazard) scale) of quantitative covariates and absence or presence of interactions between covariates. Examination of properties for the linear predictor was exemplified in Section 2.2.2. A special feature of the Cox model that needs special attention when examining goodness-of-fit is that of *proportional hazards* (no interaction between covariates and time). This is because of the non-parametric modeling of the time effect via the baseline hazard $\alpha_0(t)$. We have chosen to collect discussions of general methods for goodness-of-fit examinations for a number of different multi-state models (including the Cox model) in a separate Section 5.7. However, methods for the Cox model are also described in connection with the PBC3 example in Section 2.2.2, and in Section 3.7 examination of the proportional hazards assumption using time-dependent covariates will be discussed.

A useful extension of (3.13) when non-proportional hazards are detected, say, among the levels $j = 1,\dots,m$ of a categorical covariate, Z_0, is the *stratified Cox model*

$$\alpha(t \mid Z_0 = j, \mathbf{Z}) = \alpha_{j0}(t)\exp(\boldsymbol{\beta}^\mathsf{T}\mathbf{Z}), \quad j = 1,\dots,m. \tag{3.20}$$

In (3.20), there is an unspecified baseline hazard for each stratum, S_j given by the level of Z_0, but the same effect of $\mathbf{Z}$ in all strata ('no interaction between Z_0 and $\mathbf{Z}$', though this assumption may be relaxed). Inference still builds on the Jacod formula that now leads to a stratified Cox partial likelihood for $\boldsymbol{\beta}$

$$\text{PL}_s(\boldsymbol{\beta}) = \prod_j \prod_{i \in S_j} \prod_t \Big(\frac{Y_i(t)\exp(\boldsymbol{\beta}^\mathsf{T}\mathbf{Z}_i)}{\sum_{k \in S_j} Y_k(t)\exp(\boldsymbol{\beta}^\mathsf{T}\mathbf{Z}_k)}\Big)^{dN_i(t)} \tag{3.21}$$

and a Breslow estimator for $A_{j0}(t) = \int_0^t \alpha_{j0}(u)du$

$$\widehat{A}_{j0}(t) = \int_0^t \frac{\sum_{i \in S_j} dN_i(u)}{\sum_{i \in S_j} Y_i(u) \exp(\widehat{\boldsymbol{\beta}}^{\mathsf{T}} \mathbf{Z}_i)} \tag{3.22}$$

where, in both (3.21) and (3.22), $i \in S_j$ if $Z_{0i} = j$, see Exercise 3.4.

3.4 Piece-wise constant hazards (*)

An alternative to the non- and semi-parametric models for the hazard studied in Sections 3.2 and 3.3 is a *parametric* model for $\alpha(\cdot)$. For the special case of survival data (Figure 1.1), many parametric models have been studied in the literature, including the exponential, Weibull, and other accelerated failure time models, and the Gompertz model (e.g., Andersen et al., 1993, ch. VI). For general multi-state models, however, these parametric specifications are less frequently used, and for that reason we will in this section restrict attention to the piece-wise constant (or piece-wise exponential) hazard model (3.4) which is, indeed, useful for a general multi-state process. The situation is, as follows. We consider a single type of event and the model for the intensity process for the associated counting process $N_i(t)$ is

$$\lambda_i(t) = \alpha(t)Y_i(t)$$

with $Y_i(t) = I(B_i < t \leq X_i)$ for some entry time $B_i \geq 0$, and $\alpha(t)$ is a common hazard for all subjects. This hazard is specified as

$$\alpha(t) = \alpha_\ell \text{ for } s_{\ell-1} \leq t < s_\ell$$

for a number (L) of intervals given by $0 = s_0 < s_1 < \cdots < s_L = \infty$. Thus, in each interval, $\alpha(\cdot)$ is assumed to be constant. This model typically provides a reasonable approximation to any given hazard, it is flexible and, as we shall see shortly, inference for the model is simple. The model has the drawbacks that the cut-points (s_ℓ) need to be chosen and that the resulting hazard is not a smooth function of time. Smooth extensions of the piece-wise constant hazard model using, e.g., splines have been developed (e.g., Royston and Parmar, 2002) but will not be further discussed here. We will show the estimation details for the case $L = 2$, the general case $L > 2$ is analogous. The starting point is the Jacod formula (3.1), and the (essential part of the) associated log-likelihood (3.7) now becomes

$$\log(L) = \sum_i \left(\sum_{\ell=1,2} \left(\log(\alpha_\ell) \int_{s_{\ell-1}}^{s_\ell} dN_i(t) - \alpha_\ell \int_{s_{\ell-1}}^{s_\ell} Y_i(t)dt \right) \right).$$

With $D_\ell = \sum_i \int_{s_{\ell-1}}^{s_\ell} dN_i(t)$, the total number of observed events in interval ℓ and $Y_\ell = \sum_i \int_{s_{\ell-1}}^{s_\ell} Y_i(t)dt$, the total time at risk observed in interval ℓ, the associated score is

$$\frac{\partial}{\partial \alpha_\ell} \log(L) = \frac{D_\ell}{\alpha_\ell} - Y_\ell$$

leading to the *occurrence/exposure rate*

$$\widehat{\alpha}_\ell = \frac{D_\ell}{Y_\ell}$$

being the maximum likelihood estimator. Standard large sample likelihood inference techniques can be used to show that the pair $(\widehat{\alpha}_1, \widehat{\alpha}_2)^{\mathsf{T}}$ is asymptotically normal with the proper mean and a covariance matrix based on the derivatives of the score which is estimated by

$$\begin{pmatrix} \frac{D_1}{Y_1^2} & 0 \\ 0 & \frac{D_2}{Y_2^2} \end{pmatrix}.$$

A crucial step in this derivation is to notice that the score $\partial \log(L)/\partial \alpha_\ell$ is a martingale when evaluated at the true parameter values and considered as a process in t, when based on data in $[0,t]$ (Andersen et al., 1993, ch. VI).

By the delta-method, it is seen that the $\log(\widehat{\alpha}_\ell)$ are asymptotically normal with mean $\log(\alpha_\ell)$ and a standard deviation which may be estimated by $1/\sqrt{D_\ell}$. Furthermore, the different $\log(\widehat{\alpha}_\ell)$ are asymptotically independent. This result is used for constructing 95% confidence limits for α_ℓ which become

$$\text{from } \widehat{\alpha}_\ell \exp(-1.96/\sqrt{D_\ell}) \text{ to } \widehat{\alpha}_\ell \exp(1.96/\sqrt{D_\ell}).$$

Comparison of piece-wise constant hazards in m strata, using the same partition of time, $0 = s_0 < s_1 < \cdots < s_L = \infty$ in all strata may be performed via the *likelihood ratio test* which, under the null hypothesis of equal hazards in all strata, follows an asymptotic $\chi^2_{L(m-1)}$-distribution.

The model with a piece-wise constant hazard can also be used as baseline hazard in a multiplicative (or additive – see Section 3.5) hazard regression model. The resulting multiplicative model is

$$\alpha(t \mid \mathbf{Z}_i) = \alpha_0(t)\exp(\boldsymbol{\beta}^{\mathsf{T}}\mathbf{Z}_i),$$

with $\alpha_0(t) = \alpha_{0\ell}$ when $s_{\ell-1} \leq t < s_\ell$, $\ell = 1, \ldots, L$. Since any baseline hazard function can be approximated by a piece-wise constant function, the resulting *Poisson* or *piece-wise exponential* regression model is closely related to a Cox model, and we showed in Section 2.2.1 that (for the PBC3 data) results from the two types of model were very similar. The parameters is this model are estimated via the likelihood function (3.1), and hypothesis testing for the parameters is also based on this.

The likelihood simplifies if $\mathbf{Z}_i$ only consists of *categorical variables*, i.e., if there exist finite sets $\mathscr{C}_1, \ldots, \mathscr{C}_p$ such that $Z_{ij} \in \mathscr{C}_j$, $j = 1, \ldots, p$. In that case, $\mathbf{Z}_i$ takes values $\boldsymbol{c}$, say, in the finite set $\mathscr{C} = \mathscr{C}_1 \times \cdots \times \mathscr{C}_p$. Letting $\theta_{\boldsymbol{c}} = \exp(\boldsymbol{\beta}^{\mathsf{T}}\mathbf{Z}_i)$ when $\mathbf{Z}_i = \boldsymbol{c}$, the likelihood function becomes

$$\begin{aligned} &\prod_i \Big(\prod_t (\lambda_i(t))^{dN_i(t)}\Big) \exp\Big(-\int_0^\infty \lambda_i(t)dt\Big) \\ &= \prod_{\ell=1}^{L} \prod_{\boldsymbol{c}\in\mathscr{C}} (\alpha_{0\ell}\theta_{\boldsymbol{c}})^{N_{\ell\boldsymbol{c}}} \exp(-\alpha_{0\ell}\theta_{\boldsymbol{c}} Y_{\ell\boldsymbol{c}}) \end{aligned}$$

with

$$N_{\ell\boldsymbol{c}} = \sum_{i:\mathbf{Z}_i=\boldsymbol{c}} (N_i(s_\ell) - N_i(s_{\ell-1})), \quad Y_{\ell\boldsymbol{c}} = \sum_{i:\mathbf{Z}_i=\boldsymbol{c}} \int_{s_{\ell-1}}^{s_\ell} Y_i(t)dt,$$

the total number of events in the interval from $s_{\ell-1}$ to s_ℓ, respectively, the total time at risk in that interval among subjects with $\mathbf{Z}_i = \boldsymbol{c}$, see Exercise 3.5.

The resulting likelihood is seen to be proportional to the likelihood obtained by, formally, treating $N_{\ell \mathbf{c}}$ as independent Poisson random variables with mean $\alpha_{0\ell}\theta_{\mathbf{c}}Y_{\ell \mathbf{c}}$. This fact is the origin of the name *Poisson regression*, and it has the consequence that parameters may be estimated using software for fitting such a Poisson model. However, since there is no requirement of assuming the $N_{\ell \mathbf{c}}$ to be Poisson distributed, this name has caused some confusion and the resulting model should, perhaps, rather be called *piece-wise exponential regression* since it derives from an assumption of piece-wise constant hazards. Another consequence of this likelihood reduction is that, when fitting the model with only categorical covariates, data may first be summarized in the cross-tables $(N_{\ell \mathbf{c}}, Y_{\ell \mathbf{c}}), \ell = 1,\ldots,L, \mathbf{c} \in \mathscr{C}$. For large data sets this may be a considerable data reduction, and we will use this fact when analyzing the testis cancer incidence data (Example 1.1.3) in Section 3.6.4.

The model, as formulated here, is multiplicative in time and covariates and, thus, assumes proportional hazards. However, since the categorical time-variable and the covariates enter the model on equal footings, examination of proportional hazards can be performed by examining time$\times$covariate interactions. Other aspects of the linear predictor may be tested in the way described in Section 2.2.2.

3.5 Additive regression models (*)

Both the Cox model and the multiplicative Poisson model resulted in *hazard ratios* as measures of the association between a covariate and the hazard function. Other hazard regression models exist and may, as explained in Section 2.2.4, sometimes provide a better fit to a given data set and/or provide estimates with a more useful and direct interpretation. One such class of models is that of *additive hazard* models among which the *Aalen model* (Aalen, 1989; Andersen et al., 1993, ch. VII) is the most frequently used. In this model, the hazard function for a subject with covariates $\mathbf{Z} = (Z_1,\ldots,Z_p)^{\mathsf{T}}$ is given by

$$\alpha(t \mid \mathbf{Z}) = \alpha_0(t) + \beta_1(t)Z_1 + \cdots + \beta_p(t)Z_p. \tag{3.23}$$

Here, both the baseline hazard $\alpha_0(t)$ and the regression functions $\beta_1(t)$, ... , $\beta_p(t)$ are unspecified functions of time, t. The interpretation of the baseline hazard, like the baseline hazard $\alpha_0(t)$ in the Cox model, is the hazard function for a subject with a linear predictor equal to 0, while the value, $\beta_j(t)$, of the jth regression function is the *hazard difference* at time t for two subjects who differ by 1 unit in their values for Z_j and have identical values for the remaining covariates, see Section 2.2.4.

In this model, the likelihood is intractable and other methods of estimation are most often used (though Lu et al., 2023,studied the maximum likelihood estimator under certain constraints). To this end, we define the cumulative baseline hazard and the cumulative regression functions

$$A_0(t) = \int_0^t \alpha_0(u)du, \quad B_j(t) = \int_0^t \beta_j(u)du, \, j = 1,\ldots,p$$

with increments collected in the $(p+1)$-column vector, say,

$$d\mathbf{B}(t) = (dA_0(t), dB_1(t),\ldots,dB_p(t))^{\mathsf{T}}$$

and these functions may be estimated using multiple linear regression, as follows. For each subject $i = 1, \ldots, n$ we represent the outcome as a counting process $N_i(t)$ and collect the increments $dN_i(t)$ in the n-column vector $d\mathbf{N}(t) = (dN_1(t), \ldots, dN_n(t))^\mathsf{T}$. We, further, define the $n \times (p+1)$-matrix $\mathbf{Y}(t)$ with ith row given by $Y_i(t)(1, Z_{i1}, \ldots, Z_{ip})$, and the parameters may be estimated as solutions to the linear regression problem

$$E(d\mathbf{N}(t) \mid \mathscr{H}_{t-}) = \mathbf{Y}(t)d\mathbf{B}(t).$$

An unweighted least squares solution is

$$\widehat{\mathbf{B}}(t) = \int_0^t I(\text{rank}(\mathbf{Y}(u)) = p+1)(\mathbf{Y}(u)^\mathsf{T}\mathbf{Y}(u))^{-1}\mathbf{Y}^\mathsf{T}(u)d\mathbf{N}(u), \tag{3.24}$$

however, more efficient, weighted versions of (3.24) exist (e.g., Martinussen and Scheike, 2006, ch. 5).

Large-sample properties of the estimators may be derived using that

$$\widehat{\mathbf{B}}(t) - \int_0^t I(\text{rank}(\mathbf{Y}(u)) = p+1)d\mathbf{B}(u)$$

is a vector-valued martingale. Thereby, $\text{SD}(\widehat{B}_j(t))$ may be estimated to obtain 95% point-wise confidence limits for $B_j(t)$. Furthermore, hypothesis tests for model reductions, such as $B_j(t) = 0, 0 < t \leq \tau^*$ for a chosen τ^*, may be derived, e.g., based on $\sup_{t \leq \tau^*} |\widehat{B}_j(t)|$ (Martinussen and Scheike, 2006, ch. 5).

The Aalen model is very flexible including a completely unspecified baseline hazard and covariate effects and, as a result, the estimates from the model are entire curves. It may, thus, be of interest to simplify the model, e.g., by restricting some regression functions to be time-constant. The hypothesis of a time-constant hazard difference $\beta_j(t) = \beta_j, 0 < t \leq \tau^*$ may be tested (e.g., using a supremum-based statistic such as $\sup_{t \leq \tau^*} |\widehat{B}_j(t) - (t/\tau^*)\widehat{\beta}_j|$, where $\widehat{\beta}_j$ is an estimate under the hypothesis of a time-constant hazard difference for the jth covariate). In the resulting *semi-parametric* model where some covariate effects are time-varying and other are time-constant, parameters may be estimated as described by Martinussen and Scheike (2006, ch. 5). The ultimate model reduction leads to the additive hazard model

$$\alpha(t \mid \mathbf{Z}) = \alpha_0(t) + \beta_1 Z_1 + \cdots + \beta_p Z_p \tag{3.25}$$

with a non-parametric baseline and time-constant hazard differences, much like the Cox regression model (Lin and Ying, 1994).

The multiplicative models assuming either a non-parametric baseline (Cox) or a piece-wise constant baseline (Poisson) were quite similar in terms of their estimates and, in a similar fashion, an additive hazard model with a piece-wise constant baseline can be studied. This leads to a model like (3.25) but now with $\alpha_0(t) = \alpha_{0\ell}$ when $s_{\ell-1} \leq t < s_\ell$, $\ell = 1, \ldots, L$. This model is fully parametric and may be fitted using maximum likelihood. However, fitting algorithms may be sensitive to starting values, reflecting the general difficulty in relation to additive hazard models that estimated hazards may become negative.

Counting processes and martingales

The mathematical foundation of the models for analyzing intensities is that of *counting processes* and *martingales* (e.g., Andersen et al., 1993).

3.6 Examples

This section presents a series of worked examples to illustrate the models for rates discussed so far. We will first recap the results from the PBC3 trial (Example 1.1.1), next present extended analyses of the childhood vaccination survival data from Guinea-Bissau (Example 1.1.2), and finally discuss Examples 1.1.4 and 1.1.3.

3.6.1 PBC3 trial in liver cirrhosis

The purpose of this trial was to evaluate the effect of treatment with CyA versus placebo on the composite end-point 'failure of medical treatment'. Data from the trial were used extensively in Chapter 2 to illustrate the methods introduced there. In summary, we found in Section 2.2.1 that, unadjusted for other covariates, there was no effect of treatment. However, adjusting for the biochemical variables albumin and $\log_2$(bilirubin), both of which turned out to have less favorable values in the active group, CyA-treated patients were found to have a significantly lower hazard of the composite end-point. Further adjustment for age and sex did not change this conclusion (Section 2.4).

Even though the main scientific question in the PBC3 trial dealt with the composite end-point, further insight could be gained by separate studies of its two components: Death without liver transplantation and liver transplantation. Analyses of the two cause-specific hazards (Section 2.4) showed that, while treatment, albumin, bilirubin and sex had effects on the two outcomes going in the same direction, high age was associated with a higher rate of death and a lower rate of transplantation.

3.6.2 Guinea-Bissau childhood vaccination study

In this study, the purpose of the analyses was to assess how the mortality rate in the 6-month period between visits by a mobile team was associated with vaccinations given before the first visit. In Section 2.3, some initial analyses were presented with main focus on BCG vaccination and the choice of time-variable in the analysis. The preliminary conclusion was that BCG-vaccinated children had a lower mortality rate – both when follow-up time and current age was used as the time-variable in a Cox model – and we concluded that the latter choice of time-variable was preferable because follow-up time was not associated with the mortality rate.

We will now analyze the data to address how the mortality rate depends not only on BCG vaccination but also on vaccination with DTP. As explained in Example 1.1.2, this vaccine was given in three doses, and Table 3.1 shows the joint distribution of children according to the two vaccinations at first visit. It is seen that few children, unvaccinated with BCG, have received any dose of DTP, in other words the two covariates BCG and DTP are highly correlated. This must be borne in mind when concluding from the subsequent analyses.

Table 3.1 *Guinea-Bissau childhood vaccination study: Vaccination status at initial visit among 5,274 children.*

	DTP doses							
BCG	0		1		2		3	
Yes	1159	35.1%	1299	39.4%	582	17.6%	261	7.9%
No	1942	98.4%	19	1.0%	9	0.5%	3	0.2%
Total	3101	58.8%	1318	25.0%	591	11.2%	264	5.0%

Table 3.2 *Guinea-Bissau childhood vaccination study: Estimated coefficients (and SD) from Cox models, using age as time-variable, for vaccination status at initial visit.*

	BCG		Any DTP dose		Interaction	
Model	$\widehat{\beta}$	SD	$\widehat{\beta}$	SD	$\widehat{\beta}$	SD
Only BCG	-0.356	0.141				
Only DTP			-0.039	0.149		
Additive effects	-0.558	0.192	0.328	0.202		
Interaction	-0.576	0.202	0.125	0.718	0.221	0.743

Since relatively few children have received multiple DTP doses, we will dichotomize that variable in the following. Table 3.2 shows estimated regression coefficients from Cox models addressing the (separate and joint) effects of vaccinations on mortality. We have repeated the analysis from the previous chapter where only BCG is accounted for showing a beneficial effect of this vaccine. It it seen that, unadjusted for BCG, there is no association between any dose of DTP and mortality, while, adjusting for BCG, DTP tends to increase mortality, albeit insignificantly, while the effect of BCG seems even more beneficial than without adjustment. In this model, due to the collinearity between the two covariates, standard deviations are inflated compared to the two simple models. Finally, it is seen that there is no important interaction between the effects of the two vaccines on mortality (though the test for this has small power because few children received a DTP vaccination without a previous BCG). An explanation behind these findings could be that BCG vaccination is, indeed, beneficial as seen in the additive model, while DTP tends to be associated with an increased mortality rate. This latter effect is, however, not apparent without adjustment for BCG because most children who got the DTP had already received the 'good' BCG vaccine. Further discussions are provided by Kristensen et al. (2000).

3.6.3 PROVA trial in liver cirrhosis

In this trial, patients were randomized in a two-by-two factorial design to either propranolol, sclerotherapy, both treatments, or no treatment and followed with respect to the two competing end-points variceal bleeding or death without bleeding. The scientific question addressed was how the occurrence of these outcomes was affected by treatment. We can address this question by comparing the event rates among the treatment arms using the non-parametric four-sample logrank test and, further, discrepancies between the treatment

Table 3.3 *PROVA trial in liver cirrhosis: Cox models for the rates of variceal bleeding and death without bleeding.*

(a) *Variceal bleeding*

Covariate		$\widehat{\beta}$	SD	$\widehat{\beta}$	SD
Sclerotherapy	vs. none	0.056	0.392	0.177	0.433
Propranolol	vs. none	-0.040	0.400	0.207	0.424
Both treatments	vs. none	-0.032	0.401	0.031	0.421
Sex	male vs. female			-0.026	0.329
Coagulation factors	% of normal			-0.0207	0.0078
$\log_2$(bilirubin)	μmol/L			0.191	0.149
Medium varices	vs. small			0.741	0.415
Large varices	vs. small			1.884	0.442

(b) *Death without bleeding*

Covariate		$\widehat{\beta}$	SD	$\widehat{\beta}$	SD
Sclerotherapy	vs. none	0.599	0.450	0.826	0.459
Propranolol	vs. none	-0.431	0.570	-0.160	0.575
Both treatments	vs. none	1.015	0.419	0.910	0.420
Sex	male vs. female			0.842	0.416
Coagulation factors	% of normal			-0.0081	0.0068
$\log_2$(bilirubin)	μmol/L			0.445	0.137
Medium varices	vs. small			0.222	0.347
Large varices	vs. small			0.753	0.449

arms may be quantified via hazard ratios from Cox models for each of the two outcomes. The results from the Cox models are summarized in Table 3.3.

The four-sample logrank test statistics for the two outcomes take, respectively, the values 0.071 for bleeding and 12.85 for death without bleeding corresponding to P-values of 0.99 and 0.005. So, for bleeding there are no differences among the treatment groups, while for death without bleeding there are. Inspecting the regression parameters for this outcome in Table 3.3, it is seen that the two groups where sclerotherapy was given have higher death rates. The Cox model with separate effects in all treatment arms can be reduced to a model with additive (on the log-rate scale) effects of sclerotherapy and propranolol (LRT $= 1.63$, 1 DF, $P = 0.20$) and in the resulting model, propranolol is insignificant (LRT $= 0.35$, 1 DF, $P = 0.56$). The regression parameter for sclerotherapy in the final model is $\widehat{\beta} = 1.018\,(\text{SD} = 0.328)$.

The PROVA trial was randomized and, hence, adjustment for prognostic variables should not change these conclusions (though, in the PBC3 trial, Example 1.1.1, such an adjustment did change the estimated treatment effects considerably). Nevertheless, Table 3.3 for illustration also shows treatment effects after such adjustments. Adjustment was made for four covariates, of which two (coagulation factors and size of varices, the latter a three-category variable), are associated with bleeding and two (sex and $\log_2$(bilirubin)) with death without bleeding. However, conclusions concerning treatment effects are not changed by this

Table 3.4 *PROVA trial in liver cirrhosis: Cox model for the composite outcome bleeding-free survival.*

Covariate		$\widehat{\beta}$	SD
Sclerotherapy	vs. none	0.525	0.313
Propranolol	vs. none	0.100	0.338
Both treatments	vs. none	0.495	0.292
Sex	male vs. female	0.360	0.253
Coagulation factors	% of normal	-0.0136	0.0053
$\log_2$(bilirubin)	μmol/L	0.328	0.102
Medium varices	vs. small	0.446	0.263
Large varices	vs. small	1.333	0.301

adjustment, and the same is the case if further adjustment for age is done (results not shown). It could be noted that, from the outset of the trial, the two end-points were considered equally important and, thus, merging them into the composite end-point 'bleeding-free survival' is of interest. However, the results in Table 3.3 suggest that doing so would provide a less clear picture. Estimates are shown in Table 3.4 where the significance of treatment diminishes (LRT with 3 DF: $P = 0.19$) and, among the covariates, sex loses its significance (Wald test: $P = 0.16$), while the three remaining covariates keep their significance. Note also that Cox models for the two separate end-points are mathematically incompatible with a Cox model for the composite end-point.

3.6.4 Testis cancer incidence and maternal parity

The data set available for studying the relationship between maternal parity and sons' rates of testicular cancer (Example 1.1.3) consists of tables of testicular cancer cases (seminomas or non-seminomas) and person-years at risk according to the variables: Current age of the son, birth cohort of the son, parity, and mother's age at the birth of the son. The basis for these tables was a follow-up of sons born to women in Denmark from the birth cohorts 1935-78 and who were either alive in 1968 (when the Danish Civil Registration System was established) or born between 1968 and 1992. The sons were followed from birth or 1968, whatever came last, to a diagnosis of testicular cancer, death, emigration or end of 1992, whatever came first. Using current age as the time-variable in a Poisson regression model for the cancer rate, sons born before 1968 have left-truncated follow-up records, beginning in their age in 1968, whereas sons born after 1968 were followed from birth (age 0, no left-truncation). Figure 3.1 shows a *Lexis diagram* for this follow-up information, i.e., an age by calendar time coordinate system illustrating the combinations of age and calendar time represented in the data set (e.g., Keiding, 1998). The numbers of person-years at risk in combinations of age and birth cohort are given in the diagram showing that the majority of person-time comes from boys aged less than 15 years.

Death of the son was a competing risk for the event of interest, testis cancer. However, as explained in Section 1.1.3, the numbers of deaths in each of the categories of the tables were not part of the available data set. This has the consequence that only the rate of testicular cancer can be analyzed and not the full competing risks model including the death rate.

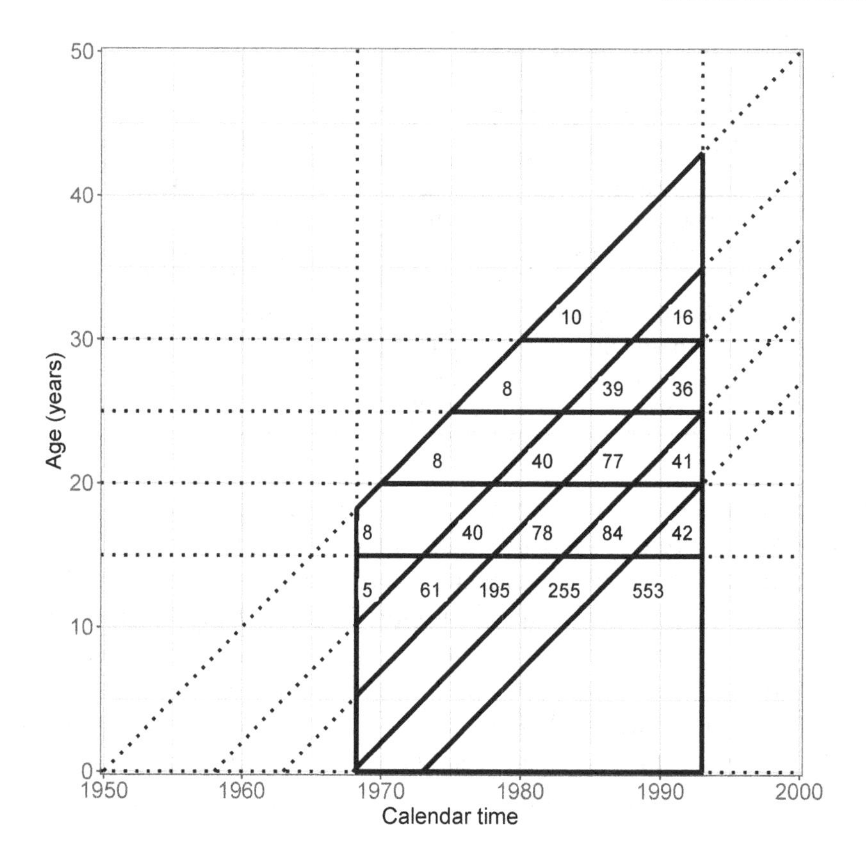

Figure 3.1 *Testis cancer incidence and maternal parity: Lexis diagram showing the numbers of person-years at risk (in units of 10,000 years) for combinations of age and birth cohort.*

Fortunately, due to the likelihood factorization (3.2), and as explained in the introduction to Chapter 2, the available data do allow a correct inference for the rate of testis cancer. Table 3.5 shows estimated log(hazard ratios) from Poisson models with age as the time-variable (in categories 0-14, 15-19, 20-24, 25-29, and 30+ years) and including either parity alone (1 vs. 2+) or parity together with birth cohort and mother's age. It is seen that, unadjusted, firstborn sons have an $\exp(0.217) = 1.24$-fold increased rate of testicular cancer and this estimate is virtually unchanged after adjustment for birth cohort and mother's age. The 95% confidence interval for the adjusted hazard ratio is (1.05, 1.50), $P = 0.01$. The rates increase markedly with age and LR tests for the adjusting factors are LRT= 2.53, 3 DF, $P = 0.47$ for mother's age and LRT= 9.22, 4 DF, $P = 0.06$ for birth cohort of son. It was also studied whether the increased rate for firstborn sons varied among age groups (i.e., a potential interaction between parity and age, non-proportional hazards). This was not the case as seen by an insignificant LR-statistic for interaction (LRT= 7.76, 4 DF, $P = 0.10$).

It was finally studied how the rates of seminomas and non-seminomas, respectively, were associated with parity. The hazard ratios (HRs) in relation to these competing end-points were remarkably similar: HR= 1.23 (0.88, 1.72) for seminomas, HR= 1.27 (1.02, 1.56) for non-seminomas.

Table 3.5 *Testis cancer incidence and maternal parity: Poisson regression models.*

Covariate		$\widehat{\beta}$	SD	$\widehat{\beta}$	SD
Parity	1 vs. 2+	0.217	0.084	0.230	0.091
Age (years)	0-14 vs. 20-24	-4.031	0.211	-4.004	0.239
	15-19 vs. 20-24	-1.171	0.119	-1.167	0.125
	25-29 vs. 20-24	0.560	0.098	0.617	0.104
	30+ vs. 20-24	0.753	0.133	0.954	0.154
Mother's age (years)	12-19 vs. 30+			0.029	0.241
	20-24 vs. 30+			0.058	0.222
	25-29 vs. 30+			-0.117	0.225
Son cohort	1950-57 vs. 1973+			-0.363	0.288
	1958-62 vs. 1973+			-0.080	0.248
	1963-67 vs. 1973+			0.124	0.237
	1968-72 vs. 1973+			0.134	0.236

3.7 Time-dependent covariates

In Chapter 2 and Sections 3.1-3.6, we have studied (multiplicative or additive) hazard regression models. Recall that the hazard function at time t, conditionally on the past, gives the instantaneous probability per time unit of having an event just after time t

$$\alpha_{hj}(t) \approx P(V(t+dt) = j \mid V(t) = h \text{ and the past for } s < t)/dt,$$

if the event is a transition from state h to state j (and $dt > 0$ is 'small'). In the model discussions and examples given so far, the past for $s < t$ only included time-fixed covariates, recorded at the time of study entry. However, one of the strengths of hazard regression models is their ability to also include covariates that are *time-dependent*. Time-dependent covariates can be quite different in nature and, in what follows, we will distinguish between *adapted* and *non-adapted* covariates and, for the latter class, between *internal* (or endogenous) and *external* (or exogenous) covariates. In later Subsections 3.7.5-3.7.8 we will, via examples, illustrate different aspects of hazard regression models with time-dependent covariates.

3.7.1 *Adapted covariates*

We will denote covariates where only aspects of the past $(V(s), s < t)$ are used as time-dependent covariates as *adapted* covariates. Thus, adapted covariates contain no randomness over and above this past (including time-fixed covariates Z, combined with time t). Examples of adapted covariates include:

- Number of events before t in a recurrent events study (e.g., Example 1.1.5).
- Time since entry into the current state h (e.g., time since bleeding in the PROVA trial, Example 1.1.4).
- Current age or current calendar time $Z(t) = Z + t$ if Z is age or calendar time at entry into the study and t time since entry.
- $Z(t) = Z \cdot f(t)$ for some specific function $f(\cdot)$ to model a non-proportional effect of Z.

Another example, not represented in any of our examples from Section 1.1, would be one where

- $Z(t)$ is a pre-specified dose (i.e., planned at time $t = 0$) given to a subject while still alive, e.g., $Z(t) = z_0$ when $0 < t \leq t_0$ and $Z(t) = z_1$ when $t_0 < t$.

Note that, in the last three examples the development of the time-dependent covariate is *deterministic* (possibly given time-fixed covariates).

3.7.2 Non-adapted covariates

Non-adapted time-dependent covariates are those that involve extra randomness, i.e., not represented by the past for $V(t)$. Examples include:

- Repeated measurements of a biochemical marker in a patient, such as repeated recordings of serum bilirubin in the PBC3 trial (Example 1.1.1). The value at time t could be the current value of the marker, lagged values (i.e., the value at time $t - \Delta$, e.g., 1 month ago), or average values over the period from $t - \Delta$ to t.
- Additional vaccinations given during follow-up in the Guinea-Bissau childhood vaccination study (Example 1.1.2).
- Complications (or improvements) that may occur during follow-up of a patient, e.g., obtaining an Absolute Neutrophil Count (ANC) above 500 cells per μL in the bone marrow transplantation study (Example 1.1.7).
- The level of air pollution in a study of the occurrence of asthma attacks.

In the first three examples, the recording of a value of the time-dependent covariate at time t requires that the subject is still alive at that time, while, in the last example, the value of time-dependent covariate exists whether or not a given subject is still alive. The former type of covariate is known as *endogenous* or *internal*, the latter as *exogenous* or *external.*

3.7.3 Inference

Inference for the parameters in the hazard model can proceed along the lines described in earlier sections. For the Cox model, the regression coefficient β can be estimated from the Cox partial log-likelihood

$$l(\beta) = \sum_{\text{event times, } X} \log \left(\frac{\exp(\beta Z_{\text{event}}(X))}{\sum_{j \text{ at risk at time } X} \exp(\beta Z_j(X))} \right)$$

where, at event time X, the covariate values at that time are used for everyone at risk at X. Similarly, the cumulative baseline hazard $A_0(t)$ is estimated by

$$\widehat{A}_0(t) = \sum_{\text{event times, } X \leq t} \frac{1}{\sum_{j \text{ at risk at time } X} \exp(\widehat{\beta} Z_j(X))}.$$

Least squares estimation in the additive Aalen model (Sections 2.2.4 and 3.5) can, likewise, be modified to include time-dependent covariates. The arguments for the Cox model depend on whether the time-dependent covariates are adapted or not. These arguments are outlined in the next section.

The ability to do *micro-simulation* (Section 5.4) depends on whether time-dependent covariates are adapted or not. Micro-simulation of models including non-adapted covariates requires *joint models* for the multi-state process and the time-dependent covariate, and in Section 7.4 we will briefly discuss joint models.

There is a related feature (to be discussed in Sections 4.1 and 5.2) that depends on whether a hazard model includes time-dependent covariates or not. This is the question of whether it is possible to estimate marginal parameters by plug-in. This will typically only be possible with time-fixed covariates, with deterministic time-dependent covariates, or with exogenous covariates, for the latter situation, see Yashin and Arjas (1988).

3.7.4 *Inference (*)*

When $\mathbf{Z}_i(t)$ is *adapted*, the arguments for deriving the likelihood follow those in Sections 3.1 and 3.3. Thus, with a Cox model, $\lambda_i(t) = Y_i(t)\alpha_0(t)\exp(\boldsymbol{\beta}^{\mathsf{T}}\mathbf{Z}_i(t))$ is still the intensity process for $N_i(t)$ with respect to the history $\mathscr{H}_t$ generated by the multi-state process and baseline covariates. This means that the Jacod formula (3.1) is applicable as the starting point for inference and, hence, that jumps in the baseline hazard, for fixed $\boldsymbol{\beta}$, are estimated by

$$\widehat{\alpha_0(t)dt} = \frac{\sum_i dN_i(t)}{\sum_i Y_i(t)\exp(\boldsymbol{\beta}^{\mathsf{T}}\mathbf{Z}_i(t))}, \tag{3.26}$$

and the essential factor in the resulting profile likelihood will still be

$$PL(\boldsymbol{\beta}) = \prod_i \prod_t \left(\frac{Y_i(t)\exp(\boldsymbol{\beta}^{\mathsf{T}}\mathbf{Z}_i(t))}{\sum_j Y_j(t)\exp(\boldsymbol{\beta}^{\mathsf{T}}\mathbf{Z}_j(t))} \right)^{dN_i(t)}, \tag{3.27}$$

which is the Cox partial likelihood with time-dependent covariates.

For *non-adapted* time-dependent covariates, the arguments for deriving a likelihood become more involved. This is because the full likelihood will also include factors resulting from the extra randomness in $\mathbf{Z}_i(t)$ and will no longer be given by the Jacod formula (3.1). It was discussed by Andersen et al. (1993; ch. III) how the likelihood given by (3.1) can still be considered a *partial* likelihood that is applicable for inference and that, for the Cox model, estimation can still be based on (3.26) and (3.27).

Estimates in the *Aalen model* may be based on (3.24) when defining the matrix $\mathbf{Y}(t)$ to have ith row equal to $Y_i(t)(1, Z_{i1}(t), \ldots, Z_{ip}(t))$. This is the case both for adapted and non-adapted covariates.

3.7.5 *Recurrent episodes in affective disorders*

The study was presented in Section 1.1.5, and in Section 2.5, where simple regression models for the association between the initial diagnosis (bipolar vs. unipolar) and the recurrence rate were studied. In an AG-model, the estimated hazard ratio between the two diagnostic groups was $\exp(0.366) = 1.442$ with 95% confidence limits from 1.198 to 1.736, see Table 2.14. In this model, recurrent episodes were assumed to be independent in the sense that the recurrence rate depended on no other aspects of the past than the initial diagnosis. One way of relaxing this assumption is to include the number of previous episodes for subject i,

Table 3.6 *Recurrent episodes in affective disorders: AG models with number of previous episodes, $N(t-)$, as time-dependent covariate.*

Covariate		$\widehat{\beta}$	SD	$\widehat{\beta}$	SD	$\widehat{\beta}$	SD
Bipolar	vs. unipolar	0.366	0.094	0.318	0.095	0.067	0.097
$N(t-)$				0.126	0.0087	0.425	0.032
$N(t-)^2$						-0.0136	0.0016

$N_i(t-)$ as a time-dependent covariate. This is an adapted variable. Since its effect appears highly non-linear ($P < 0.001$ for linearity in a model including both $N_i(t-)$ and $N_i(t-)^2$), we quote the hazard ratio for diagnosis from the model including the quadratic term which is $\exp(0.067) = 1.069$ with 95% confidence limits from 0.884 to 1.293. In a model including only $N_i(t-)$, it is seen that the recurrence rate increases with the number of previous episodes and so does the rate in the quadratic model where the maximum of the estimated parabola for $N_i(t-)$ is $-0.425/(-2 \cdot 0.0136) = 15.6$. Table 3.6 summarizes the results. Including $N_i(t-)$ as a categorical variable instead gives the hazard ratio $\exp(0.090) = 1.094$ (0.904, 1.324). In both cases, a substantial reduction of the hazard ratio is seen when comparing to the value 1.442 from the model without adjustment for $N_i(t-)$. The explanation was given in Section 2.5 in connection with the PWP model where adjustment for previous episodes was carried out using (time-dependent) stratification, namely that the occurrence of repeated episodes is itself affected by the initial diagnosis, and previous episodes, therefore, serve as an *intermediate variable* between the baseline covariate and recurrent episodes. The AG model including functions of $N_i(t-)$ is, thus, more satisfactory in the sense that the independence assumption is relaxed; however, it is less clear if it answers the basic question of comparison between the two diagnostic groups. To answer this question, models for the marginal mean number of events over time may be better suited (Sections 4.2.3 and 5.5.4).

Another time-dependent covariate which may affect the estimate is current calendar period. This is also an adapted covariate for given value of the baseline covariate calendar time at diagnosis. The number of available beds in psychiatric hospitals has been decreasing over time, and if access to hospital varies between the two diagnostic groups, then adjustment for period may give rise to a different hazard ratio for diagnosis. We, therefore, created a covariate by categorizing current period, i.e., calendar time at diagnosis + follow-up time, into the intervals: Before 1965, 1966-70, 1971-75, 1976-80, 1981+, and adjusted for this in the simple AG model including only diagnosis. As seen in Table 3.7, the resulting hazard ratio for diagnosis is only slightly smaller than without adjustment ($\exp(0.361) = 1.435$, 95% c.i. from 1.193 to 1.728). It is also seen that the recurrence rate tends to decrease with calendar period.

3.7.6 *PROVA trial in liver cirrhosis*

The study was presented in Section 1.1.4, and in Section 3.6.3 analyses addressing the main question in this trial were presented. Thus, the rates of variceal bleeding and death without bleeding were related to the given treatment, concluding that treatment arms involving sclerotherapy had a higher mortality rate whereas no effect on the bleeding rate was seen. The course of the patients after onset of the primary end-point, bleeding, was not part of

Table 3.7 *Recurrent episodes in affective disorders: AG model with current calendar period as time-dependent covariate.*

Covariate			$\widehat{\beta}$	SD
Bipolar	vs. unipolar		0.361	0.095
Period	1966-70	vs. 1959-65	-0.251	0.208
	1971-75	vs. 1959-65	-0.179	0.331
	1976-80	vs. 1959-65	-0.367	0.439
	1981+	vs. 1959-65	-1.331	0.554

the basic trial question; however, it is of clinical interest to study the $1 \to 2$ transition rate in the illness-death model of Figure 1.3. For this purpose, a choice of time-variable in the model for the rate $\alpha_{12}(\cdot)$ is needed. For the two-state model for survival analysis and for the competing risks model, a single time origin was assumed and all intensities depended on time t since that origin. The same is the case with the rates $\alpha_{01}(t)$ and $\alpha_{02}(t)$ in the illness-death model. However, for the rate $\alpha_{12}(\cdot)$, both the time-variable t (time since randomization) and time since entry into state 1, duration $d = d(t) = t - T_1$, may play a role. Note the similarity with the choice of time-variable for models for transition intensities in models for recurrent events (Section 2.5). If $\alpha_{12}(\cdot)$ only depends on t, then the multi-state process is said to be *Markovian*; if it depends on d, then it is *semi-Markovian*; see Section 1.4. In the Markovian case, inference for $\alpha_{12}(t)$ needs to take delayed entry into account; if $\alpha_{12}(\cdot)$ only depends on d, then this is not the case.

Results from Cox regression analyses are displayed in Table 3.8. It should be kept in mind when interpreting these results that this is a small data set with 50 patients and 29 deaths (Table 1.2). We first fitted a (Markov) model with t as the baseline time-variable including treatment, sex, and $\log_2$(bilirubin) (top panel (a), first column in Table 3.8). The latter two covariates were not strongly associated with the rate and their coefficients are not shown – the same holds for subsequent models. In this model, the treatment effect was significant (LRT= 11.12, 3 DF, $P = 0.01$) and there was an interaction between propranolol and sclerotherapy (LRT= 6.78, 1 DF, $P = 0.009$). The combined treatment group seems to have a high mortality rate and the group receiving only sclerotherapy a low rate, however, it should be kept in mind that one is no longer comparing randomized groups because the patients entering the analysis are selected by still being alive and having experienced a bleeding – features that may, themselves, be affected by treatment. The Breslow estimate of the cumulative baseline hazard is shown in Figure 3.2 and is seen to increase sharply for small values of t. It should be kept in mind that there is delayed entry and, as a consequence, few patients at risk at early failure times: For the first five failure times the numbers at risk were, respectively, 2, 3, 5, 6, and 6. The Markov assumption corresponds to no effect of duration since bleeding on the mortality rate after bleeding, and this hypothesis may be investigated using adapted time-dependent covariates. Defining $d_i(t) = t - T_{1i}$ where T_{1i} is the time of entry into state 1 for patient i, i.e., his or her time of bleeding, the following two covariates were added to the first model

$$
\begin{aligned}
Z_{i1}(t) &= I(d_i(t) < 5 \text{ days}) \\
Z_{i2}(t) &= I(5 \text{ days} \leq d_i(t) < 10 \text{ days}),
\end{aligned}
$$

Table 3.8 *PROVA trial in liver cirrhosis: Cox models for the rate of death after bleeding using different baseline time-variables.*

(a) *Time since randomization*

Covariate		$\widehat{\beta}$	SD	$\widehat{\beta}$	SD
Sclerotherapy	vs. none	-1.413	0.679	-1.156	0.684
Propranolol	vs. none	-0.115	0.595	-0.024	0.631
Both treatments	vs. none	0.733	0.544	0.425	0.611
$d(t)$ <5 days	vs. $\geq$10 days			2.943	0.739
5 days$\leq d(t)$ <10 days	vs. $\geq$10 days			2.345	0.803

(b) *Duration*

Covariate		$\widehat{\beta}$	SD	$\widehat{\beta}$	SD
Sclerotherapy	vs. none	-0.997	0.643	-1.019	0.650
Propranolol	vs. none	-0.300	0.596	-0.312	0.601
Both treatments	vs. none	0.871	0.514	0.847	0.524
t <1 year	vs. t $\geq$2 years			-0.172	0.910
1 year$\leq t$ <2 years	vs. t $\geq$2 years			-0.221	0.886

see top panel (a), second column of Table 3.8. These covariates were strongly associated with the mortality rate ($P < 0.001$), and the Markov assumption is clearly rejected. We can also see that the treatment effect changes somewhat and it is no longer statistically significant ($P = 0.29$). The same conclusion is arrived at if, instead, the time-dependent covariate $d_i(t)$ is included with an assumed linear effect on the log(rate) (not shown).

The coefficients for the time-dependent covariates show that the mortality rate is very high shortly after the bleeding episode and instead of attempting to model this effect parametrically, using time-dependent covariates, an alternative would be to use duration since bleeding as the baseline time-variable in a Cox model. Results from such models are shown in the lower panel (b) of Table 3.8. In the model including treatment (together with sex and $\log_2$(bilirubin) – coefficients not shown) this is statistically significant ($P = 0.02$), and there is a tendency that the combined treatment group has the highest mortality rate and the group receiving only sclerotherapy the lowest. The Breslow estimate of the cumulative baseline hazard is shown in Figure 3.3 and is seen to increase sharply for small values of duration. With this time-variable there is no left-truncation and the estimator does not have a particularly large variability for small values of duration.

To this model, one may add functions of t as time-dependent covariates to investigate whether time since randomization affects the mortality rate. Neither a piece-wise constant effect (Table 3.8, lower panel (b), second column) nor a linear effect (not shown) suggested any importance and their inclusion has little impact on the estimated treatment effects.

We will prefer the model with duration as the baseline time-variable because, of the two time-variables, duration seems to have the strongest effect on the mortality rate and by using this in the Cox baseline hazard, one avoids making parametric assumptions about the way in which it affects the rate.

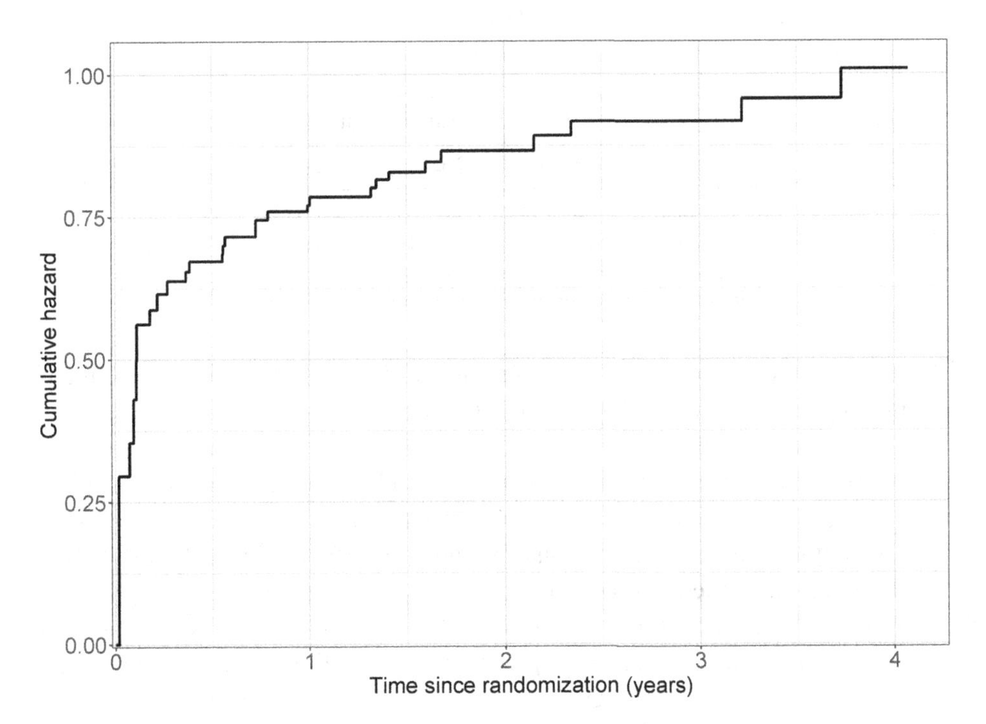

Figure 3.2 *PROVA trial in liver cirrhosis: Breslow estimate for the cumulative baseline hazard in a Cox model for the* $1 \to 2$ *transition rate as a function of time since randomization.*

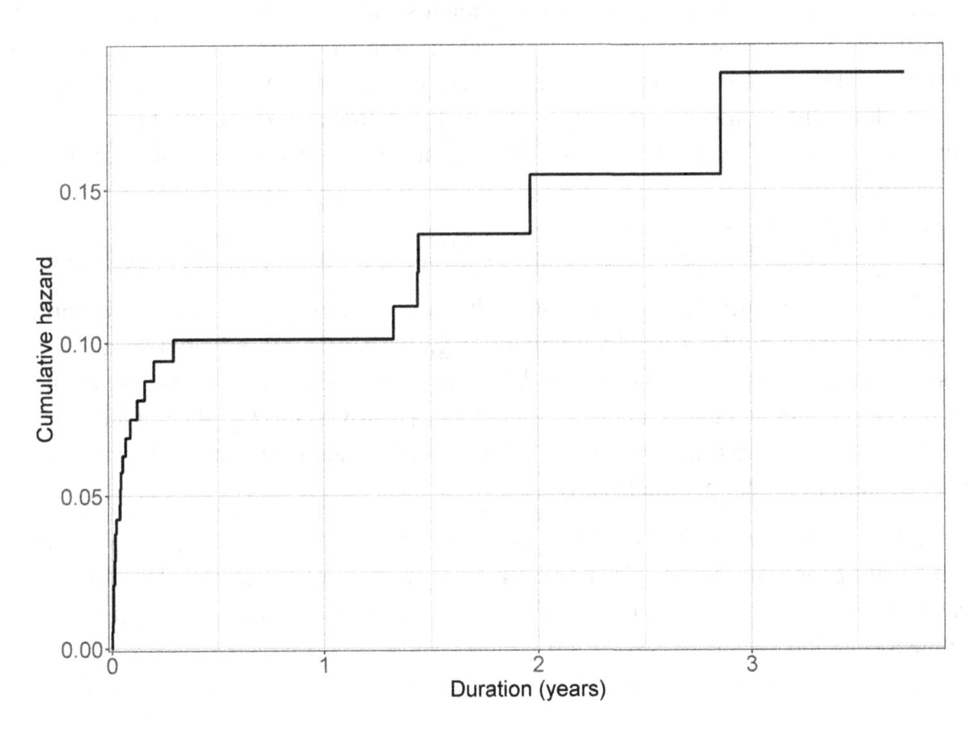

Figure 3.3 *PROVA trial in liver cirrhosis: Breslow estimate for the cumulative baseline hazard in a Cox model for the* $1 \to 2$ *transition rate as a function of time since bleeding (duration).*

Table 3.9 *PROVA trial in liver cirrhosis: Deaths after bleeding/person-years at risk according to duration and time since randomization.*

	Time since randomization		
Duration	0-1 year	1-2 years	2+ years
0-4 days	8/0.463	2/0.131	0/0.014
5-9 days	2/0.386	1/0.116	0/0.014
10+ days	7/12.430	5/19.882	4/21.565

The choice between duration of bleeding and time since randomization as baseline time-variable may be entirely avoided by, instead of using a Cox regression model, analyzing the data using a Poisson regression model and splitting the follow-up time (since bleeding) according to both time-variables. This will require a choice of cut-points for both time-variables. Using the same intervals as in the two Cox models (i.e., 1 and 2 years for t and 5 and 10 days for $d(t)$) gives the distribution of deaths and person-years at risk as shown in Table 3.9. It is seen that no patients with a bleeding episode after 2 years since randomization died within the first 10 days after bleeding, and one is also reminded of the fact that this is a small data set. Having split the follow-up time after bleeding according to the two time-variables, Poisson regression models including either time-variable or both may be fitted. The results from these models, also including the insignificant variables sex and $\log_2$(bilirubin), are presented in Table 3.10. It can be noticed (cf. Section 2.2) that results from similar Cox or Poisson models tend to be very close and, furthermore, that it is not crucial for the estimated regression coefficients for treatment which time-variable(s) we adjust for. This is also the case in a model allowing for an interaction between the two time-variables (not shown). However, the fact that in the model including both time-variables additively, duration is strongly associated with the rate ($P < 0.001$) but time since randomization is not ($P = 0.53$) suggests that it is most important to account for duration since bleeding.

Time-variable and time-dependent covariates

The Cox model requires a specification of the time-variable for the baseline hazard. If several time-variables affect the intensity (e.g., both time on study and duration in a state), there is a choice to be made: Which time-variable should be baseline, and which can be handled using adapted time-dependent covariates? A general advice is to choose as baseline time-variable, one that has a marked effect on the hazard that may be hard to model parametrically.

For a Poisson model, several time-variables can be handled in parallel (i.e., without pin-pointing one as 'baseline'); however, in that case all time-variables must be categorized with an assumption that they have a piece-wise constant effect on the intensity.

Table 3.10 *PROVA trial in liver cirrhosis: Poisson regression models for the rate of death after bleeding accounting for duration since bleeding (a), time since randomization (b), or both (c).*

(a)

Covariate		$\widehat{\beta}$	SD
Sclerotherapy	vs. none	-1.130	0.642
Propranolol	vs. none	-0.314	0.589
Both treatments	vs. none	0.967	0.516
$d(t)$ <5 days	vs. $\geq$10 days	3.602	0.439
5$\leq d(t)$ <10 days	vs. $\geq$10 days	2.844	0.637

(b)

Covariate		$\widehat{\beta}$	SD
Sclerotherapy	vs. none	-1.281	0.652
Propranolol	vs. none	-0.432	0.566
Both treatments	vs. none	0.801	0.525
t <1 year	vs. t $\geq$2 years	1.507	0.579
1$\leq t$ <2 years	vs. t $\geq$2 years	0.430	0.648

(c)

Covariate		$\widehat{\beta}$	SD
Sclerotherapy	vs. none	-1.110	0.648
Propranolol	vs. none	-0.318	0.579
Both treatments	vs. none	0.826	0.514
$d(t)$ <5 days	vs. $\geq$10 days	3.350	0.464
5$\leq d(t)$ <10 days	vs. $\geq$10 days	2.583	0.659
t <1 year	vs. t $\geq$2 years	0.733	0.618
1$\leq t$ <2 years	vs. t $\geq$2 years	0.189	0.655

3.7.7 PBC3 trial in liver cirrhosis

In Section 2.2 we saw how the assumption of proportional hazards could be checked in a Poisson regression model by introducing an interaction between covariates and the categorized time-variable. By using adapted time-dependent covariates, the same idea applies for the Cox regression model, and we will illustrate this using the PBC3 trial as example. To the model including treatment, albumin, and $\log_2$(bilirubin), a time-dependent covariate of the form

$$Z_i(t) = Z_i \cdot f(t)$$

for some function $f(\cdot)$ of time t was added. Here Z_i is one of the time-fixed covariates. To do this, a choice of $f(t)$ must be made and choosing a monotone $f(\cdot)$, a test of proportionality versus an alternative of a monotone hazard ratio is obtained. Typical choices of $f(\cdot)$ are the identity $f(t) = t$, $\log(t)$ or $I(t > t_0)$ for some threshold t_0. Table 3.11 shows results from such tests in the form of the estimated coefficient for $Z_i(t)$ and its SD. It is seen that, in all cases the estimated coefficient is rather small compared to its SD resulting nowhere in

Table 3.11 *PBC3 trial in liver cirrhosis: Examination of proportional hazards in a Cox model including treatment, albumin, and* $\log_2$*(bilirubin).*

	Treatment		Albumin		$\log_2$(bilirubin)	
Function	$\widehat{\beta}$	SD	$\widehat{\beta}$	SD	$\widehat{\beta}$	SD
$f(t) = t$	0.045	0.178	0.023	0.019	-0.063	0.062
$f(t) = \log(t)$	0.108	0.254	0.033	0.026	-0.077	0.088
$f(t) = I(t > 2 \text{ years})$	0.032	0.434	0.057	0.043	-0.182	0.149

any evidence against proportionality. It is also seen that, for each of the three covariates, the coefficient has the same sign for all choices of $f(t)$. The tendencies for treatment and albumin are that the hazard ratio increases over time while, for bilirubin, it decreases.

3.7.8 *Bone marrow transplantation in acute leukemia*

The four-state model for the bone marrow transplantation study was shown in Figure 1.6. We will first present Nelson-Aalen estimates for the cumulative transition rates. Figure 3.4 shows the estimates for the cumulative mortality rates $A_{03}(t)$, without relapse or GvHD, $A_{13}(t)$, without relapse and with GvHD, and $A_{23}(t)$, after relapse. The latter two are estimated taking the delayed entry into states 1 or 2 into account. Here, time t denotes time since bone marrow transplantation (BMT). It is seen that occurrence of a relapse markedly increases the mortality rate and also that the direct mortality rate for patients after experiencing GvHD is higher than without. Figure 3.5 shows the estimated cumulative GvHD rate, $A_{01}(t)$, and shows that the rate is high shortly after BMT and thereafter it decreases. Finally, Figure 3.6 shows the estimated cumulative relapse rates, $A_{02}(t)$ without GvHD and $A_{12}(t)$ after GvHD, the latter estimated using delayed entry. It is seen that the GvHD event decreases the relapse rate.

For these estimates, no assumptions are made in relation to how the two different relapse rates and the three death rates are connected. More parsimonious models could build on various proportional hazards assumptions, e.g., $\alpha_{12}(t) = \exp(\beta_2)\alpha_{02}(t)$ for the relapse rates and $\alpha_{13}(t) = \exp(\beta_3)\alpha_{03}(t)$ for the death rates with or without GvHD. Such models can be fitted by including GvHD as a *time-dependent covariate* in separate models for the relapse and death rates. When GvHD is a *state* in the multi-state model, the covariate

$$Z_i(t) = I(i \text{ had GvHD before time } t)$$

is an adapted time-dependent covariate. For relapse, the estimated hazard ratio for GvHD is $\exp(\widehat{\beta}_2) = 0.858$ with 95% confidence limits from 0.663 to 1.112 ($P = 0.25$). The proportional hazards assumption was evaluated by including the time-dependent covariate $Z_i(t)\log(t+1)$ which gives a P-value of 0.35. A graphical evaluation, following the lines from the stratified Cox model in Section 2.2, can be performed by plotting $\widehat{A}_{12}(t)$ against $\widehat{A}_{02}(t)$, see Figure 3.7. Under proportional hazards, the resulting curve should be a straight line through the point (0,0), with slope equal to $\exp(\widehat{\beta}_2) = 0.858$ and this is seen to be a good approximation. For death, the similar analyses yield $\exp(\widehat{\beta}_3) = 3.113$ with 95% confidence limits from 2.577 to 3.760. Addition of an interaction between GvHD and $\log(t+1)$

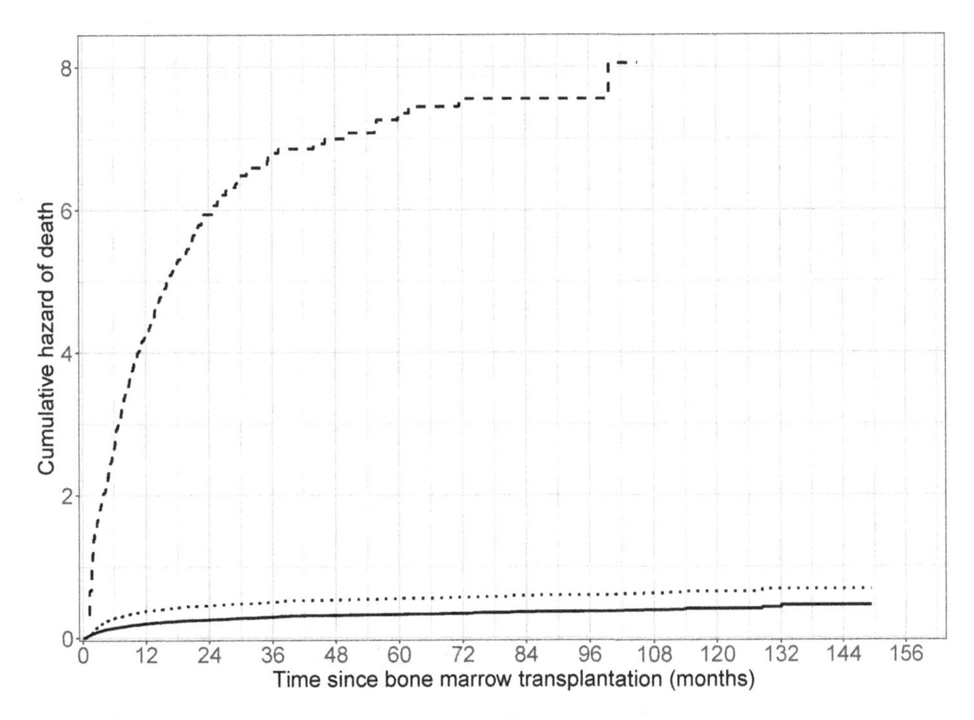

Figure 3.4 *Bone marrow transplantation in acute leukemia: Cumulative mortality rate after relapse (dashed line); cumulative mortality rate after GvHD (dotted line); cumulative mortality rate without relapse or GvHD (solid line) (GvHD: Graft versus host disease).*

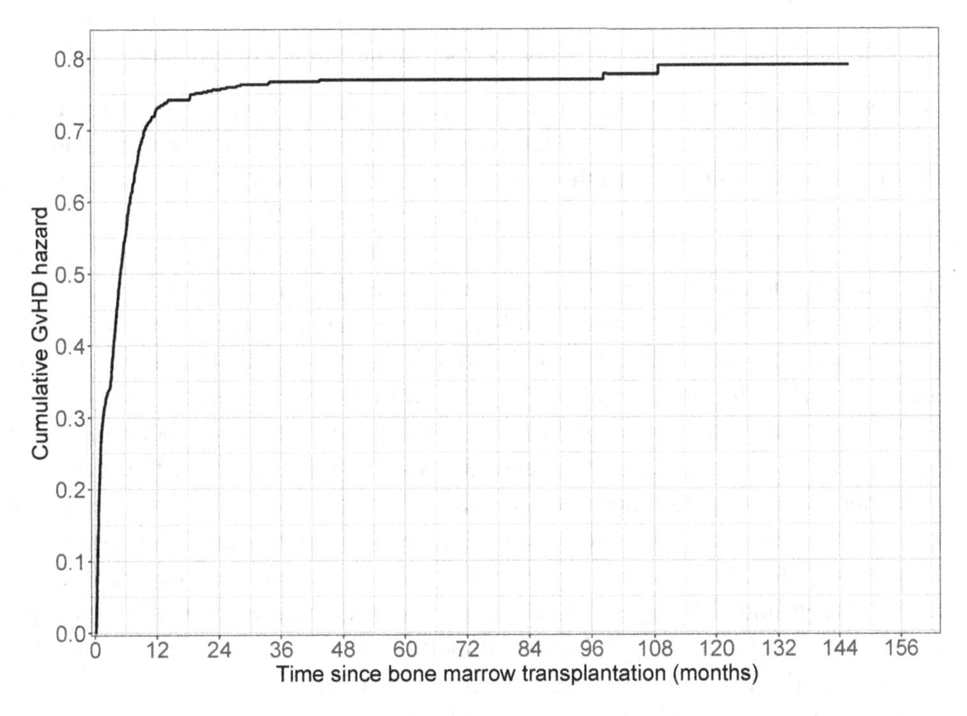

Figure 3.5 *Bone marrow transplantation in acute leukemia: Cumulative rate of GvHD (Graft versus host disease).*

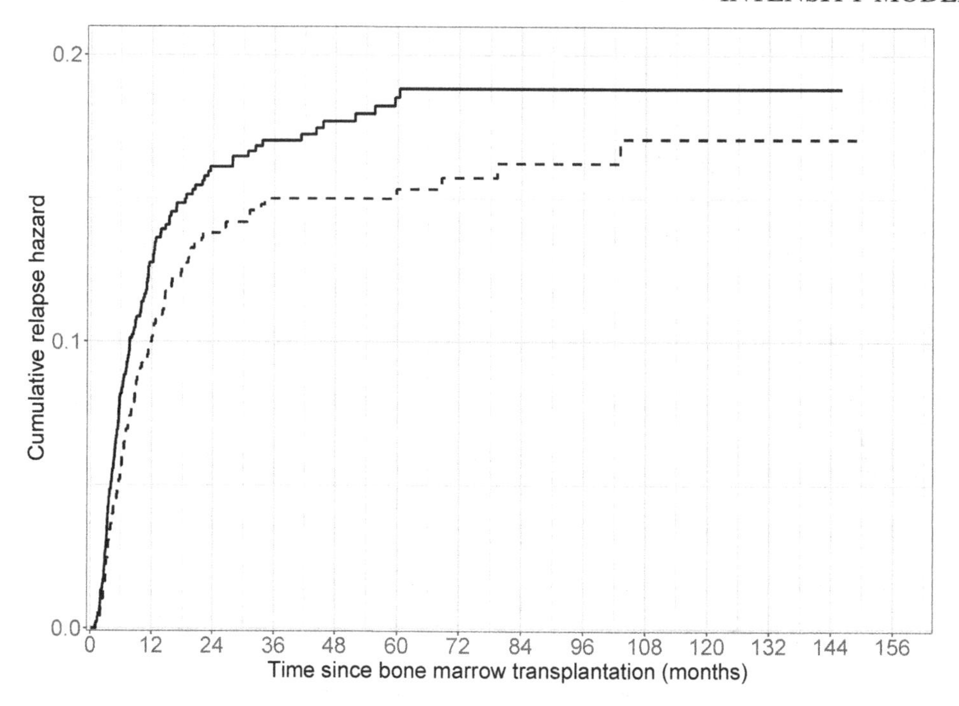

Figure 3.6 *Bone marrow transplantation in acute leukemia: Cumulative relapse rate after GvHD (dashed line); cumulative relapse rate without GvHD (solid line) (GvHD: Graft versus host disease).*

gives $P = 0.11$, and the goodness-of-fit plot is seen in Figure 3.8. This figure does suggest some deviations from proportional hazards apparently caused by a too low hazard ratio early on (convex shape of the curve); however, the formal test is insignificant.

Bone marrow transplantation studies often aim at studying the two adverse end-points relapse and death without relapse, the latter often termed death in remission or treatment-related mortality, both signaling that the treatment with BMT is no longer effective. In such a situation, a relevant multi-state model to use would be the competing risks model, Figure 1.2, i.e., the disease course after relapse is not studied, and GvHD is no longer considered a separate state in the model. However, in an analysis of the rates of relapse and death in remission, it would still be of interest to study how these may be affected by occurrence of GvHD over time. This can be done as just described, i.e., by including the time-dependent GvHD covariate $Z_i(t)$ in the Cox models for the two rates. However, for the competing risks model, this will no longer be an adapted time-dependent covariate but rather a non-adapted, *internal* or *endogenous* time-dependent covariate, because the past history at time t in the competing risks model does not contain information on GvHD and because the existence of $Z_i(t)$ requires subject i to be alive. At this point, the distinction between these two situations may look rather academic but when, later in the book (Section 5.2.4), we go beyond rate models and also target marginal parameters, such as the probability of experiencing a relapse, the distinction will become important. We will conclude this example by presenting results from analyses of these two event rates taking both of the time-dependent covariates GvHD and ANC500 into account, the latter taking the value 1 at time t if, at that time, the Absolute Neutrophil Count is above 500 cells per μL, based on repeated blood

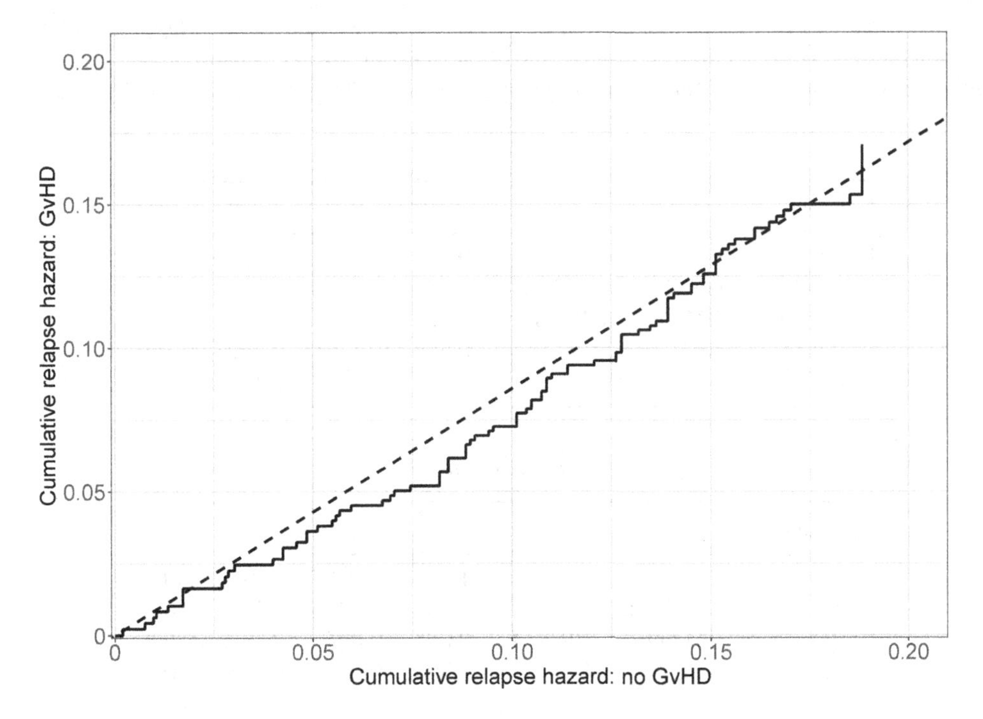

Figure 3.7 *Bone marrow transplantation in acute leukemia: Cumulative relapse rate with GvHD plotted against cumulative relapse rate without GvHD. The dashed straight line has slope equal to* $\exp(\widehat{\beta}_2) = 0.858$ *(GvHD: Graft versus host disease).*

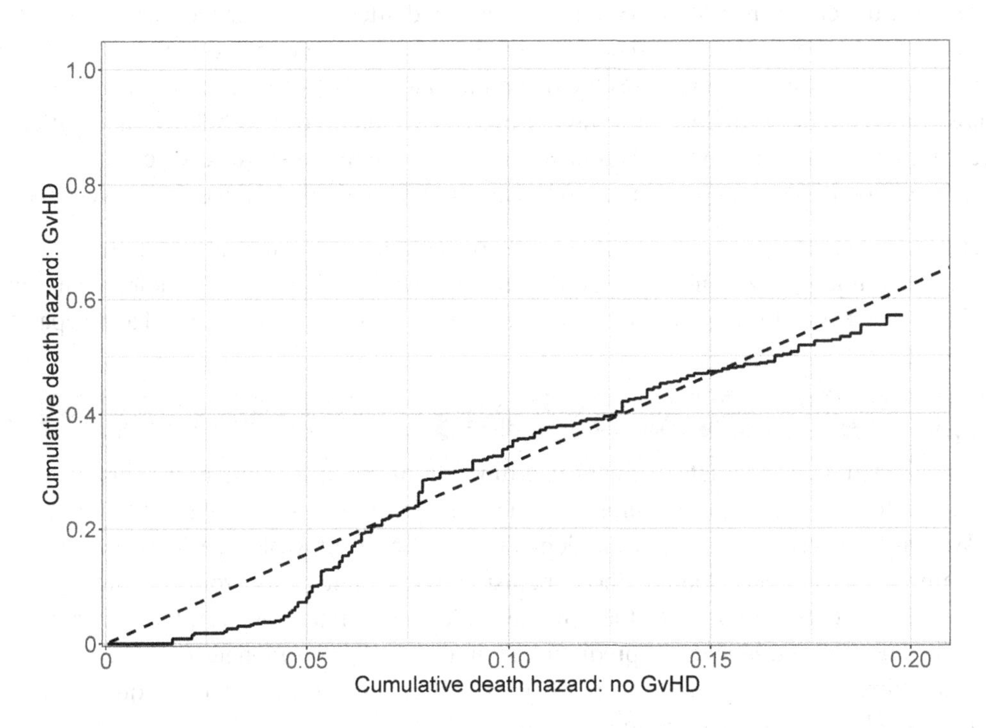

Figure 3.8 *Bone marrow transplantation in acute leukemia: Cumulative death rate with GvHD plotted against cumulative death rate without GvHD. The dashed straight line has slope equal to* $\exp(\widehat{\beta}_3) = 3.113$ *(GvHD: Graft versus host disease).*

Table 3.12 *Bone marrow transplantation in acute leukemia: Cox models for relapse and death in remission (GvHD: Graft versus host disease, BM: Bone marrow, PB: Peripheral blood, AML: Acute myelogenous leukemia, ALL: Acute lymphoblastic leukemia, ANC: Absolute neutrophil count)*

(a) *Relapse*

Covariate		$\widehat{\beta}$	SD	$\widehat{\beta}$	SD
GvHD(t)		-0.184	0.134	-0.188	0.134
Age	per 10 years	-0.040	0.045	-0.039	0.045
Graft type	BM only vs. PB/BM	-0.125	0.135	-0.130	0.135
Disease	ALL vs. AML	0.563	0.130	0.562	0.130
ANC500(t)				-2.138	1.077

(b) *Death in remission*

Covariate		$\widehat{\beta}$	SD	$\widehat{\beta}$	SD
GvHD(t)		1.041	0.098	1.040	0.099
Age	per 10 years	0.263	0.033	0.263	0.033
Graft type	BM only vs. PB/BM	-0.085	0.096	-0.140	0.096
Disease	ALL vs. AML	0.334	0.098	0.336	0.098
ANC500(t)				-2.228	0.305

samples taken during follow-up, and equal to 0 otherwise. Table 3.12 shows the results. It is seen that the earlier results for GvHD are sustained after adjustment for age, graft type, disease (and ANC500), i.e., GvHD tends to reduce the relapse rate and increase the death rate. Higher age markedly increases the death rate and tends to be associated with a lower relapse rate. Patients with ALL have higher event rates than those with AML, and patients receiving only bone marrow tend to have lower event rates than those who also receive peripheral blood. Finally, obtaining an ANC above 500 markedly reduces both event rates.

In principle, one could go a step further and concentrate on the mortality rate (cf. Figure 1.1), treating relapse as an internal time-dependent covariate. However, as indicated in Figure 3.4, occurrence of relapse is such a serious event that it is typically treated as a separate end-point.

Time-dependent covariates or state

An adapted time-covariate at time t is a function of the past of the multi-state process before that time, for example whether a given state has been visited before t. We have seen that the same time-dependent covariate may also be included in a simplified multi-state model without inclusion of that state. This avoids the need for modeling the development of the time-dependent covariate; however, the covariate is then no longer adapted and prediction from the model is no longer feasible.
The ultimate choice between these options will depend of the scientific questions the model is aimed at addressing.

3.7.9 *Additional issues*

We have seen that incorporating time-dependent covariates into a hazard regression model is, in principle, straightforward. The same methods of estimation apply as for models including only time-fixed covariates, e.g., the Cox partial likelihood. Furthermore, the interpretation of $\exp(\beta)$ is the ratio between the instantaneous event risks per time unit for a one-unit change in Z and for given values of the remaining explanatory variables in the model. However, this simplicity is somewhat deceptive and models including time-dependent covariates are considerably more involved than models without. In the following, we will discuss this in more detail, see also Fisher and Lin (1999).

Data availability; measurement errors

To set up the Cox partial likelihood or the estimating equations for the Aalen model, values for all covariates at all event times are needed. For internal covariates, this may entail some difficulties if covariates are based on repeated measurements of some marker, such as the ANC in the bone marrow transplantation study. Thus, at any given event time, T, what will typically be known is the last value recorded before time T, and in order to assess the covariate value at T, some extrapolation or modeling of the repeated measurements of the marker is needed. Frequently, *last observation carried forward* is used, though studies have shown that this may not be optimal and better ways of extrapolating are preferable (e.g., Andersen and Liestøl, 2003). Note that *inter*polation between values before and after T is not recommended since later values are only available for the selected group of subjects who survive until the time of the next measurement, thereby one is 'conditioning on the future' when including this information. The fact that the marker varies over time will also imply that it is measured with some error, so that accounting for measurement error may be needed (e.g., Bycott and Taylor, 1998). In the bone marrow transplantation study, what was recorded was the first time for which ANC exceeded 500 cells per μL and the true time of crossing this threshold would have been at an earlier point in time. A solution that is sometimes used to such problems is *joint modeling* of the marker and the multi-state process. We will briefly discuss such techniques in Section 7.4.

A related problem occurs when time-dependent covariates are used in situations with delayed entry. Here, the information collected for a subject at the time of entry into the study may not suffice for calculating the value of a covariate such as the duration since some earlier event, and care must be exercised when trying to include such covariates. Another problem with delayed entry and internal time-dependent covariates may occur when modeling rates in terms of age rather than time on study. This is because the risk set at any given age may consist of subjects with strongly varying times since the latest measurement of the marker and, thereby, possibly with differential measurement error (e.g., Andersen and Liestøl, 2003).

Interpretation of covariate effects

In a multiplicative model, $\exp(\beta)$ is the ratio between event rates for a one-unit change in Z and for given values of the remaining explanatory variables in the model. This means that if a model includes both a time-fixed covariate, Z_1, such as initial diagnosis or randomized treatment group, and a non-deterministic time-dependent covariate, $Z_2(t)$ such as occurrence of later events, then $\exp(\beta_1)$ is the hazard ratio for Z_1 for given value of $Z_2(t)$. This may mask some of the true effect of Z_1 because some of this effect may be mediated via changes in $Z_2(t)$. An example of this was presented in Section 3.7.5 when analyzing the data on repeated episodes in affective disorders and where the hazard ratio for the initial diagnosis, bipolar vs. unipolar, was quite different without or with adjusting for $N(t-)$, the number of episodes before time t. A similar problem was observed when, in the PROVA trial (Section 3.7.6), the mortality rate after a bleeding episode was modeled in relation to the randomized treatment and it was noticed that the patients contributing to this analysis may be selected differently in the treatment groups due to the way in which treatment affects the occurrence of a bleeding and the mortality rate without bleeding. The desired effect of a baseline covariate on the event occurrence may, in such situations, be better studied in terms of the cumulative risk of the event or the expected number of recurrent events. We will return to this problem in Sections 4.2.3 and 5.5.4.

An additional problem with interpretation that we will also return to in later chapters is that a hazard ratio cannot necessarily be interpreted as reflecting a risk ratio. This is the case for both time-fixed and for time-dependent covariates and is due to the fact that the same covariate may affect other transition hazards in the multi-state model (e.g., in the case of competing risks, see Sections 4.1.2 and 4.2.2).

Immortal time bias

A time-dependent covariate is sometimes, mistakenly, considered to be time-fixed and this will lead to *immortal time bias* (e.g., Suissa, 2007; Andersen et al., 2021). The name of this bias is explained below. In the bone marrow transplantation study, one might try to compare rates of relapse among those who do or do not ever experience GvHD, or, in the study of vaccinations and mortality in Guinea-Bissau, one may wish to compare the mortality rates between children who do or do not receive additional vaccinations in the period between the two visits by the mobile team. This entails conditioning on the future and leads to bias because those who live long enough to obtain GvHD or receive additional vaccinations without experiencing the event of interest will appear to have a longer time until event occurrence. The correct way to handle the situation is to treat GvHD or additional vaccinations as time-dependent covariates (e.g., Jensen et al., 2007).

We will illustrate this bias via the bone marrow transplantation study. Recall that when, correctly, treating GvHD as a time-dependent covariate, the estimated hazard ratio for GvHD is $\exp(\widehat{\beta}_2) = 0.858$ for relapse and $\exp(\widehat{\beta}_3) = 3.113$ for death in remission (Section 3.7.8). If one, naively, includes instead the variable 'GvHD: yes/no' in the model as a time-fixed covariate, the hazard ratio in relation to relapse is 0.544, and for death in remission it is 1.576. Thus, for both outcomes GvHD appears more beneficial in the incorrect analyses. This can be explained, as follows. There are 3,938.61 person-years at risk in the initial state (0) in the multi-state model of Figure 1.6 and 3,255.19 in the GvHD state (1). However, out of the person-years spent in state 0, 283.42 are from patients who later develop GvHD. These person-years are in state 0 and should not be credited the GvHD group. However, when treating GvHD as time-fixed, these immortal years are included in the risk time among GvHD patients which leads to the bias described.

3.8 Models with shared parameters

In most of the examples so far, different transitions in a multi-state model were modeled separately, i.e., models for different transition intensities had no parameters in common. Thus, in Section 3.6.3, separate models for the rate of bleeding and the rate of death without bleeding in the PROVA trial (Example 1.1.4) were studied and, in Section 3.7.8, the rates of relapse and that of death in remission in the bone marrow transplantation study (Example 1.1.7) were also modeled separately. Since, in these examples the rates of completely different events were studied, having parameters in common for the different transition intensities seems quite unnatural. However, in the case of recurrent events (Section 2.5), examples were given (in the form of PWP or AG models) where the initial diagnosis of bipolar versus unipolar disorder in the study of recurrent episodes in affective disorders (Example 1.1.5) had the same multiplicative effect on the rates of first, second, third, ... recurrence. Also in the PROVA trial and in the bone marrow transplantation study, models with common regression coefficients could be envisaged. Thus, in the former case, variables such as sex and age could have the same effects on the mortality rates with or without a bleeding episode, and in the latter, variables such as disease and graft type could have the same effect on the rate of death in remission with or without GvHD.

In the present section, we will study models for several transition intensities where some covariates may have common effects across different transitions. We will provide a detailed study of the illness-death model (Figure 1.3); however, the ideas presented for this model carry over to more complex multi-state models. As we shall see, the concepts of *type-specific covariates* and *time-dependent strata* are crucial in this discussion.

We address simultaneous (Cox) modeling of the rates $\alpha_{02}(t)$ and $\alpha_{12}(t)$ in an illness-death model, and for each covariate Z there is a choice whether it has the same or different effects on these two rates. Furthermore, $\alpha_{02}(t)$ and $\alpha_{12}(t)$ may or may not be proportional. The modeling combination that we have mostly focused on so far is when all covariates have different effects and where the two rates are not proportional, in which case the two rates are modeled separately. However, we shall see in what follows that all modeling combinations may be obtained by fitting one common model for the two hazards to a *duplicated data set* where either Z is used directly or replaced by two *type-specific covariates*, and the model is either *stratified* (time-dependently) by the starting state (0 or 1) or that state is used as a *time-dependent covariate*.

3.8.1 Duplicated data set

The starting point for such an analysis is the two separate data sets for, respectively, the $0 \to 2$ and $1 \to 2$ transitions, see Table 1.10 for a discussion in relation to the PROVA trial where 'disease' corresponds to bleeding. The data set for the $0 \to 2$ transition has variables (`Start, Stop, Status`) where `Start` = 0, `Stop` = time last seen in state 0, and `Status` = 1, if at time `Stop`, a $0 \to 2$ transition (death without the disease) was observed and = 0 otherwise. Suppose that the data set, additionally, contains a (numerical) covariate Z. The data set for the $1 \to 2$ transition has `Start` = time of entry into state 1 (time of disease), `Stop` = time last seen in state 1, and `Status` = 1, if at time `Stop`, a $1 \to 2$ transition was observed (death with the disease) and = 0 otherwise. Suppose that this data set also contains the covariate Z. The duplicated data set should have one or two records for each subject, one if the disease was not observed and two if it was observed. Each record should have the following variables:

- `(Start, Stop, Status)` copied from the original data sets.
- Z copied from the original data sets.
- A *stratum* variable, say `Type`, that is 0, if the record came from the $0 \to 2$ data set and 1 if the record came from the $1 \to 2$ data set.
- Two *type-specific covariates* $(Z_0, Z_1) = (Z, 0)$ if `Type=0`, and $(Z_0, Z_1) = (0, Z)$ if `Type=1`, i.e., the type-specific covariate corresponding to 'the other value of `Type`' is set to 0.

The models with different combinations of common versus different effects of Z and proportional versus non-proportional baseline $0 \to 2$ and $1 \to 2$ transition rates can now be fitted to the duplicated data set, as follows, where, in all cases, the `(Start, Stop, Status)` triple is used as response variable.

- Different covariate effects and non-proportional hazards

$$\alpha_{02}(t \mid Z) = \alpha_{02,0}(t) \exp(\beta_0 Z), \quad \alpha_{12}(t \mid Z) = \alpha_{12,0}(t) \exp(\beta_1 Z). \tag{3.28}$$

  Include the type-specific covariates (Z_0, Z_1) and stratify by `Type` (time-dependent strata). This is equivalent to fitting separate models for the two transition rates.

- Same covariate effect and non-proportional hazards

$$\alpha_{02}(t \mid Z) = \alpha_{02,0}(t) \exp(\beta Z), \quad \alpha_{12}(t \mid Z) = \alpha_{12,0}(t) \exp(\beta Z). \tag{3.29}$$

  Include the original covariate Z and stratify by `Type` (time-dependent strata).

- Different covariate effects and proportional hazards

$$\alpha_{02}(t \mid Z) = \alpha_0(t) \exp(\beta_0 Z), \quad \alpha_{12}(t \mid Z) = \alpha_0(t) \exp(\beta_1 Z + \gamma). \tag{3.30}$$

  Include the type-specific covariates (Z_0, Z_1) and use `Type` as a (time-dependent) covariate – the latter yielding the hazard ratio $\exp(\gamma)$.

- Same covariate effect and proportional hazards

$$\alpha_{02}(t \mid Z) = \alpha_0(t) \exp(\beta Z), \quad \alpha_{12}(t \mid Z) = \alpha_0(t) \exp(\beta Z + \gamma). \tag{3.31}$$

  Include the original covariate Z and use `Type` as a (time-dependent) covariate – the latter yielding the hazard ratio $\exp(\gamma)$.

Models (3.28) and (3.29), respectively (3.30) and (3.31), may be compared using likelihood ratio tests, i.e., it can be examined whether the regression coefficients for Z can be taken to be the same for the two transition types. Comparing models (3.28) and (3.30), respectively (3.29) and (3.31), corresponds to an examination of proportional hazards as exemplified, e.g., in Sections 2.2.2 and 3.7.8. Multiple regression models, i.e., including more covariates with combinations of identical and different effects on the two rates, can be set up along these lines and will be exemplified below. For all models, model-based SD can be applied.

The models (3.28) and (3.30) with type-specific covariates correspond to inclusion of an interaction between Type and Z. This observation suggests how joint Poisson models for the two mortality rates may also be fitted. This will require a duplicated data set including cases of death and person-years at risk, both before and after disease occurrence, and where interactions between some covariates and Type may be included.

The data duplication trick for the illness-death model works in a similar fashion for other multi-state models, including the competing risks model (Figure 1.2). Thus, for the PBC3 trial, models with common covariate effects on the rate of death without transplantation and the rate of transplantation could be fitted as well as models with proportional cause-specific hazards. However, since the interpretation of such common effects on rates of quite different events is not attractive, we will not illustrate this feature in the following.

3.8.2 PROVA trial in liver cirrhosis

We will study the PROVA trial (Example 1.1.4) and joint models for the mortality rates with or without a previous bleeding. For simplicity, we drop the treatment variables and study models including the covariates sex and $\log_2$(bilirubin) (see also Section 3.7.6). We will also (for illustration, and in spite of the fact that better-fitting models were identified in Section 3.7.6) study Markov models, i.e., the time-variable t used for both the mortality rate without bleeding $\alpha_{02}(t)$ and that after bleeding, $\alpha_{12}(t)$ is time since randomization. Table 3.13 shows the estimated coefficients for the two covariates obtained by fitting a number of joint models, and Figure 3.9 shows the estimated cumulative baseline hazards for the model (3.28) where proportionality is not assumed.

In model (3.28) (Table 3.13a), it is seen that the coefficients for sex are close, while those for bilirubin are not (likelihood ratio tests for the corresponding model reductions are, respectively, 0.04 and 9.8 both with 1 DF). In the model with identical sex effects (Table 3.13b), it is seen that bilirubin only affects the mortality rate without bleeding and the coefficient for $\log_2$(bilirubin) in the model for $\alpha_{12}(t)$ can be set to 0 (Table 3.13c). Note the gain in efficiency for the coefficient for sex when based on both mortality rates (SD in (b), (c) vs. (a)).

Judged from Figure 3.9, it seems that the baseline rates are far from proportional and this is supported by a test for proportionality in the model where Type is a time-dependent covariate – here the coefficient for Type$\cdot\log(t)$ is strongly significant ($P < 0.001$). For this reason, no results from models assuming proportional mortality rates with and without bleeding are presented.

3.8.3 Bone marrow transplantation in acute leukemia

In Section 3.7.8, we studied models for the rates of relapse and death in remission in the bone marrow transplantation study (Example 1.1.7) using GvHD as a time-dependent covariate. In this section, we follow up on that example and illustrate joint modeling of the rates $\alpha_{03}(t)$ and $\alpha_{13}(t)$ in Figure 1.6, i.e., the rates of death in remission without or with GvHD. We will include the two covariates age and disease (ALL vs. AML), and Table 3.14 shows estimated regression coefficients from a series of such models. The model (a) corresponds to fitting separate Cox models for the two transition rates, and it is seen that

Table 3.13 *PROVA trial in liver cirrhosis: Joint Cox models for the mortality rates without or with bleeding (sex: Males vs. females).*

(a)

	Sex		$\log_2$(bilirubin)	
Event type	$\widehat{\beta}$	SD	$\widehat{\beta}$	SD
Death without bleeding	1.041	0.411	0.528	0.115
Death after bleeding	0.910	0.481	-0.162	0.183
Both				

(b)

	Sex		$\log_2$(bilirubin)	
Event type	$\widehat{\beta}$	SD	$\widehat{\beta}$	SD
Death without bleeding			0.527	0.115
Death after bleeding			-0.169	0.179
Both	0.987	0.312		

(c)

	Sex		$\log_2$(bilirubin)	
Event type	$\widehat{\beta}$	SD	$\widehat{\beta}$	SD
Death without bleeding			0.526	0.115
Death after bleeding			0	
Both	0.942	0.307		

both covariates have quite similar effects. The cumulative baseline rates from this model are shown in Figure 3.10 and do not contra-indicate proportionality, so, model (b) in the table shows results from a model assuming $\alpha_{13}(t) = \exp(\gamma)\alpha_{03}(t)$, corresponding to treating `Type` (GvHD) as a time-dependent covariate. Here, $\widehat{\gamma} = 0.842$ (SD $= 0.269$). Proportionality is also supported by including the covariate GvHD$(t)\log(t)$ for which the likelihood ratio test is LRT=2.08 with 1 DF. The resulting model is on the form (3.30) with different coefficients and proportional hazards. In model (c) of the table, coefficients are shown for the model where the two type-specific covariates for age and disease are replaced by common covariates leading to a model of the form (3.31). The LRT for common coefficients is 0.74 with 2 DF, supporting the simpler model in which the gain in efficiency for the regression coefficients can be noticed. In model (c), $\widehat{\gamma} = 1.049$ (SD $= 0.098$).

3.8.4 Joint likelihood (*)

The estimators in joint Cox models for all transition intensities $\alpha_{hj}(t)$ in a multi-state model are related to those based on the stratified Cox partial likelihood (3.21) and the corresponding Breslow estimators (3.22). That is, there is a single p-vector of regression coefficients $\boldsymbol{\beta}$ and a number, K of unspecified baseline hazards $\alpha_{v0}(t)$, $v = 1, \ldots, K$. To realize this, it

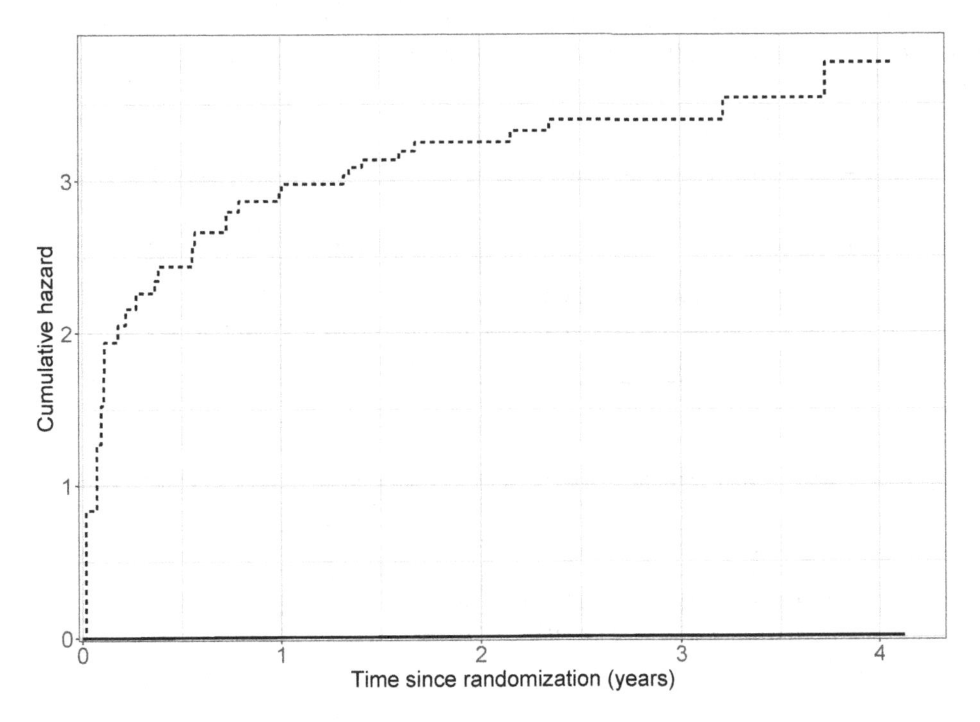

Figure 3.9 *PROVA trial in liver cirrhosis: Breslow estimates for the cumulative baseline mortality rates in a joint Cox model for the* $0 \rightarrow 2$ *(solid line) and* $1 \rightarrow 2$ *transition rates (dashed line).*

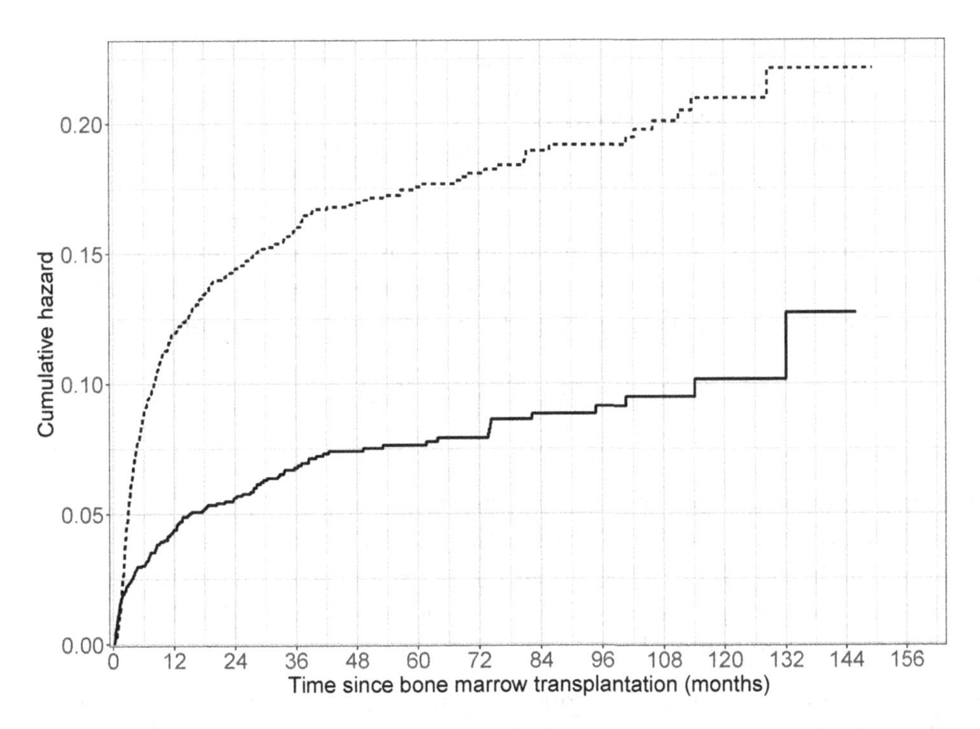

Figure 3.10 *Bone marrow transplantation in acute leukemia: Breslow estimates for the cumulative baseline rates of death in remission in a joint Cox model for the mortality rates with (dashed line) or without GvHD (solid line) (GvHD: graft versus host disease).*

Table 3.14 *Bone marrow transplantation in acute leukemia: Joint Cox models for the rates of death in remission without or with GvHD (GvHD: Graft versus host disease, disease: Acute myelogenous leukemia (AML) vs. acute lymphoblastic leukemia (ALL)).*

(a)

	Disease		Age (years)	
Event type	$\widehat{\beta}$	SD	$\widehat{\beta}$	SD
Death in remission without GvHD	0.267	0.159	0.256	0.048
Death in remission with GvHD	0.360	0.125	0.287	0.041
Both				

(b)

	Disease		Age (years)	
Event type	$\widehat{\beta}$	SD	$\widehat{\beta}$	SD
Death in remission without GvHD	0.262	0.159	0.246	0.048
Death in remission with GvHD	0.379	0.125	0.292	0.041
Both				

(c)

	Disease		Age (years)	
Event type	$\widehat{\beta}$	SD	$\widehat{\beta}$	SD
Death in remission without GvHD				
Death in remission with GvHD				
Both	0.332	0.098	0.272	0.031

has to be argued that all the models studied in Section 3.8.1, i.e., with covariates having either different or common regression coefficients for the different transition hazards and with separate or proportional baseline hazards, may be written in the form

$$\begin{aligned}\alpha_{hji}(t) &= \alpha_{v0}(t)\exp(\boldsymbol{\beta}^{\mathsf{T}}\mathbf{Z}_{hji}(t)) \\ &= \alpha_{\phi(h,j)0}(t)\exp(\boldsymbol{\beta}^{\mathsf{T}}\mathbf{Z}_{hji}(t))\end{aligned} \tag{3.32}$$

with proper definition of (possibly time-dependent) p-vectors of *type-specific* covariates $\mathbf{Z}_{hji}(t) = (Z_{hj1i}(t), \ldots, Z_{hjpi}(t))$, i.e., there is one set of p covariates for each type, $h \to j$, of transition in the model. In (3.32), $v = \phi(h, j)$ is the function of the transition type that takes the same value for pairs of states, h, j for which the corresponding baseline hazards $\alpha_{hj0}(t)$ are assumed proportional.

We will now illustrate how this idea works using the three-state illness-death model as an example (Figure 1.3). Assume that we will fit the following models for the three transition intensities for subject i

$$\begin{aligned}\alpha_{01i}(t) &= \alpha_{01,0}(t)\exp(\beta_0 Z_{1i}),\\ \alpha_{02i}(t) &= \alpha_{02,0}(t)\exp(\beta_1 Z_{1i}+\beta_2 Z_{2i}),\\ \alpha_{12i}(t) &= \alpha_{02,0}(t)\exp(\gamma)\exp(\beta_1 Z_{1i}+\beta_2' Z_{2i}).\end{aligned}$$

Thus, there are two time-fixed covariates: Z_1 that influences all three hazards and influences the two death intensities in the same way (β_1), and Z_2 that does not influence the disease rate $\alpha_{01}(t)$ and influences the mortality rates without or with the disease in different ways (β_2 and β_2'). Furthermore, the mortality rates with and without the disease are proportional. It is seen that, in this case, there are $p = 5$ unknown regression coefficients $\boldsymbol{\beta} = (\beta_0, \beta_1, \beta_2, \beta_2', \gamma)$ and $K = 2$ unspecified baseline hazards $\alpha_{01,0}(t)$ and $\alpha_{02,0}(t)$. The models may be written in the form (3.32) with this p-vector of regression coefficients and these K baseline hazards by defining the type-specific covariates,

$$\begin{aligned}\mathbf{Z}_{01i}(t) &= (Z_{1i},0,0,0,0)\\ \mathbf{Z}_{02i}(t) &= (0,Z_{1i},Z_{2i},0,0,0)\\ \mathbf{Z}_{12i}(t) &= (0,Z_{1i},0,Z_{2i},1).\end{aligned}$$

The mapping ϕ is $\phi(0,1) = 1$, $\phi(0,2) = \phi(1,2) = 2$ corresponding to $\alpha_{02,0}(t)$ and $\alpha_{12,0}(t)$ being proportional.

We now take the Jacod likelihood (3.1) as the staring point which, for the current model, becomes

$$\begin{aligned}L &= \prod_i \prod_{(h,j)} \{\exp\Big(-\int_0^\infty Y_{hi}(t)\alpha_{\phi(h,j)0}(t)\exp(\boldsymbol{\beta}^\mathsf{T}\mathbf{Z}_{hji}(t))dt\Big)\\ &\times \prod_t \Big(Y_{hi}(t)\alpha_{\phi(h,j)0}(t)\exp(\boldsymbol{\beta}^\mathsf{T}\mathbf{Z}_{hji}(t))\Big)^{dN_{hji}(t)}\}. \qquad (3.33)\end{aligned}$$

Note that (3.33) *no longer factorizes* over types, v since the same $\boldsymbol{\beta}$ appears for all types. Transforming by the logarithm, differentiating with respect to a single $\alpha_{v0}(t)$, and solving for $\alpha_{v0}(t)$ as in Section 3.3 leads, for fixed $\boldsymbol{\beta}$, to the estimate

$$\widehat{\alpha_{v0}(t)dt} = \frac{\sum_i \sum_{\phi(h,j)=v} dN_{hji}(t)}{\sum_i \sum_{\phi(h,j)=v} Y_{hi}(t)\exp(\boldsymbol{\beta}^\mathsf{T}\mathbf{Z}_{hji}(t))} = \frac{dN_v(t)}{S_{0v}(\boldsymbol{\beta},t)}, \qquad (3.34)$$

say. Inserting this into (3.33) leads to the relevant version of the stratified Cox partial likelihood

$$\mathrm{PL}(\boldsymbol{\beta}) = \prod_i \prod_v \prod_{\phi(h,j)=v} \prod_t \Big(\frac{Y_{hi}(t)\exp(\boldsymbol{\beta}^\mathsf{T}\mathbf{Z}_{hji}(t))}{S_{0v}(\boldsymbol{\beta},t)}\Big)^{dN_{hji}(t)},$$

and the Breslow estimator becomes

$$\widehat{A}_{v0}(t) \int_0^t \frac{dN_v(u)}{S_{0v}(\widehat{\boldsymbol{\beta}},u)}$$

with notation as in (3.34), see Exercise 3.6. Since the resulting estimators are likelihood-based, model-based SD may be obtained from the second derivative of the log-likelihood, and likelihood ratio tests are also available.

3.9 Frailty models

In Chapter 2 and in previous sections of the present chapter, we have shown several examples of regression models for a single transition rate in a multi-state model and unless there were parameters that were shared among several transitions (see Section 3.8), these intensities could be analyzed separately. In all these examples, an assumption of *independence among observational units* (typically among subjects/patients) was reasonable. In the present section, we will study situations where the independence assumption is not necessarily met. First of all, correlated event history data may be a consequence of subjects 'coming in clusters', such as members of the same family, patients attending the same general physician or medical center, or inhabitants in the same community. In these situations it is likely that the event occurrences for subjects from the same cluster are more alike than those among clusters. An example is the bone marrow transplantation study (Example 1.1.7) where the 2,009 patients were treated in one of 255 different medical centers and where patients attending the same center may share some common traits and, thereby, be more alike than patients from different centers. As a quite different situation, one may be interested in the distribution of times to entry into different states in a multi-state model, e.g., time to event no. $h = 1, 2, \ldots$ in a recurrent events situation (e.g., Figure 1.5), or time to relapse or to GvHD in the model for the bone marrow transplantation data (Example 1.1.7, Figure 1.6). Here, within any given patient, these times will likely be dependent.

A classical way of modeling dependence among observational units in statistics is to use *random effects* to represent unobserved common traits, and in event history analysis random effects models are known as *frailty* models (e.g., Hougaard, 2000). In the present section, we will discuss frailty models, first presenting some of the more technical inference details (Section 3.9.1) and, next, focusing on two major examples of using frailty models. Thus, in Section 3.9.2, we will study shared frailty models for clustered data, while Section 3.9.3 presents frailty models for recurrent events, possibly jointly with mortality. In Sections 4.3 and 5.6, we will return to the problem of dependent event history data and discuss *marginal hazard models*, and in Section 7.2 we will summarize our discussions.

3.9.1 *Inference (*)*

Frailty models may be set up quite generally for both situations giving rise to dependent data. If data come in (independent) clusters $i = 1, \ldots, n$ with n_i observations in cluster i, then the intensity model for subject h in cluster i could be

$$\alpha_{ih}(t \mid \mathbf{Z}_{ih}, \mathbf{A}_i) = A_{ih}\alpha^c_{ih}(t \mid \mathbf{Z}_{ih}), \; h = 1, \ldots, n_i. \tag{3.35}$$

In the situation where, for each subject $i = 1, \ldots, n$ (assumed independent), we have a multi-state model with possible transition types $\nu = 1, \ldots, K$ (Section 3.1), the frailty model for the type ν transition could be

$$\alpha_{\nu i}(t \mid \mathbf{Z}_i, \mathbf{A}_i) = A_{\nu i}\alpha^c_{\nu i}(t \mid \mathbf{Z}_i) \tag{3.36}$$

where the independent frailties $\mathbf{A}_i = (A_{1i}, \ldots, A_{Ki}), i = 1, \ldots, n$ follow some K-variate distribution across the population. In both cases, inference for parameters in the baseline intensities $\alpha^c_{\nu i}(t \mid \mathbf{Z}_i)$, respectively $\alpha^c_{ih}(t \mid \mathbf{Z}_{ih})$ (the conditional hazards given covariates for a frailty of 1) and in the frailty distribution may, in principle, be performed using the likelihood approach as described in Section 3.1. For given frailty, obervations are independent with the likelihood given by the Jacod formula (3.1), and the likelihood for the observed data is obtained by integrating out the frailty. This may entail technical and numerical challenges and, furthermore, for this approach to work, the assumption that censoring is independent of the frailty must be imposed (Nielsen et al., 1992). This is because the full likelihood, as explained in Section 3.1, in addition to the factors arising from (3.1), also involves factors reflecting the censoring distribution, and if these factors depend on the frailty, then integration of the likelihood over the frailty distribution may become intractable.

Models like (3.35) and (3.36) were discussed by Putter and van Houwelingen (2015) and by Balan and Putter (2020). Even though models of both types may, in principle, be analyzed, these authors concluded that frailty models are most useful for *clustered data* and for *recurrent events* (without or with competing risks). As a side remark, we can mention that frailty models may also be studied for univariate survival data to explain effects of omitted covariates (e.g., Aalen et al., 2008, ch. 6). However, as discussed by Putter and van Houwelingen (2015) and Balan and Putter (2020), this may become quite speculative because information on effects of missing covariates comes from deviations from proportional hazards and a proportional hazards model with a missing covariate and a non-proportional hazards model will be virtually indistinguishable. Following this we will, in what follows, concentrate on frailty models for clustered data and for recurrent events.

3.9.2 *Clustered data*

The set-up is as follows. Data come in independent clusters $i = 1, \ldots, n$ with n_i observations in cluster i and we will assume that the intensity for subject h in cluster i is given by

$$\alpha_{ih}(t \mid Z_{ih}, A_i) = A_i \alpha^c_{ih}(t \mid Z_{ih}), \; h = 1, \ldots, n_i. \tag{3.37}$$

Here, the Z_{ih} are observed individual level covariates and $A_1, \ldots, A_n$ are independent and identically distributed random frailties representing unobserved covariates shared by members of cluster i. We will assume that their distribution is independent of the observed covariates. Standard choices for the frailty distribution include the gamma distribution with mean $E(A) = 1$ and an unknown standard deviation $\sigma = \mathrm{SD}(A)$ to be estimated, and the log-normal distribution with $E(\log(A)) = 0$ and $\mathrm{SD}(\log(A)) = \sigma$. The parameter σ quantifies the unobserved heterogeneity among clusters and, at the same time, the intra-cluster association. We will exemplify this below. The baseline hazard could be of the Cox-form

$$\alpha^c_{ih}(t \mid Z_{ih}) = \alpha_0(t) \exp(\mathrm{LP}_{ih}),$$

with an unspecified $\alpha_0(t)$, possibly stratified, or $\alpha_0(t)$ could be piece-wise constant. The regression parameters β_ℓ in the linear predictor $\mathrm{LP}_{ih} = \beta_1 Z_{ih1} + \cdots + \beta_p Z_{ihp}$ have a *within-cluster* interpretation, $\exp(\beta_\ell)$ giving the hazard ratio for a one-unit difference in covariate $Z_{ih\ell}$ for given values of the remaining observed covariates, cf. Section 2.2.1, *and for given frailty*.

Table 3.15 *Bone marrow transplantation in acute leukemia: Frailty models for relapse-free survival taking clustering by medical center into account (BM: Bone marrow, PB: Peripheral blood, AML: Acute myelogenous leukemia, ALL: Acute lymphoblastic leukemia).*

		Gamma frailty		Log-normal frailty	
Covariate		$\widehat{\beta}$	SD	$\widehat{\beta}$	SD
Graft type	BM only vs. BM/PB	-0.175	0.087	-0.177	0.088
Disease	ALL vs. AML	0.472	0.080	0.474	0.080
Age	per 10 years	0.187	0.028	0.189	0.028
Frailty SD^2		0.117		0.139	

Bone marrow transplantation in acute leukemia

We return to the bone marrow transplantation study where patients were treated in one of 255 different medical centers. These centers had strongly varying sizes, contributing between 1 and 110 subjects. Shared frailty Cox models for relapse-free survival were fitted with, first, a gamma distributed random effect representing unobserved factors shared among patients from the same center and, next, a log-normal frailty. Table 3.15 shows the results. The two sets of coefficients using the two different frailty distributions are seen to be similar. The interpretation of, e.g., the effect of graft type is that two patients from the same center – one receiving bone marrow only, the other bone marrow or peripheral blood and having identical values for disease and age – have a ratio between their hazards for relapse-free survival of $\exp(-0.175) = 0.839$ with 95% confidence limits from 0.708 to 0.995. The estimated SD^2 of the gamma frailty distribution was $\widehat{\sigma}^2 = 0.117$. This corresponds to a value of Kendall's τ-coefficient of concordance of $0.117/(0.117+2) = 0.055$ (Hougaard, 2000, ch. 4) – a fairly small value reflecting that the within-cluster association is low. For the log-normal frailty, $\widehat{\sigma}^2 = 0.139$.

A standard Cox model stratified by center results in regression coefficients with the same within-center interpretation as the frailty models. Estimates from this model were, as follows (with SD in brackets): graft type: -0.197 (0.114), disease: -0.471 (0.091), age 0.210 (0.035). They are seen to be close to those from the frailty models, however, with somewhat larger SD. This is because many of the small centers contribute with little information to the stratified model.

3.9.3 *Recurrent events*

In the previous section, we studied the use of a *shared* frailty model for the analysis of clustered data. A similar model may be applicable for recurrent events without competing risks (Figures 1.4 or 1.5 without the absorbing state 2, respectively D). Here, an AG-type shared frailty model for the recurrent events intensity would be

$$\alpha_i(t \mid A_i) = A_i \alpha_0(t) \exp(\text{LP}_i(t)),$$

allowing for a linear predictor including time-dependent covariates. Here, there is an individual-level frailty, A_i, that is assumed to follow some distribution (typically a gamma distribution with mean 1 and SD$= \sigma$) across the population. As was the case for clustered data, the interpretation of a regression parameter in the linear predictor is a *within-subject* effect and, for that reason, the model may be questionable for time-fixed covariates. Thus, for a randomized study such as the LEADER trial (Example 1.1.6), the treatment effect would be the ratio between intensities for the same subject under, respectively, treatment and control and, in any given study, both situations would not be observed and model estimates build on extrapolations beyond the observed data. With time-dependent covariates, however, the model may be more directly applicable, a possible example being the number of previous episodes in the study of recurrence in affective disorders – see Example 1.1.5, where one of the questions addressed was whether the disease course was deteriorating. A deteriorating disease course is suggested if, for given frailty, there is a tendency for the re-admission intensity to increase with the number of previous episodes. This was studied by Kessing et al. (1999, 2004); however, under the assumption that the discharge intensity (and censoring) was independent of frailty. This assumption may be unrealistic since more severely ill patients (i.e., with a high frailty) are likely to also spend more time in the hospital than patients with a lower frailty. An extended model to address this problem was proposed by O'Keefe et al. (2018) where the discharge intensity has a frailty factor of $1/A_i$.

In many situations, there will be a competing risk in the form of a mortality that needs to be addressed. In Kessing et al. (1999, 2004), the approximation that mortality is independent of frailty was imposed – an assumption that is likely to be violated since more severely ill patients with a high frailty may also have a higher mortality rate. Having a frailty effect that is shared between the re-admission intensity and the mortality rate (e.g., Huang and Wang, 2004) may be more satisfactory; however, as discussed in Section 3.8, models with the same effect of covariates (observed or unobserved) on different transitions are not easy to interpret. More general models for recurrent events with competing risks were discussed by Cook and Lawless (2007, ch. 6) who considered a bivariate frailty (A_{i1}, A_{i2}) and the model

$$\alpha_i(t \mid A_{i1}) = A_{i1}\alpha_0(t)\exp(\mathrm{LP}_{i1}(t))$$

for the recurrent events intensity and

$$\alpha_{Di}(t \mid A_{i2}) = A_{i2}\alpha_{D0}(t)\exp(\mathrm{LP}_{i2}(t))$$

for the mortality rate. A more parsimonious model with $A_{i2} = A_{i1}^{\gamma}$, i.e.,

$$\begin{aligned} \alpha_i(t \mid A_i) &= A_i\alpha_0(t)\exp(\mathrm{LP}_{i1}(t)) \\ \alpha_{Di}(t \mid A_i) &= A_i^{\gamma}\alpha_{D0}(t)\exp(\mathrm{LP}_{i2}(t)) \end{aligned} \tag{3.38}$$

was studied by Liu et al. (2004) and Rondeau et al. (2007). Here, γ is an additional parameter to be estimated and A_i follows a gamma distribution with mean 1. In the former paper, inference was based on the EM algorithm while, in the latter paper, a penalized likelihood approach was used.

LEADER cardiovascular trial in type 2 diabetes

We will illustrate the joint frailty model by quoting results from analyzing the LEADER trial (Example 1.1.6) by Furberg et al. (2022). We will do this even though a frailty model may not be the best choice for analyzing trial data due to its within-subject interpretation of

Table 3.16 *LEADER cardiovascular trial in type 2 diabetes: Frailty models for recurrent myocardial infarctions (MI) with a gamma frailty distribution.*

	Piece-wise constant		Cox-type	
	$\widehat{\beta}$	SD	$\widehat{\beta}$	SD
Liraglutide vs. placebo	-0.177	0.088	-0.177	0.088
Frailty SD	2.38		2.39	

coefficients. The recurrent event under study is recurrent myocardial infarctions (MI) and the competing event is all-cause death. The joint frailty model with Cox baseline hazards, $\alpha_0(t), \alpha_{D0}(t)$, did not converge when using the penalized likelihood approach of Rondeau et al. (2007), so instead, models with piece-wise constant baseline hazards were studied. Analyses of frailty models with Cox-type or piece-wise constant baseline hazards for the recurrent events process alone (i.e., assuming that frailty does not affect mortality) yielded quite similar results. This is seen in Table 3.16, where the effect of treatment (log(rate ratio) for liraglutide vs. placebo) on the recurrent MI rate is $\widehat{\beta} = -0.177\,(\text{SD} = 0.088)$ for both the piece-wise constant and the Cox-type model. The estimated frailty SD ($\widehat{\sigma}$) in the two models (assuming a gamma distributed frailty) was 2.38 and 2.39, respectively.

The similar estimates from the joint frailty model, Equation (3.38), were $\widehat{\beta} = -0.186\,(\text{SD} = 0.068)$ and $\widehat{\sigma} = 0.947\,(\text{SD} = 0.031)$, see Table 3.17. In this model, the estimated effect of treatment on mortality was $\widehat{\beta}_D = -0.211\,(\text{SD} = 0.078)$ and the association parameter linking the frailties for recurrent events and mortality was estimated to be $\widehat{\gamma} = 1.860\,(\text{SD} = 0.115)$. The interpretation is that, for any given patient (i.e., for given frailty), treatment with liraglutide reduces the MI rate by a factor of $\exp(-0.186) = 0.830$ and reduces the mortality rate by $\exp(-0.211) = 0.809$. Furthermore, patients at a high rate of an MI (high frailty) also have a high mortality rate ($\widehat{\gamma} > 0$). Heterogeneity among patients (frailty SD, $\widehat{\sigma}$) appears to be considerably higher when not accounting for mortality.

Table 3.17 *LEADER cardiovascular trial in type 2 diabetes: Joint frailty model for recurrent myocardial infarctions (MI) and all-cause mortality – piece-wise constant baseline hazards and gamma frailty distribution.*

	Recurrent MI		All-cause death	
	$\widehat{\beta}$	SD	$\widehat{\beta}$	SD
Liraglutide vs. placebo	-0.186	0.068	-0.211	0.078
Frailty SD ($\widehat{\sigma}$)		0.947		
Association ($\widehat{\gamma}$)		1.86		

3.10 Exercises

Exercise 3.1 (*) Show that, under the null hypothesis $H_0 : A_0(t) = A_1(t)$, the test statistic $\int_0^t K(u)\big(d\widehat{A}_1(u) - d\widehat{A}_0(u)\big)$ is a martingale (Section 3.2.2).

Exercise 3.2 (*) Show that, when evaluated at the true parameter vector $\boldsymbol{\beta}_0$, the Cox partial likelihood score

$$\int_0^t \sum_i (\mathbf{Z}_i - \frac{\sum_j Y_j(u)\mathbf{Z}_j \exp(\boldsymbol{\beta}_0^\mathsf{T}\mathbf{Z}_j)}{\sum_j Y_j(u)\exp(\boldsymbol{\beta}_0^\mathsf{T}\mathbf{Z}_j)})dN_i(u)$$

is a martingale (Section 3.3).

Exercise 3.3 (*) Show that, for a Cox model with a single binary covariate, the score test for the hypothesis $\beta = 0$ based on the first and second derivative of $\log \text{PL}(\beta)$ (Equation (3.16)) is equal to the logrank test.

Exercise 3.4 (*) Show that, for the stratified Cox model (3.20), the profile likelihood is given by (3.21) and the resulting Breslow estimator by (3.22).

Exercise 3.5 (*) Consider the situation in Section 3.4 with categorical covariates and show that the likelihood is given by

$$\prod_{\ell=1}^{L}\prod_{c\in\mathscr{C}}(\alpha_{0\ell}\theta_c)^{N_{\ell c}}\exp(-\alpha_{0\ell}\theta_c Y_{\ell c}).$$

Exercise 3.6 (*) Derive the estimating equations for the model studied in Section 3.8.4.

Exercise 3.7 Consider the Cox model for stroke-free survival in the Copenhagen Holter study including the covariates ESVEA, sex, age, and systolic blood pressure (Exercise 2.4). Test, using time-dependent covariates, whether the effects of these covariates may be described as time-constant hazard ratios.

Exercise 3.8 Consider the Cox model for stroke-free survival in the Copenhagen Holter study including the covariates ESVEA, sex, age, and systolic blood pressure. Add to that model the time-dependent covariate $I(\text{AF} \leq t)$. How does this affect the effect of ESVEA?

Exercise 3.9 Consider the Cox model for stroke-free survival in the Copenhagen Holter study including the covariates ESVEA, sex, age, and systolic blood pressure. Add to that model, incorrectly, the covariate AF – now considered as *time-fixed*. How does this affect the AF-effect?

Exercise 3.10 Consider an illness-death model for the Copenhagen Holter study with states '0: Alive without AF or stroke', '1: Alive with AF and no stroke', '2: Dead or stroke', see Figures 1.3 and 1.7.

1. Fit separate Cox models for the rates of the composite end-point for subjects without or with AF, i.e., for the $0 \to 2$ and $1 \to 2$ transitions including the covariates ESVEA, sex, age, and systolic blood pressure. The time-variable in both models should be time since recruitment.
2. Examine to what extent a combined model for the two intensities (i.e., possibly with common regression coefficients and/or proportional hazards between the $0 \to 2$ and $1 \to 2$ transition rates) may be fitted.

Exercise 3.11 Consider the data on repeated episodes in affective disorder, Example 1.1.5.

1. Fit separate gamma frailty models for unipolar and bipolar patients including the covariate 'number of previous events $N(t-)$' assuming (not quite satisfactorily!) that the mortality rate is independent of frailty.
2. Do the recurrence rates tend to increase with number of previous episodes?

Exercise 3.12 Consider the data on mortality in relation to childhood vaccinations in Guinea-Bissau, Example 1.1.2.

1. Fit a gamma frailty model with a random effect of cluster ('village') including binary variables for BCG and DTP vaccination and adjusting for age at recruitment (i.e., using time since recruitment as time-variable). Compare the results with those in Table 2.12.
2. Fit a Cox model stratified on cluster, including binary variables for BCG and DTP vaccination and adjusting for age at recruitment. Compare with the results from the frailty model.

Chapter 4

Intuition for marginal models

In this chapter, we will give a less technical introduction to the different models for risks and other marginal parameters to be discussed in more mathematical details in Chapter 5. Along with the introduction of the models, examples will be given to illustrate how results from analysis of these models can be interpreted. In Chapter 2, we gave an intuitive introduction to models for the basic parameter in multi-state models, the transition *intensity*. As explained in Section 1.2, knowing all rates in principle enables calculation of marginal model parameters such as the probability (risk), $Q_h(t)$ of occupying state h at time t. In some multi-state models it is possible, mathematically, to describe this relationship. This is the case for the two-state survival model of Figure 1.1, the competing risks model (Figure 1.2), and the progressive illness-death model (Figure 1.3). This means that if estimates are given for all transition rates, e.g., via a regression model, then the marginal parameters may be estimated (for given covariates) by *plug-in*. Plug-in refers to the idea of estimating a given function, say $g(\theta)$, of the parameter θ, by first getting an estimate $\widehat{\theta}$ of θ, and then using $g(\widehat{\theta})$ as the plug-in estimate of $g(\theta)$. This is the topic of Section 4.1. As we shall see there, in a regression situation this activity does not provide parameters that directly describe how the marginal parameters are associated with the covariates. Therefore, it may be of interest to set up regression models where marginal parameters are linked directly to covariates. This is the topic of Section 4.2. The *direct model* approach has the additional advantage that while plug-in builds on correctly specified models for all intensities (and, thereby, a risk of *model misspecification* is run), only a single directly specified model for the marginal parameter needs to be correct. In Section 4.3, we introduce *marginal hazard* models that may be applicable in situations where an independence assumption need not be justified and in Section 4.4, we return to a discussion of the concept of *independent censoring*, including ways of studying whether censoring is affected by observed covariates.

4.1 Plug-in methods

4.1.1 Two-state model

In this model (Figure 1.11) there is only one transition rate, $\alpha_{01}(t) = \alpha(t)$, the hazard function for the distribution of the survival time T. It has the interpretation that $\alpha(t)dt$ is (approximately) the conditional probability of dying before time $t + dt$ (for a small $dt > 0$) given survival till time t

$$\alpha(t)dt \approx P(T \leq t + dt \mid T > t). \tag{4.1}$$

From the hazard function, the *survival function*, a marginal parameter, $S(t) = Q_0(t)$, is derived as follows. Divide the interval from 0 to t into small intervals all of length Δ. From Equation (4.1), the probability of surviving the next little time interval (from u to $u+\Delta$) given survival till u is $(1-\alpha(u)\Delta)$ as illustrated in Figure 4.1.

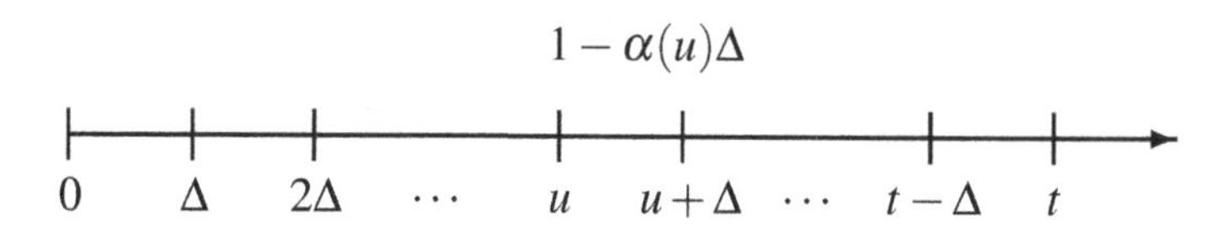

Figure 4.1 *The probability of surviving the time interval from u to $u+\Delta$ given survival till u is* $(1-\alpha(u)\Delta)$.

The marginal probability, $S(t)$, is the product of such factors (conditional probabilities) for $u < t$

$$S(t) = (1-\alpha(\Delta)\Delta)\cdot(1-\alpha(2\Delta)\Delta)\cdots(1-\alpha(u)\Delta)\cdots(1-\alpha(t-\Delta)\Delta). \quad (4.2)$$

This observation leads to the *Kaplan-Meier estimator* for the survival function (Kaplan and Meier, 1958), as follows. We follow the arguments in Section 2.1.1 leading to the Nelson-Aalen estimator for the cumulative hazard, i.e., estimating $\alpha(u)\Delta$ by the fraction

$$\frac{\text{No. of patients with an event in } (u, u+\Delta)}{\text{No. of patients at risk of an event just before time } u} = \frac{dN(u)}{Y(u)},$$

where $dN(u)$, the observed number of failures at time u, is typically 0 or 1. The survival function is then estimated by plugging-in this fraction into the product-representation for $S(t)$. Hereby, the Kaplan-Meier estimator is obtained

$$\begin{aligned} \widehat{S}(t) &= (1-\frac{dN(X_1)}{Y(X_1)})\cdots(1-\frac{dN(X_k)}{Y(X_k)}) \\ &= \prod_{\text{event times, } X \leq t} (1-\frac{1}{\text{No. at risk at } X}). \end{aligned} \quad (4.3)$$

In Equation (4.3), $X_1, \ldots, X_k$ are the individual observation times before time t (some are event times, others are censoring times) and the second line of the equation uses the 'product symbol' $\prod$ which is similar to the 'summation symbol' $\sum$ used previously. Note that, for times $u < t$ with no observed event, $dN(u)$ is 0, and the factor $1 - dN(u)/Y(u)$ becomes 1, so the plug-in estimator effectively becomes a product over observed event times before time t as seen in Equation (4.3). The standard deviation of $\widehat{S}(t)$ can be estimated using the *Greenwood formula* whereby confidence limits around $\widehat{S}(t)$ may be obtained (Kaplan and Meier, 1958). This is typically done by taking as starting point symmetric confidence limits for $\log(A(t))$, see Section 2.1.1. This amounts to 95% confidence limits for $S(t)$ obtained by raising $\widehat{S}(t)$ to the powers $\exp(\pm 1.96 \cdot \text{SD}/(\widehat{S}(t)\widehat{A}(t)))$ where SD is the Greenwood estimate.

As an example, Figure 4.2 shows the Kaplan-Meier estimates for the two treatment groups in the PBC3 trial (Example 1.1.1) for the time to the composite end-point 'failure of medical

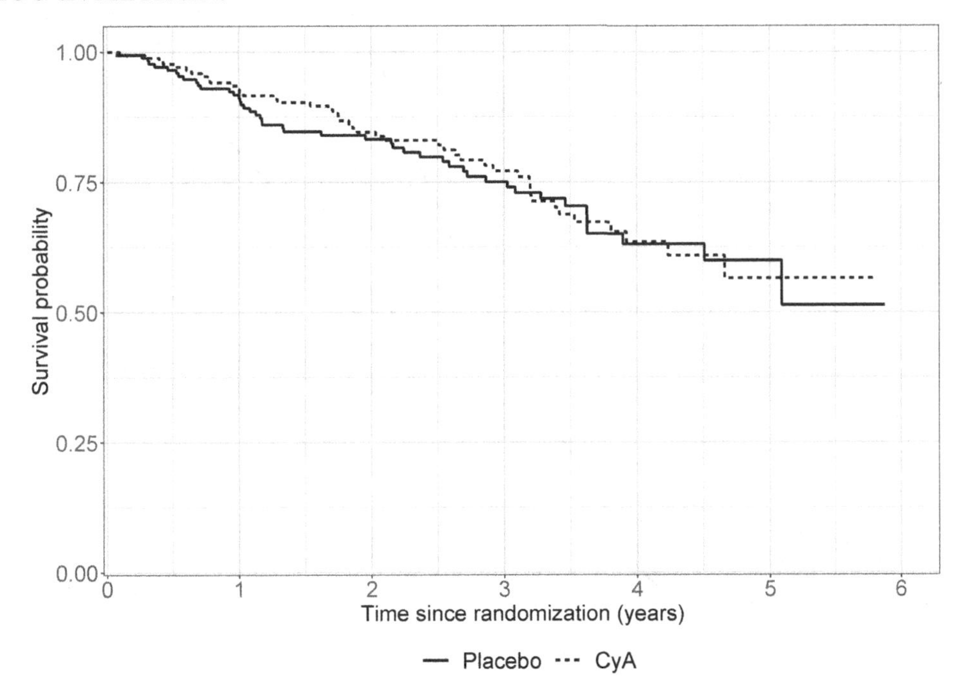

Figure 4.2 *PBC3 trial in liver cirrhosis: Kaplan-Meier estimates by treatment.*

treatment'. Similar to Figure 2.1, displaying the Nelson-Aalen estimates, the figure suggests that, unadjusted for prognostic variables, the survival probability is unaffected by treatment. This should be no surprise since the two estimators build on exactly the same information and, therefore, are in *one-to-one correspondence* with each other. One important difference, however, lies in the interpretation of the values on the vertical axis. In Figure 2.1, the cumulative hazard was estimated and, as discussed there, the interpretation of this quantity is not so direct. Figure 4.2, on the other hand, depicts the fraction of patients that, over time, is still event-free. At 2 years, the estimates are 0.846 in the CyA group and 0.832 for placebo with values of the Greenwood SD equal to 0.029, respectively 0.030. The resulting 95% confidence interval for $S(2 \text{ years})$ is then (0.766, 0.882) for the placebo group and (0.800, 0.894) for CyA. To enhance readability, confidence limits have not been added to Figure 4.2.

Let us have a closer look at the product representation Equation (4.2) for $S(t)$ to understand this one-to-one correspondence. From the product representation we get a sum representation by using the logarithm

$$-\log(S(t)) = \sum_{u<t} -\log(1-\alpha(u)\Delta) \approx \sum_{u\leq t} \alpha(u)\Delta,$$

where for the approximation we have used that $-\log(1-x) \approx x$ which holds for small positive values of x as seen in Figure 4.3. It then follows that

$$S(t) = \exp(-\int_0^t \alpha(u)du) \tag{4.4}$$

because, for small Δ, the sum $\sum_{u\leq t} \alpha(u)\Delta$ will be equal to the integral in Equation (4.4).

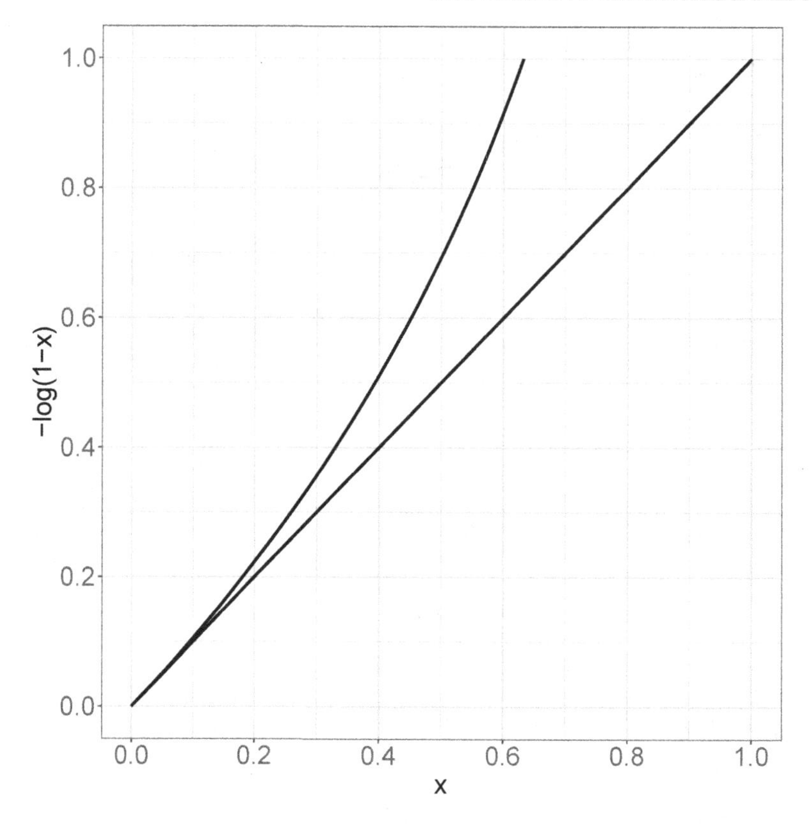

Figure 4.3 *The functions* $y = -\log(1-x)$ *and* $y = x$ *for* $0 < x < 1$. *Note that the two functions almost coincide for small values of* $x > 0$.

This equation was already given in Equation (1.2). This formula expresses the one-to-one correspondence between the survival function and the hazard. Knowing the hazard is knowing the survival function, and vice versa.

Suppose now that in the PBC3 example we have estimated the hazard by assuming it to be piece-wise constant as in Table 2.1. We can then estimate $S(t)$ by plug-in

$$\widehat{S}(t) = \exp(-\int_0^t \widehat{\alpha}(u)du).$$

Figure 4.4 shows the Kaplan-Meier estimator for the PBC3 placebo group together with the plug-in estimator using a piece-wise constant hazard model for $\alpha(t)$ and, just like in Figure 2.3, it is seen that the two models give quite similar results.

Having the one-to-one correspondence, Equation (4.4), a regression model for $\alpha(t)$ induces a regression model for $S(t)$. Assume a Cox regression model for $\alpha(t)$

$$\alpha(t) = \alpha_0(t)\exp(\text{LP}),$$

where the linear predictor LP is given by Equation (1.4). The survival function is then given by

$$S(t \mid Z) = \exp\big(-\int_0^t \alpha_0(u)\exp(\text{LP})du\big) = \exp\big(-A_0(t)\exp(\text{LP})\big), \tag{4.5}$$

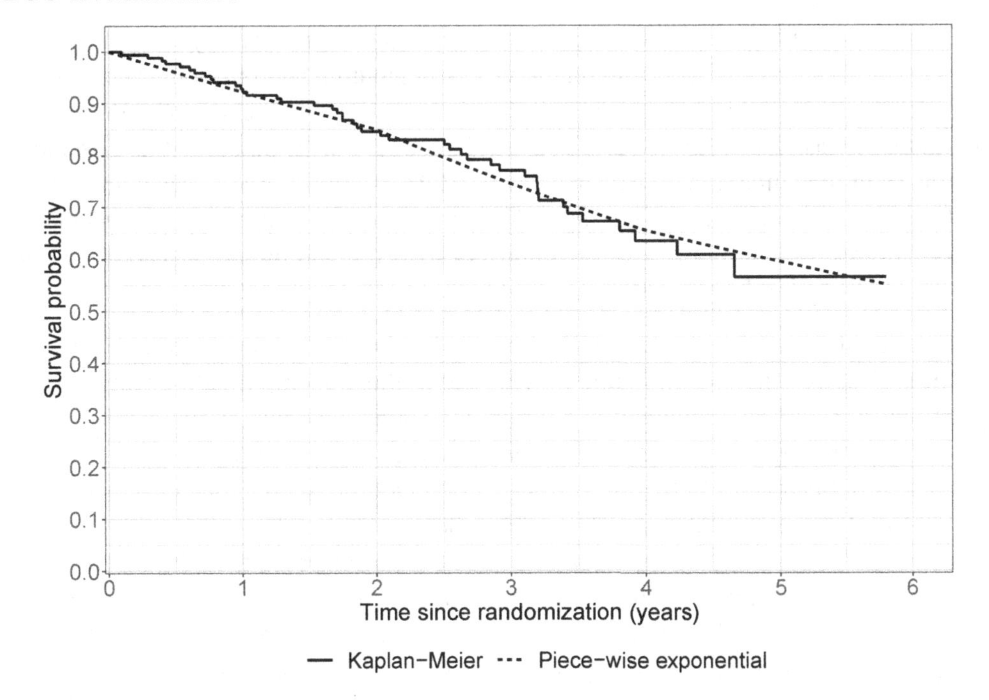

Figure 4.4 *PBC3 trial in liver cirrhosis: Estimated survival curves for the placebo group.*

where $A_0(t) = \int_0^t \alpha_0(u)du$. Using the complementary log-log transformation *cloglog* of the distribution function $F(t)$

$$\text{cloglog}(F(t)) = \log(-\log(1-F(t))) = \log(-\log(S(t))),$$

we get the regression model in the cloglog scale

$$\log(-\log(S(t \mid Z)) = \log(A_0(t)) + \text{LP}. \tag{4.6}$$

The cloglog function is the *link function* which takes us from the marginal parameter to the linear predictor. As an example, let us consider the models for the PBC3 trial presented in Table 2.7. Here, the survival function at time t for a CyA treated subject ($Z_1 = 1$) with biochemical values albumin $= Z_2$, bilirubin $= Z_3$ may (based on the Cox model) be estimated by

$$\widehat{S}(t \mid Z_1 = 1, Z_2, Z_3) = \exp\bigl(-\widehat{A}_0(t)\exp(-0.574 - 0.091Z_2 + 0.665\log_2(Z_3))\bigr)$$

while, for a placebo treated patient with the same values of the biochemical variables, the estimated survival function at time t is

$$\widehat{S}(t \mid Z_1 = 0, Z_2, Z_3) = \exp\bigl(-\widehat{A}_0(t)\exp(-0.091Z_2 + 0.665\log_2(Z_3))\bigr).$$

Figure 4.5 shows the estimated survival curves for albumin = 38g/L and bilirubin = 45μmol/L – values close to the observed average values among all patients. We can see that, on the probability scale Equation (4.5), the treatment effect is time-dependent, while on the cloglog scale Equation (4.6), the effect is time-constant as assumed in the Cox and

Poisson models, see Figure 4.6. It is also the case that, on the probability scale, the difference between the curves will depend on the values of albumin and bilirubin. The fact that, on the cloglog scale, vertical distances between survival curves are constant under proportional hazards was classically used to construct goodness-of-fit plots for the Cox model based on a stratified model (e.g., Andersen et al., 1993, Section VII.3) – a technique that is still offered by standard software packages. However, since we find that these plots may be hard to interpret, we will not provide examples of their use and prefer, instead, plots such as those exemplified in Figure 2.10.

If a single set of covariate-adjusted survival curves for the two treatment groups is desired, then this may be obtained by averaging curves such as those exemplified over the observed distribution of Z_2, Z_3. As explained in Section 1.2.5, this is known as the *g-formula* and works by performing two predictions for each subject, i, one setting treatment to CyA and one setting treatment to placebo, and in both predictions keeping the observed values (Z_{2i}, Z_{3i}) for albumin and bilirubin. The predictions for each value of treatment are then averaged over $i = 1, \dots, n$ (see Equation (1.5)). The g-formula results in the curves shown in Figure 4.7. Note that, if randomization in the PBC3 trial had been more successful, then these curves would resemble the Kaplan-Meier estimates in Figure 4.2. Using the curves obtained based on the g-formula, it is possible to visualize the treatment effect on the probability scale after covariate-adjustment using plug-in. At 2 years, the values of the curves in Figure 4.7 are 0.799 for placebo and 0.867 for CyA with estimated SD, respectively, 0.025 and 0.019 – slightly smaller than what is obtained based on 1,000 bootstrap replications, namely SD values of 0.028 and 0.022. The treatment effect (risk difference at 2 years) is thus $0.867 - 0.799 = 0.068$, and it has an estimated SD of 0.026 close to that based on 1,000 bootstrap replications which is 0.027.

On a technical note, one may wonder why one does typically not estimate $S(t)$ non-parametrically by plugging-in the Nelson-Aalen estimator into Equation (4.4). The answer is that for a distribution with jumps as the one estimated by Nelson-Aalen, the relationship between the cumulative hazard and the survival function is given by the product-representation rather than by Equation (4.4). Having said this, it should be mentioned that computer packages often offer the estimator 'exp($-$Nelson-Aalen)' as an alternative to Kaplan-Meier (and '$-\log$(Kaplan-Meier)' as an alternative to Nelson-Aalen) and that this in practice makes little difference. Following this remark, the survival function for given covariates based on the Cox model could, alternatively, have been estimated by a product-representation based on the Breslow estimator for the cumulative baseline hazard. Figure 4.8 shows the result of using these, alternative, estimators when predicting the survival function in the two treatment groups for albumin $= 38$ and bilirubin $= 45$ and, as we can see, this makes virtually no difference compared to Figure 4.5.

Restricted mean life time

Having estimated $Q_0(t) = S(t)$, estimates of the marginal parameter τ-restricted mean life time $\varepsilon_0(\tau)$, i.e., the expected time spent in state 0 before time τ, may be obtained by plug-in. This is because

$$\varepsilon_0(\tau) = \int_0^\tau S(t)dt, \tag{4.7}$$

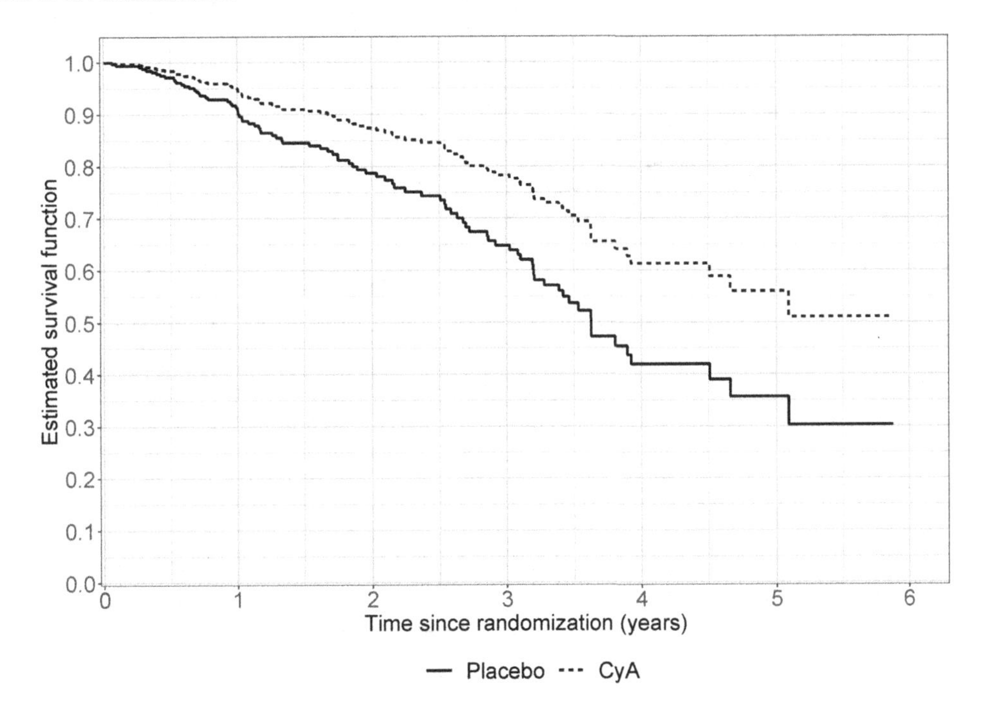

Figure 4.5 *PBC3 trial in liver cirrhosis: Estimated survival curves for a patient with albumin = 38 g/L and bilirubin = 45 µmol/L based on a Cox model. There is one curve for each value of treatment.*

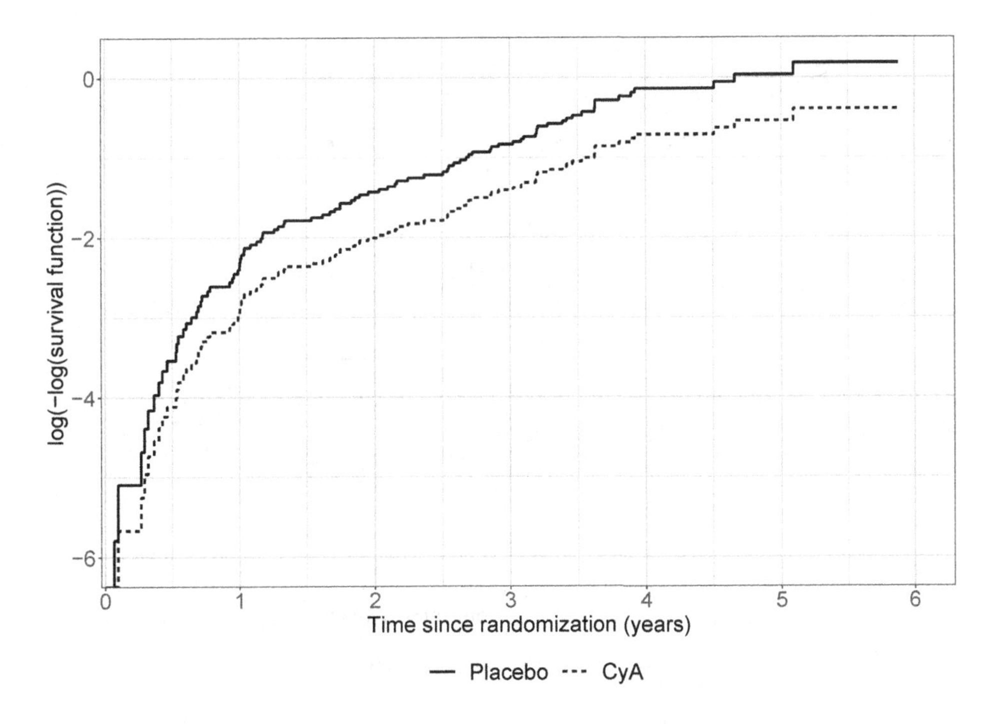

Figure 4.6 *PBC3 trial in liver cirrhosis: Estimated survival curves for a patient with albumin = 38 g/L and bilirubin = 45 µmol/L based on a Cox model. The vertical scale is cloglog-transformed, and there is one curve for each value of treatment.*

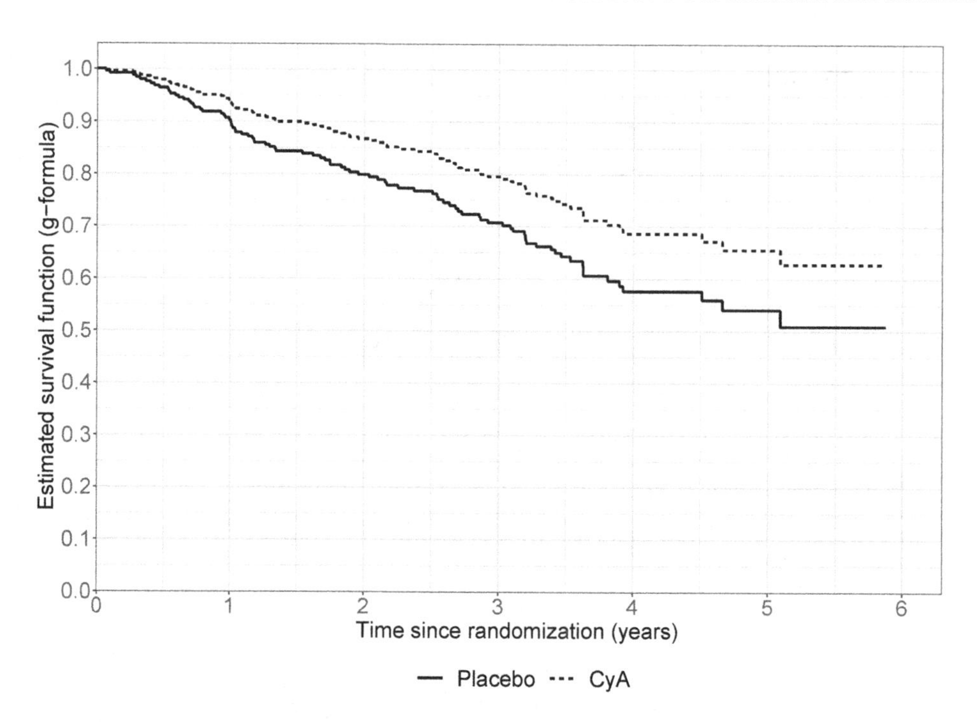

Figure 4.7 *PBC3 trial in liver cirrhosis: Estimated survival curve in the two treatment groups based on the g-formula.*

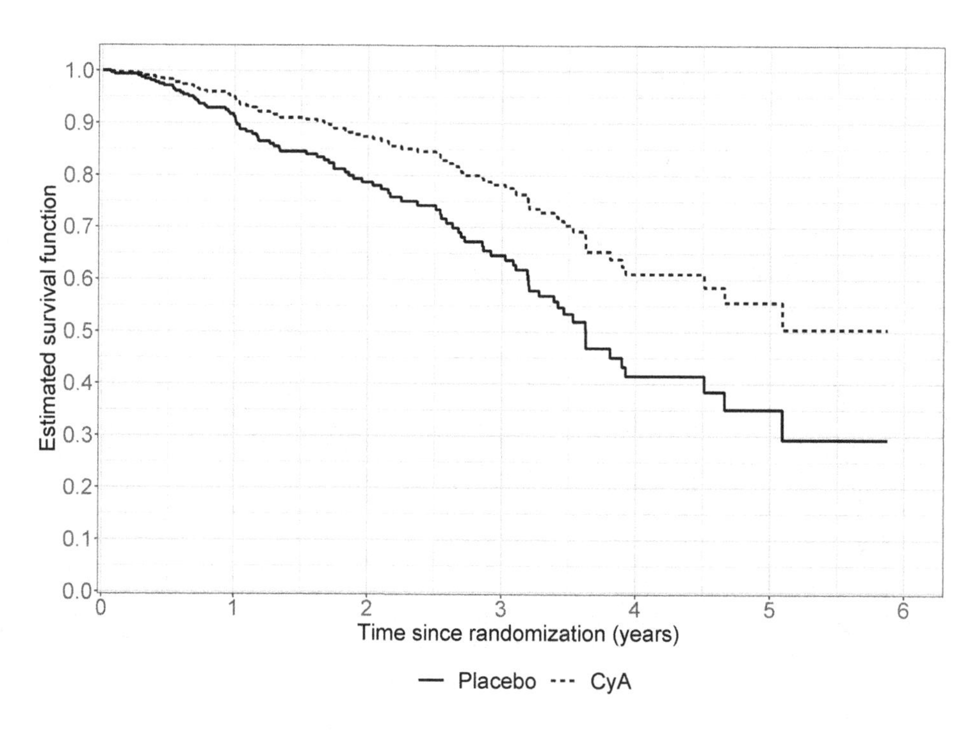

Figure 4.8 *PBC3 trial in liver cirrhosis: Estimated survival curves for a patient with albumin = 38 g/L and bilirubin = 45 μmol/L based on a Cox model. There is one curve for each value of treatment and estimates are based on the product-formula.*

Table 4.1 *PBC3 trial in liver cirrhosis: Estimated 3-year restricted means (and SD) by treatment group. *: Based on 1,000 bootstrap replications.*

Model	(albumin, bilirubin)	Placebo $\widehat{\varepsilon}_0(3)$	Placebo SD	CyA $\widehat{\varepsilon}_0(3)$	CyA SD
Non-parametric		2.61	0.064	2.68	0.057
			0.064*		0.058*
Cox	(38, 45)	2.53	0.068*	2.72	0.054*
	(20, 90)	0.96	0.268*	1.38	0.279*
	g-formula	2.55	0.060*	2.71	0.046*

i.e., it is the area under the survival function (Figure 1.10). The equation can be derived in the following (perhaps less intuitive) way. If T is the life time, then $\varepsilon_0(\tau)$ is the expected value of the minimum, $\min(T,\tau)$, of T and the threshold τ. This random variable may be written as

$$\min(T,\tau) = \int_0^{\min(T,\tau)} 1dt = \int_0^{\tau} I(T > t)dt$$

and, therefore, $\varepsilon_0(\tau)$, the expected value of this is

$$E(\int_0^{\tau} (I(T > t)dt) = \int_0^{\tau} E(I(T > t))dt = \int_0^{\tau} P(T > t)dt = \int_0^{\tau} S(t)dt$$

which is exactly the right-hand side of Equation (4.7). A non-parametric estimator for $\varepsilon_0(\tau)$ is obtained by plugging-in the Kaplan-Meier estimator for $S(t)$ into Equation (4.7) while a regression model for $\varepsilon_0(\tau)$ may be obtained by plugging-in, e.g., a Cox model-based estimator for $S(t \mid Z)$ into the equation.

The method is illustrated using the PBC3 data, and Table 4.1 shows the results. It is seen that, unadjusted, the 3-year restricted means do not differ between the two treatment groups. If based on a Cox model, the restricted means differ according to the chosen covariate pattern and single, adjusted values may be obtained using the g-formula. Most of the SD values in the table are based on 1,000 bootstrap replications, and it is seen that the scenario with albumin and bilirubin values of 20 and 90 provides larger SD – the explanation being that these values are more extreme compared to the observed distributions of the two biochemical variables.

4.1.2 *Competing risks*

In the competing risks model (Figure 1.12), there is one transition hazard for each absorbing state h, the cause-specific hazard $\alpha_{0h}(t) = \alpha_h(t)$. We discussed in Chapter 2 how each single hazard can be analyzed using, e.g., the Nelson-Aalen estimator or a Cox regression model and an important point was that modeling of $\alpha_1(t)$ and $\alpha_2(t)$ could be done separately. An intuitive argument for that was given, and in Section 2.4 the cause-specific hazards for the two competing events 'transplantation' and 'death without transplantation' in the PBC3 trial (Example 1.1.1) were analyzed.

The situation is different when we want to estimate cumulative probabilities over time, i.e., when we wish to go from the two cause-specific hazards to the three state occupation probabilities: The overall (survival) probability of no event $S(t) = Q_0(t)$, the probability (or risk) of transplantation, the cumulative incidence $F_1(t) = Q_1(t)$, and the probability of death without transplantation, the cumulative incidence $F_2(t) = Q_2(t)$.

The overall survival probability $S(t)$ has the same relation to the total transition hazard out of the initial state 0 as the one we saw in the case of overall survival data, Equation (4.4), i.e.,

$$S(t) = \exp(-\int_0^t (\alpha_1(u) + \alpha_2(u))du).$$

To derive the expression for the cause h cumulative incidence ($h = 1,2$), the following argument applies. Recall the definition of the cause-specific hazard

$$\alpha_h(u)du \approx P(\text{in state } h \text{ at time } u + du \mid \text{in state } 0 \text{ at time } u)$$

(Section 1.2.3). The cumulative incidence at time t is the probability that a cause h event has happened between time 0 and time t. The probability that it happens in the little time interval from u to $u + du$ (with $0 < u \leq t$) is the probability $S(u)$ of no events before time u (being in state 0 at time u) times the conditional probability $\alpha_h(u)du$ of cause h happening in that little interval given no previous events as illustrated in Figure 4.9.

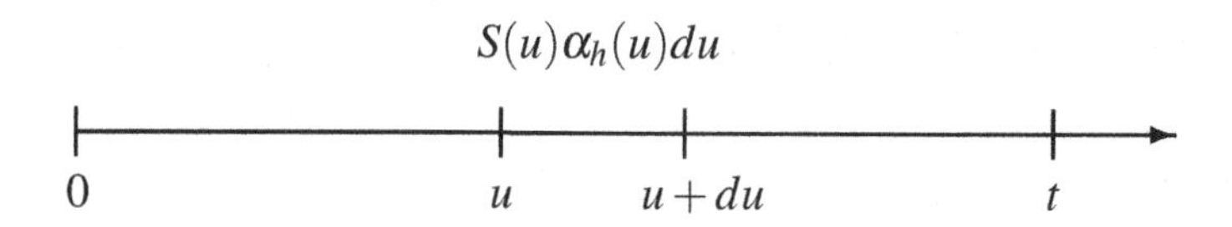

Figure 4.9 *The probability that a cause h event happens in the little time interval from u to u + du is the probability S(u) of no events before time u times the conditional probability $\alpha_h(u)du$.*

Since, for different values of u, the events 'cause h happens in the interval from u to $u + du$' are exclusive, their total probability is the sum (integral) of the separate probabilities from 0 to t, i.e.,

$$F_h(t) = \int_0^t S(u)\alpha_h(u)du, \tag{4.8}$$

an equation that was already given in (1.3).

Estimating $S(t)$ by the overall Kaplan-Meier estimator and the cumulative cause h specific hazard by the Nelson-Aalen estimator, plug-in into Equation (4.8) leads to the non-parametric *Aalen-Johansen* estimator

$$\widehat{F}_h(t) = \sum_{\text{cause } h \text{ event times } X \leq t} \widehat{S}(X-)\frac{1}{Y(X)} \tag{4.9}$$

of the cause h cumulative incidence. In Equation (4.9), $\widehat{S}(X-)$ is the Kaplan-Meier value just before the event time, X, i.e., the jump at that time is not included. Confidence limits around $\widehat{F}_h(t)$ may also be computed, preferably based on a symmetric confidence interval for cloglog$(F_h(t))$ in the same way as for the Kaplan-Meier estimator (Section 4.1.1).

In a similar fashion, Cox models (i.e., regression coefficients $(\widehat{\beta}_1, \widehat{\beta}_2)$ and Breslow estimates $(\widehat{A}_{10}, \widehat{A}_{20})$) for each of the cause-specific hazards may be plugged-in into Equation (4.8) to obtain estimates of $F_h(t \mid Z)$ for given values of covariates Z.

It is very important to notice that, via the factor $S(\cdot)$, the cumulative incidence for cause 1 *depends on both of the cause-specific hazards* $\alpha_1(\cdot)$ and $\alpha_2(\cdot)$. This means that, in spite of the fact that inference for $\alpha_1(\cdot)$ could be carried out by, formally, censoring for cause 2 events, both causes must be taken into account when estimating the cumulative risk of cause 1 events. An estimator for $F_1(t)$ obtained as '$1 - \widehat{S}_1(t)$' where $\widehat{S}_1(t)$ is a Kaplan-Meier estimator counting only cause 1 events as events (and cause 2 events as censorings) will be a *biased* estimator. This Kaplan-Meier curve estimates $\exp(-\int_0^t \alpha_1(u)du)$ – a quantity that does not possess a probability interpretation in the population where both causes are operating (the population for which we wish to make inference, cf. Section 1.3). The incorrect cumulative incidence estimator $1 - \widehat{S}_1(t)$ will be *upwards biased* because $F_1(t) \leq 1 - \exp(-\int_0^t \alpha_1(u)du)$, intuitively because, by counting cause 2 events as censorings, we pretend that, had these subjects not been 'censored', then they would still be at risk for the event of interest (i.e., cause 1), see, e.g., Andersen et al. (2012). In other words, the one-to-one correspondence between a single rate $(\alpha_h(t))$ and the risk $(F_h(t))$ that we saw for the two-state model (Section 4.1.1) does not exist in the competing risks model: To compute the cause h risk, $F_h(t)$, not only the rate for cause h is needed, but also the rates for the competing cause(s) (and vice versa).

We will illustrate cumulative incidences and the bias incurred when using the incorrect estimator using the PBC3 data. Figures 4.10 and 4.11 show, for the placebo group, a *stacked plot* of cumulative incidences and overall survival function computed, respectively, correctly by using the Aalen-Johansen estimator, Equation (4.9), and incorrectly using '1– Kaplan-Meier' based on a single cause. More specifically, $\widehat{F}_1(t)$, $\widehat{F}_1(t) + \widehat{F}_2(t)$ and $\widehat{F}_1(t) + \widehat{F}_2(t) + \widehat{S}(t)$ are plotted against t. In Figure 4.10, the latter curve is, correctly, equal to 1 while, in Figure 4.11, this sum exceeds 1 because '1– Kaplan-Meier' is an upwards biased estimator of the cumulative incidence. Note that (in the correct Figure 4.10), the values of the vertical axis have simple interpretations as the fractions of patients who, over time, are expected to experience the various events.

We will also illustrate predicted cumulative incidences from cause-specific Cox models. This can be done for a given pattern for the covariates that enter into the Cox models. Figure 4.12 shows predicted stacked cumulative incidences and overall survival for placebo treated female patients with, respectively (age, albumin, bilirubin) equal to (40, 38, 45), (40, 20, 90), and (60, 38, 45). It is seen that, for the second pattern, both cumulative risks are considerably larger than for the first while, comparing the first and the last pattern, it is seen that the older patient has a much higher risk of death and a lower risk of transplantation.

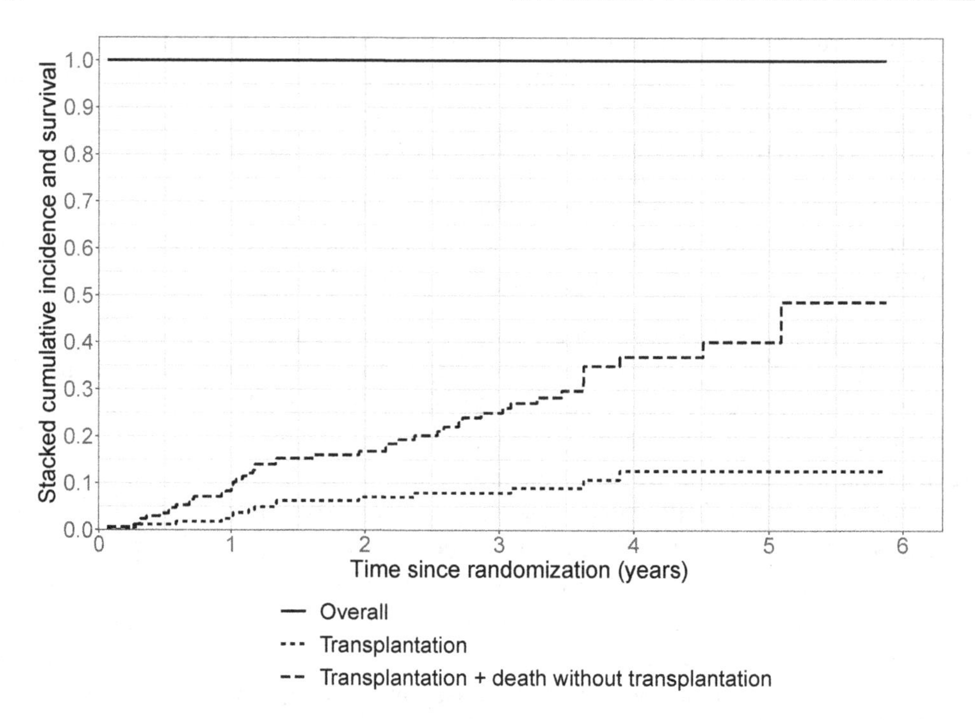

Figure 4.10 *PBC3 trial in liver cirrhosis: Stacked cumulative incidence and survival curves for the placebo group. Cumulative incidences are, correctly, estimated using the Aalen-Johansen estimator, Equation (4.9).*

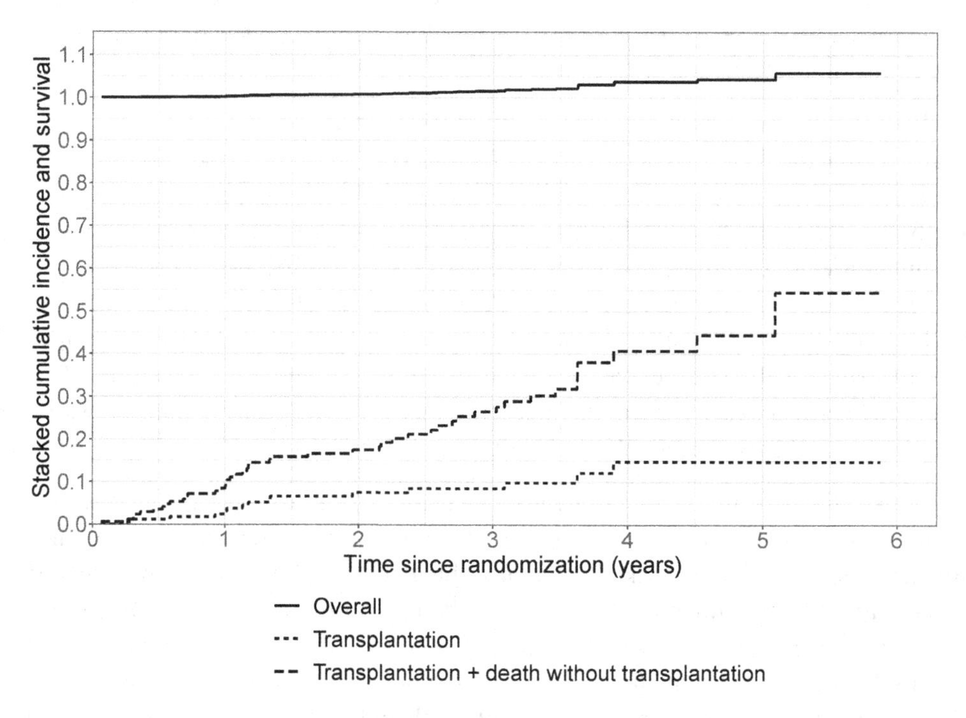

Figure 4.11 *PBC3 trial in liver cirrhosis: Stacked cumulative incidence and survival curves for the placebo group. Cumulative incidences are, incorrectly, estimated using the '1−Kaplan-Meier' estimators based on single causes.*

(a) *40 years old, albumin = 38 g/L, bilirubin = 45 μmol/L*

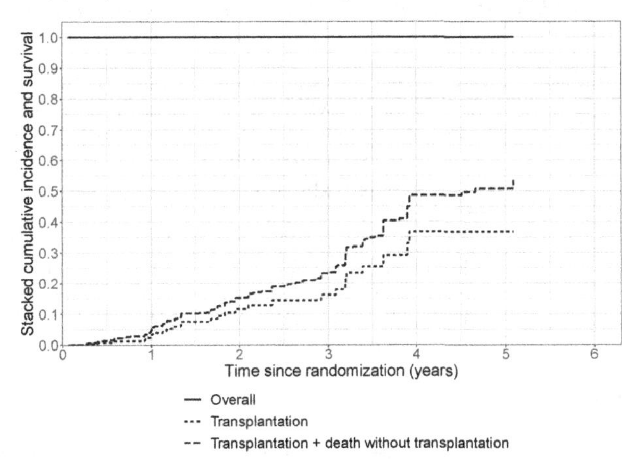

(b) *40 years old, albumin = 20 g/L, bilirubin = 90 μmol/L*

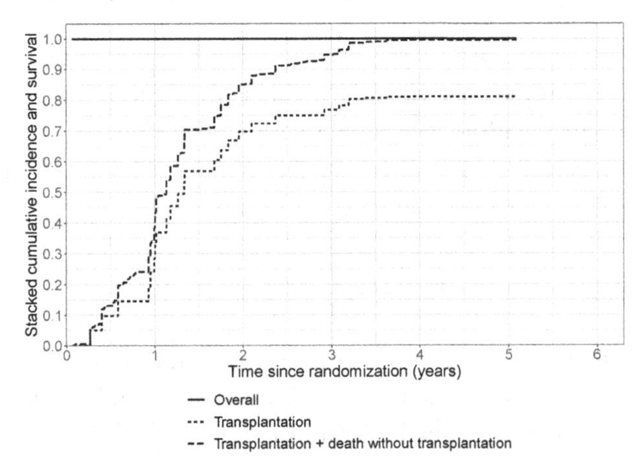

(c) *60 years old, albumin = 38 g/L, bilirubin = 45 μmol/L*

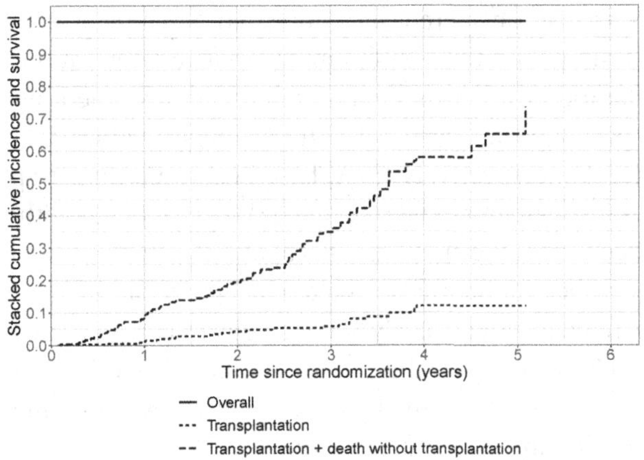

Figure 4.12 *PBC3 trial in liver cirrhosis: Predicted, stacked cumulative incidence and survival curves for three women in the placebo group based on cause-specific Cox models.*

Table 4.2 *PBC3 trial in liver cirrhosis: Estimated time lost (in years) before 3 years due to transplantation (T) and death without transplantation (D) by treatment group in four scenarios (F: female).*

Scenario	Event type	Sex	Age	Albumin	Bilirubin	Placebo	CyA
1	T			No adjustment		0.142	0.086
	D			No adjustment		0.250	0.235
2	T	F	40	38	45	0.220	0.117
	D	F	40	38	45	0.090	0.061
3	D	F	40	20	90	1.377	0.967
	D	F	40	20	90	0.364	0.302
4	T	F	60	38	45	0.080	0.043
	D	F	60	38	45	0.373	0.256

Cause-specific time lost

For overall survival data we saw how the τ-restricted mean survival time, $\varepsilon_0(\tau)$ could be estimated as the area under the survival curve. This is the expected time lived before time τ. The maximum time lived before τ is equal to τ and, therefore, $\tau - \varepsilon_0(\tau) = \int_0^\tau F(t)dt$ is the expected *time lost* before time τ. We have that $F(t) = F_1(t) + F_2(t)$ is the sum of the cause-specific cumulative incidences and

$$\varepsilon_h(\tau) = \int_0^\tau F_h(t)dt$$

is the *expected time lost 'due to cause h' before time* τ (Andersen, 2013). This may be estimated as the area under the Aalen-Johansen estimator (or under a model-predicted cumulative incidence estimator) over the interval from 0 to τ.

Table 4.2 shows the estimated expected time lost (in years) before $\tau = 3$ years in each treatment group of the PBC3 trial due to transplantation and death without transplantation, respectively. The scenarios considered are either: No adjustment or adjustment for sex, age, albumin, and bilirubin with three different covariate configurations. In the unadjusted situation, note that the total time lost equals 3 minus the restricted means (2.61 and 2.68 years, respectively) presented in Table 4.1. The estimated values for the SD in the two treatment groups: Placebo, respectively CyA, based on 1,000 bootstrap replications, were 0.040 and 0.030 for transplantation, and 0.053 and 0.050 for death without transplantation. For the other three scenarios, the numbers in Table 4.2 (for the placebo group) provide, for each cause, one-number summaries of the tendencies seen in Figure 4.12. In this scale we can notice the beneficial effect of CyA after adjustment. For these values, estimates of SD could also be obtained using the bootstrap (though we have not illustrated this). Single values for each treatment group could be obtained by averaging over the observed covariate distribution using the g-formula (Section 1.2.5). We will illustrate this based on a direct regression model for $\varepsilon_h(\tau)$ in Section 4.2.2.

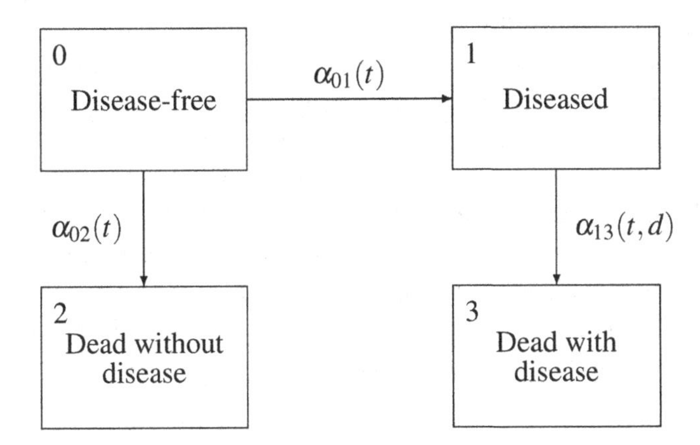

Figure 4.13 *States and transitions in the modified illness-death model without recovery.*

Estimation of cumulative incidence

The cumulative incidence for a given cause h depends on the cause-specific hazards for *all* causes. This means that, in spite of the fact that inference for separate $\alpha_h(\cdot)$ can be carried out one at a time, all causes must be taken into account when estimating the cumulative risk of cause h events. This is done using the Aalen-Johansen estimator. An estimator for $F_h(t)$ obtained as '$1 - \widehat{S}_h(t)$' where $\widehat{S}_h(t)$ is a Kaplan-Meier estimator counting only cause h events as events (and other events as censorings) will be a biased estimator. It will be *upwards biased* because, by counting other events as censorings, we pretend that, had these subjects not been 'censored', then they would still be at risk for the event of interest (i.e., cause h). The one-to-one correspondence between a single rate ($\alpha_h(t)$) and the risk ($F_h(t)$) that is present in the two-state model does not exist in the competing risks model.

4.1.3 *Illness-death models*

The illness-death model without recovery is illustrated in Figure 1.3. There are three states: '0: Disease-free', '1: Diseased', and '2: Dead'; and three transition rates: $\alpha_{02}(\cdot)$, the mortality rate without the disease, $\alpha_{01}(\cdot)$, the disease rate, and $\alpha_{12}(\cdot)$, the mortality rate with the disease. Some discussions of this model simplify if state 2 is split into two: '2: Dead without the disease' and '3: Dead with the disease' and renaming the rate $\alpha_{12}(\cdot)$ as $\alpha_{13}(\cdot)$, and we will do so in what follows, see Figure 4.13.

If $\alpha_{13}(\cdot) = 0$, then the model is the competing risks model. This means that the probabilities $Q_0(t)$ and $Q_2(t)$ may be obtained from the rates $\alpha_{01}(\cdot)$ and $\alpha_{02}(\cdot)$ as described in Section 4.1.2 and the same holds for the expected time, $\varepsilon_0(\tau)$, spent in state 0 before time τ, i.e., the restricted disease-free mean life time. In the general case, i.e., $\alpha_{13}(\cdot) \geq 0$, the probability $Q_1(t) + Q_3(t)$ of having experienced the disease before time t (and either being alive with the disease, state 1, or dead with the disease, state 3 at that time) is the cumulative incidence $F_1(t) = \int_0^t Q_0(u)\alpha_{01}(u)du$ in that competing risks model. The novelty compared to the competing risks model lies in distinguishing between occupancy of the states 1 and 3. The state 1 occupation probability, $Q_1(t)$ will also depend on the $1 \to 3$ transition intensity $\alpha_{13}(\cdot)$. Additionally, as discussed in Section 3.7, a complication arises in connection with the choice of time origin for the $1 \to 3$ transition intensity $\alpha_{13}(\cdot)$. For this rate both the time-variable t and duration since entry into state 1, say $d = d(t) = t - T_1$, may play a role. If $\alpha_{13}(\cdot)$ only depends on t then, as mentioned previously, the multi-state process is *Markovian*; if it depends on d, then it is *semi-Markovian* (Sections 1.2.3 and 1.4). In the Markovian case, inference for $\alpha_{13}(t)$ needs to take delayed entry into account, if $\alpha_{13}(\cdot)$ only depends on d, then this is not the case. The case where $\alpha_{13}(\cdot)$ depends on both t and d is more complex and models for this situation were discussed in Section 3.7. For the Markovian case, a direct argument for $Q_1(t)$ is possible, as follows.

To occupy state 1 at time t, a $0 \to 1$ transition must have occurred at some earlier time point $u < t$ and, subsequently, no $1 \to 3$ transition has occurred. The probability of a $0 \to 1$ transition between u and $u + du$ is (following the lines of argument leading to the expression for the cumulative incidence, Section 4.1.2) $Q_0(u)\alpha_{01}(u)du$ and the conditional probability of staying in state 1 between times u and t is $\exp(-\int_u^t \alpha_{13}(x)dx)$ as illustrated in Figure 4.14. Adding up these probabilities for the different $u < t$ (i.e., integrating over u from 0 to t)

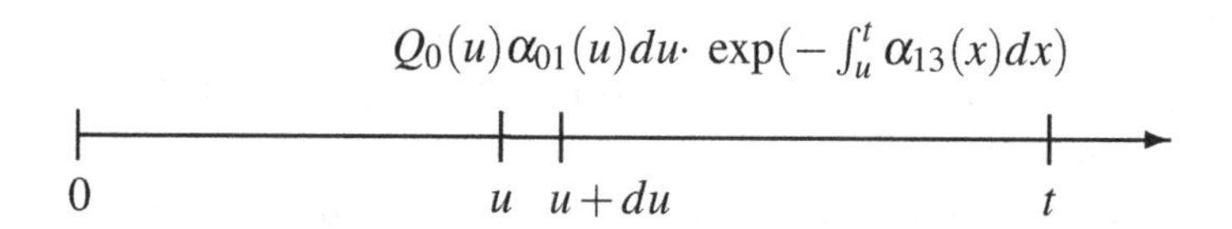

Figure 4.14 *The probability that a $0 \to 1$ transition happens in the small time interval from u to $u + du$ is the probability $Q_0(u)$ of no transition out of state 0 before time u times the conditional probability $\alpha_{01}(u)du$. The probability of no $1 \to 3$ transition from u to t given in state 1 at time u is $\exp(-\int_u^t \alpha_{13}(x)dx)$.*

leads to the desired expression

$$Q_1(t) = \int_0^t Q_0(u)\alpha_{01}(u)\exp(-\int_u^t \alpha_{13}(x)dx)du$$

for the state 1 occupation probability. The semi-Markovian case is similar; however, now the probability of staying in state 1 from entry at time u and until t is $\exp(-\int_u^t \alpha_{13}(x, x - u)dx)$.

The probability $Q_3(t)$ equals $F_1(t) - Q_1(t)$ and, for all h, plug-in is applicable for estimation of $Q_h(t)$. The latter approach for estimation of $Q_3(t)$ is related to an idea of Pepe (1991), as follows. For both the Markovian and the semi-Markovian situation, the probability $Q_0(t) + Q_1(t)$ of being alive with or without the disease, may be estimated by the Kaplan-Meier estimator, say $\widehat{S}(t)$, counting all deaths and disregarding disease occurrences. This leads to the alternative *Pepe estimator*

$$\widehat{Q}_1(t) = \widehat{S}(t) - \widehat{Q}_0(t).$$

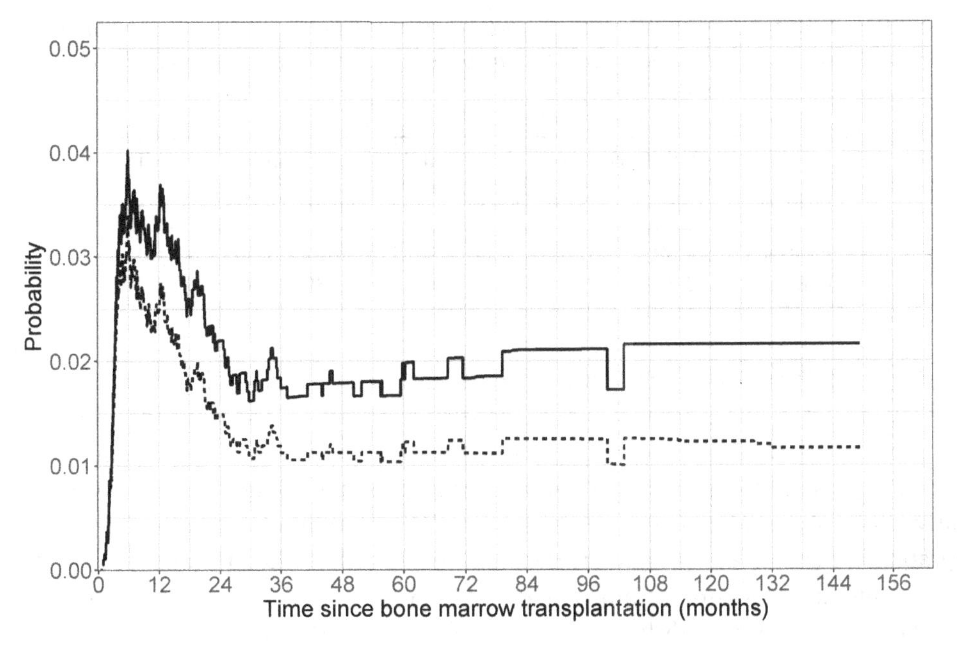

Figure 4.15 *Bone marrow transplantation in acute leukemia: State occupation probability and prevalence for the relapse state.*

Another quantity of interest in this model is the *disease prevalence* at time t

$$\frac{Q_1(t)}{Q_0(t)+Q_1(t)},$$

i.e., the conditional probability of being diseased at time t given alive at time t. Finally, the expected time lived with the disease before time τ is

$$\varepsilon_1(\tau)=\int_0^\tau Q_1(u)du.$$

To illustrate this, we use a simplified version of the model for the bone marrow transplantation data (Example 1.1.7) where graft versus host disease is not accounted for, see Figure 1.6. Figure 4.15 shows both the estimated probability, $\widehat{Q}_1(t)$ of being alive with relapse and the estimated prevalence. As a consequence of the high mortality rate with relapse (Figure 3.4), both probabilities are rather low. From $\widehat{Q}_1(t)$, the expected time lived with relapse before time τ can be estimated, and with $\tau = 120$ months the estimate is $\widehat{\varepsilon}_1(\tau) = 1.62\,(\text{SD} = 0.28)$ months (with SD in brackets based on 1,000 bootstrap samples). For state 0, the expected time spent alive without relapse before $\tau = 120$ months is $\widehat{\varepsilon}_0(\tau) = 75.78\,(\text{SD} = 1.25)$ months while the expected time lost due to death before $\tau = 120$ months is 42.61 (SD $= 1.23$) months ($\widehat{\varepsilon}_2(\tau) = 29.13, \text{SD} = 1.13$ months lost without relapse and $\widehat{\varepsilon}_3(\tau) = 13.48, \text{SD} = 0.80$ months lost after relapse).

So, in the illness-death model, the use of plug-in becomes cumbersome though still doable and the same is the case for more complicated *irreversible* ('forward-going') models, i.e.,

Table 4.3 *Recurrent episodes in affective disorders: Estimated numbers of years spent in and out of hospital, and lost due to death before $\tau = 15$ years for patients with unipolar or bipolar disease (SD based on 1,000 bootstrap replications).*

	Out of hospital		In hospital		Dead		All
Disease	Years	SD	Years	SD	Years	SD	(τ)
Unipolar	9.59	0.51	2.20	0.23	3.21	0.48	15.00
Bipolar	12.33	0.77	1.87	0.34	0.80	0.63	15.00

those for which transitions back into previous states are not possible (e.g., Figures 1.5-1.6). However, it seems clear that a more general technique – also covering models with back-transitions – would be preferable, and for Markov processes such a technique exists. This technique, based on *product-integration* of the transition intensities, is, however, not as intuitive as those described in the present section. We will return to a discussion of product-integration in Section 5.1 and skip the details here. We will rather illustrate results from an analysis using the illness-death model *with recovery*, i.e., the model for recurrent episodes (recurrent events with periods between times at which subjects are at risk for a new event), Figure 1.4, as an example.

We study Example 1.1.5 on recurrent episodes in affective disorders and refer to an ongoing episode as 'being in hospital'. As for the illness-death model without recovery, we may be interested in the state occupation probabilities, $Q_0(t)$, the probability of being out of the hospital t years after the initial diagnosis, $Q_1(t)$, the probability of being in the hospital at time t, and $Q_2(t)$, the probability of being dead at time t. Likewise, the average times, $\varepsilon_0(\tau), \varepsilon_1(\tau), \varepsilon_2(\tau)$, spent in each of the states until some threshold τ may be of interest. Figure 4.16 shows the stacked estimates of the state occupation probabilities for patients with unipolar or bipolar disorder. It is seen that bipolar patients spend more time out of the hospital, and unipolar patients have a higher mortality – an observation that is emphasized by computing the one-number summaries $\widehat{\varepsilon}_h(15 \text{ years}), h = 0, 1, 2$, see Table 4.3. Note that, as explained in Section 1.2.2, the estimates add up to $\tau = 15$ years.

Plug-in

Plug-in is a technique that enables estimation of a marginal parameter based on a specification of all intensities in a multi-state model, at least whenever a mathematical expression is available that expresses this dependence. It has the advantage that the censoring distribution needs not be specified (except from the fact that it should be considered which covariates affect censoring). However, plug-in (1) requires a correct specification of all intensities in the multi-state model, and (2) does not provide parameters that directly link the marginal parameter to covariates.

4.2 Direct models

This section discusses models where a marginal parameter in a multi-state model is directly linked to covariates. This will be done on a case-by-case basis for specific multi-state diagrams as exemplified in Chapter 1. In each case the method involves setting up a set

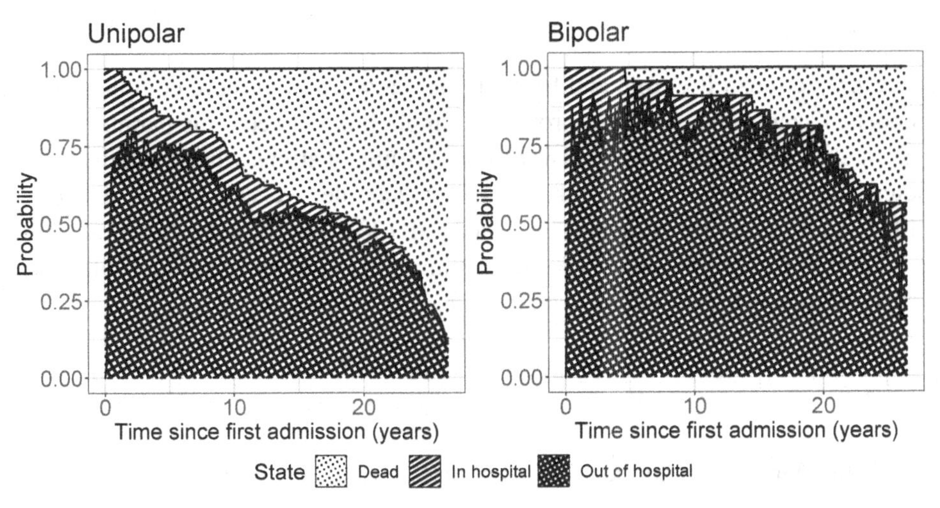

Figure 4.16 *Recurrent episodes in affective disorders: Estimated stacked state occupation probabilities for patients with unipolar or bipolar disorder.*

of *generalized estimating equations* (GEEs) whose solutions are the regression parameters giving this direct link. Mathematical details will be described in Section 5.5 where we will see that this approach typically also involves estimation of the *distribution of censoring times* (see also Section 4.4.1).

4.2.1 Two-state model

In this model, there is only one transition intensity, the hazard function $\alpha(t)$ and, as seen in Section 4.1.1, a regression model for this hazard also induces a regression model for the survival probability $S(t) = Q_0(t)$ (and at the same time for the failure probability $F(t) = Q_1(t)$). For a multiplicative hazard model such as the Cox model, the survival function is

$$S(t \mid Z) = \exp(-A_0(t)\exp(\mathrm{LP}))$$

and, therefore,

$$\log(-\log(S(t \mid Z))) = \log(-\log(1 - F(t \mid Z))) = \log(A_0(t)) + \mathrm{LP} \tag{4.10}$$

where $\mathrm{LP} = \beta_1 Z_1 + \cdots + \beta_p Z_p$ is the linear predictor, see Section 2.2.1 and Equation (4.6).

For an additive hazard model with constant hazard differences, $\alpha(t \mid Z) = \alpha_0(t) + \mathrm{LP}$, we have that

$$-\log(S(t \mid Z))/t = A_0(t)/t + \mathrm{LP}.$$

In both cases, a certain transformation, the *link function*, of the marginal parameter (here $S(t)$) gives the *linear predictor* and, therefore, the regression parameters can be interpreted in the scale given by this link function, see Section 1.2.5. In the case of a multiplicative hazard, the link function is the cloglog function, corresponding to $\exp(\beta)$ being hazard ratios (see Equation (4.6)), and in the additive case the link function is $-\log(\cdot)$ (note the minus sign) and the $\exp(\beta)$ coefficients correspond to ratios between survival probabilities.

Restricted mean life time

For the restricted mean life time and for marginal parameters in more complicated multi-state models, hazard models do not provide a simple link between this parameter and covariates. In such situations, a way forward is to set up equations whose solutions provide parameters that establish a direct link. As a first example, we will look at the restricted mean life time $\varepsilon_0(\tau)$. For this parameter, Tian et al. (2014) proposed estimating equations where some transformation, such as the logarithm or the identity function of the restricted mean survival time is linear in the covariates, i.e.,

$$\begin{aligned}\log(\varepsilon_0(\tau)) &= \beta_0 + \text{LP}\\ \varepsilon_0(\tau) &= \beta_0 + \text{LP}.\end{aligned}$$

We provide more mathematical details in Section 5.5.2 and here illustrate the method using the PBC3 data.

Table 4.4 shows the estimated coefficients in a linear model for the 3-year restricted mean including treatment (Z_1), albumin (Z_2), and bilirubin (Z_3)

$$\varepsilon_0(3 \mid Z) = \beta_0 + \beta_1 Z_1 + \beta_2 Z_2 + \beta_3 \log_2(Z_3),$$

i.e., the link function is the identity function, meaning that it is the restricted mean itself that is given by the linear predictor. In the estimation, it has been assumed that censoring does not depend on covariates, see Section 4.4.1 for more details. To estimate the variability of the estimators, *robust* or *sandwich* estimators of the SD are used (we will be using both names in what follows). The use of the word 'sandwich' stems from the fact that the mathematical expression for the collection of standard deviations and correlations consists of two identical parts, the 'bread', with something different, the 'meat', in between. The coefficients β_1, β_2, and β_3 have attractive interpretations. For given values of albumin and bilirubin, a CyA-treated patient, on average lives 0.168 years longer without transplantation during the first 3 years after randomization compared to a placebo-treated patient, for each extra 10 g/L of albumin the average years lived without transplantation during 3 years increases by 0.31 years, and for each doubling of bilirubin the average years lived without transplantation decrease by 0.214 years. The intercept β_0 is the 3-year restricted mean when the covariates all take the value 0 – an enhanced interpretability would be obtained by centering the quantitative covariates (Section 2.2.2). However, more informative absolute values for the restricted mean can be obtained using the *g-formula* (Section 1.2.5). This gives the values 2.559 (0.065) years for placebo and 2.729 (0.053) for CyA with estimated SD values based on 1,000 bootstrap replications in brackets. Since the model is *linear*, the difference between these two restricted means is the coefficient (β_1) for treatment and based on the bootstrap procedure, the estimated value is 0.170 year with a bootstrap SD of 0.079, almost as in Table 4.4.

4.2.2 *Competing risks*

For the competing risks model it is of interest to relate the cumulative incidence to covariates, and this may by done using the Fine-Gray model (Fine and Gray, 1999). In this model, the cumulative incidence $F_h(t)$ for cause h is linked to covariates in the same way

Table 4.4 *PBC3 trial in liver cirrhosis: Direct linear model (identity as link function) for the 3-year restricted mean.*

Covariate		$\widehat{\beta}$	SD
Intercept		2.376	0.381
Treatment	CyA vs. placebo	0.168	0.078
Albumin	per 1 g/L	0.031	0.008
$\log_2$(Bilirubin)	per doubling	-0.214	0.034

Table 4.5 *PBC3 trial in liver cirrhosis: Estimated coefficients (and SD) from Fine-Gray models for death without transplantation and transplantation.*

		Death without transplantation		Transplantation	
Covariate		$\widehat{\beta}$	SD	$\widehat{\beta}$	SD
Treatment	CyA vs. placebo	-0.353	0.260	-0.409	0.368
Albumin	per 1 g/L	-0.061	0.031	-0.070	0.033
$\log_2$(Bilirubin)	per doubling	0.616	0.089	0.619	0.101
Sex	male vs. female	-0.415	0.317	-0.092	0.580
Age	per year	0.087	0.016	-0.075	0.017

as $F(t) = 1 - S(t)$ is linked to covariates in the Cox model for survival data, i.e., via the cloglog function – see Equation (4.6). The model is

$$\log(-\log(1 - F_h(t \mid Z))) = \log(\widetilde{A}_{0h}(t)) + \text{LP}_h$$

with linear predictor $\text{LP}_h = \beta_{1h}Z_1 + \cdots + \beta_{ph}Z_p$. Estimating equations for the β-parameters were proposed together with an estimator for $\widetilde{A}_{0h}(t)$, from which $F_h(t \mid Z)$ may be predicted. Robust estimators for the associated SD were also given. We will give more details in Section 5.5.3 and here illustrate the model using the PBC3 data.

Table 4.5 shows the estimated coefficients when fitting Fine-Gray models to the cumulative incidences for the two competing events transplantation and death without transplantation. In the estimation, it has been assumed that censoring does not depend on covariates. From the negative signs of the coefficients for treatment, sex and albumin, it appears that CyA treatment, male sex and high albumin all decrease the risks of both end-points. High bilirubin increases the risk of both end-points, while advanced age increases the death risk and decreases the risk of transplantation. These results are qualitatively well in line with those obtained when analyzing the cause-specific hazards. It is important to realize that the two sets of models target different parameters and, as we have seen in Section 4.1.2, each cumulative incidence depends on both cause-specific hazards and, therefore, a coefficient in a Fine-Gray model depends on how the corresponding covariate is associated with both cause-specific hazards. It follows that a situation can occur where a covariate is associated with, e.g., an increased cause-specific hazard for cause 1 but not associated with that for cause 2, in which case that covariate could affect (decrease) the cumulative incidence for cause 2. This is because a high cause 1 risk ‘leaves fewer subjects to experience cause

2'. An example of this situation was provided by Andersen et al. (2012). Similar mechanisms also explain differences between the coefficients from the cause-specific Cox models (Table 2.13) and the Fine-Gray models (Table 4.5). For those covariates where the cause-specific Cox coefficients have the same sign for both events (i.e., treatment, sex, albumin and bilirubin), the Fine-Gray coefficients are numerically smaller while, for age where the Cox coefficients have opposite signs, the Fine-Gray coefficients are numerically larger.

One may wonder, what is the exact interpretation of the Fine-Gray coefficients (except from being risk differences on the cloglog scale)? When applying the $\text{cloglog}(x) = \log(-\log(1-x))$ transformation to the risk function $F = 1 - S$ in the case of no competing risks, the result is the cumulative hazard, i.e., its slope (the hazard $\alpha(\cdot)$) has the interpretation

$$\alpha(t)dt \approx P(\text{event in } (t, t+dt) \mid \text{no event } < t).$$

However, application of the cloglog function to the cause h cumulative incidence in the presence of competing risks results in a function whose slope (say, $\widetilde{\alpha}_h(t)$) has the following interpretation

$$\widetilde{\alpha}_h(t)dt \approx P(\text{cause } h \text{ event in } (t, t+dt) \mid \text{no cause } h \textit{ or } \text{a competing event } < t).$$

This function is known as the cause h *sub-distribution hazard* and its interpretation is not very appealing: It gives the cause h event rate among those who have either not yet had a cause h event or have experienced a competing event. This awkward 'risk set' has caused some debate (e.g., Putter et al., 2020), but it is also the basis for the equations from which the regression coefficients are estimated, see Section 5.5.3 for more details. In conclusion, the Fine-Gray model has the nice feature that it provides a direct link between a cumulative incidence and covariates, but this association, $\exp(\beta)$, is expressed on the not-so-nice scale of sub-distribution hazard ratios. The use of other link functions will be discussed in Sections 5.5.5 and 6.1.4 and, as we shall see there, these choices may entail other difficulties. An appealing feature of the Fine-Gray model is the ease with which the cause h cumulative incidence can be predicted for given covariates by combining the estimated regression coefficients for that cause with the estimate of the baseline cumulative sub-distribution hazard $\widetilde{A}_{0h}(t)$. Recall that, in order to predict a cause h cumulative incidence based on cause-specific hazard models, both regression coefficients and cumulative baseline hazards for all causes are needed. Prediction based on Fine-Gray models is exemplified in Figure 4.17 where the cumulative incidences for both treatments and both events in the PBC3 trial are estimated for a 40-year old woman with albumin equal to 38 g/L and bilirubin equal to 45 μmol/L. The curves for placebo treatment can be compared with those in Figure 4.12a and are seen to be close to those presented there. It should be noted that cause-specific Cox proportional hazards models and Fine-Gray (proportional sub-distribution hazards) models are mathematically incompatible.

To illustrate how single predictions across the patient population may be obtained, we use the g-formula (Section 1.2.5) to estimate cumulative incidences at 2 years based on the Fine-Gray models. To estimate the SD of the resulting risks of transplantation or death without transplantation, 1,000 bootstrap samples were drawn. For transplantation, the g-formula gives a 2-year risk of 0.069 for placebo and 0.049 for CyA. Based on the bootstrap, the corresponding values (SD) are, respectively, 0.113 (0.198) and 0.092 (0.202); however, excluding samples with degenerate estimates of 1 (47 samples), the bootstrap-based values are 0.069 (0.020) for placebo and 0.048 (0.013) for CyA – in line with the risk estimates from the original data. The estimated treatment effect (risk difference at 2 years) is 0.021 with a bootstrap SD of 0.020. For death without transplantation, the estimated 2-year risks are 0.117 for placebo and 0.088 for CyA in accordance with the corresponding bootstrap values (SD) of, respectively, 0.117 (0.021) and 0.088 (0.019) leading to an estimated risk difference at 2 years of 0.030 (0.023).

It is sometimes advocated that the Fine-Gray model is used only for the 'cause of interest' (say, cause 1). However, we believe that, in the competing risks model, all causes should be analyzed because the association between the cause 1 cumulative incidence and a covariate may be a result of an association between that covariate and the cause-specific hazard for a competing cause. Therefore, Latouche et al. (2013) argued that, to get an overview, all cause-specific hazards and all cumulative incidences should be studied. Another item to pay attention to is the fact that separate Fine-Gray models for two competing causes may be mathematically incompatible (and certainly incompatible with a Cox model for overall survival) and may provide overall risk estimates exceeding 1.

Cause-specific time lost

In Section 4.1.2, we estimated the transplant-free time lost from either death or transplantation in the PBC3 trial (Example 1.1.1) by plug-in and in Section 4.2.1, a direct regression model was set up for the restricted mean life time following Tian et al. (2014). We will now illustrate how a similar direct regression model may be studied for the *cause-specific time lost* in the competing risks model (Conner and Trinquart, 2021). We will return to a more detailed discussion of the resulting estimating equations in Section 5.5.3 and here illustrate results from an analysis of the competing risks data from the PBC3 trial (assuming that censoring does not depend on covariates).

The parameter of interest is $\varepsilon_h(\tau)$, and we wish to relate this to covariates via a linear predictor LP_h. This is typically done by assuming either of the models

$$\begin{aligned}\log(\varepsilon_h(\tau)) &= \beta_{h0} + \text{LP}_h \\ \varepsilon_h(\tau) &= \beta_{h0} + \text{LP}_h,\end{aligned}$$

though other link functions may be applied. In the PBC3 example we use the identity link and look at linear models for $\varepsilon_h(\tau), h = 1, 2$, i.e., either transplantation or death without transplantation, for $\tau = 3$ years. Table 4.6 shows results from fitting three sets of models for these two end-points. We first fit a model including only the treatment variable (CyA vs. placebo, first column) yielding, for transplantation, an intercept of $\widehat{\beta}_{10} = 0.138$ years (corresponding to placebo) and a difference of $\widehat{\beta}_{11} = -0.049$ years when comparing CyA with placebo. For death without transplantation, the corresponding estimates are $\widehat{\beta}_{20} = 0.244$ years and $\widehat{\beta}_{21} = 0.000$ years, respectively. Note that these numbers are close to those found in Table 4.2: 0.142 years and $0.086 - 0.142 = -0.056$ years for transplantation and 0.250 years and $0.235 - 0.250 = 0.015$ years for death without transplantation. Next, we adjust for sex, age, albumin, and $\log_2$(bilirubin), second column. In Table 4.2, we estimated the time lost in each treatment group for different sets of fixed values for these covariates, yielding different treatment effects for the different sets ($0.117 - 0.220 = -0.103$,

(a) *Death without transplantion*

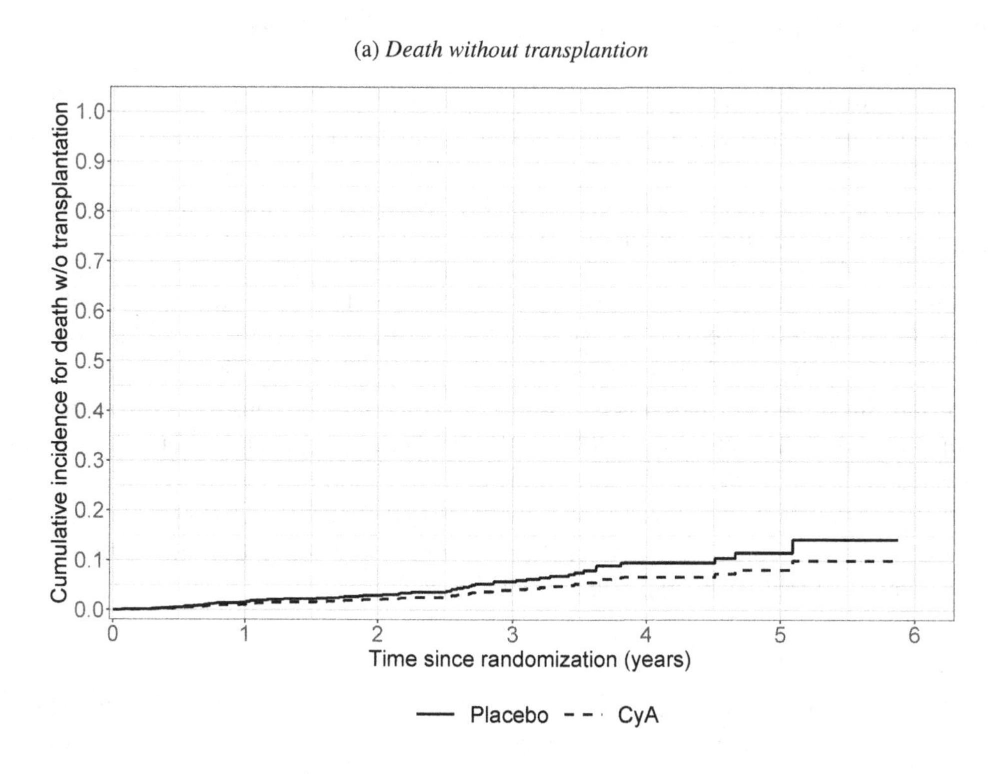

(b) *Transplantation*

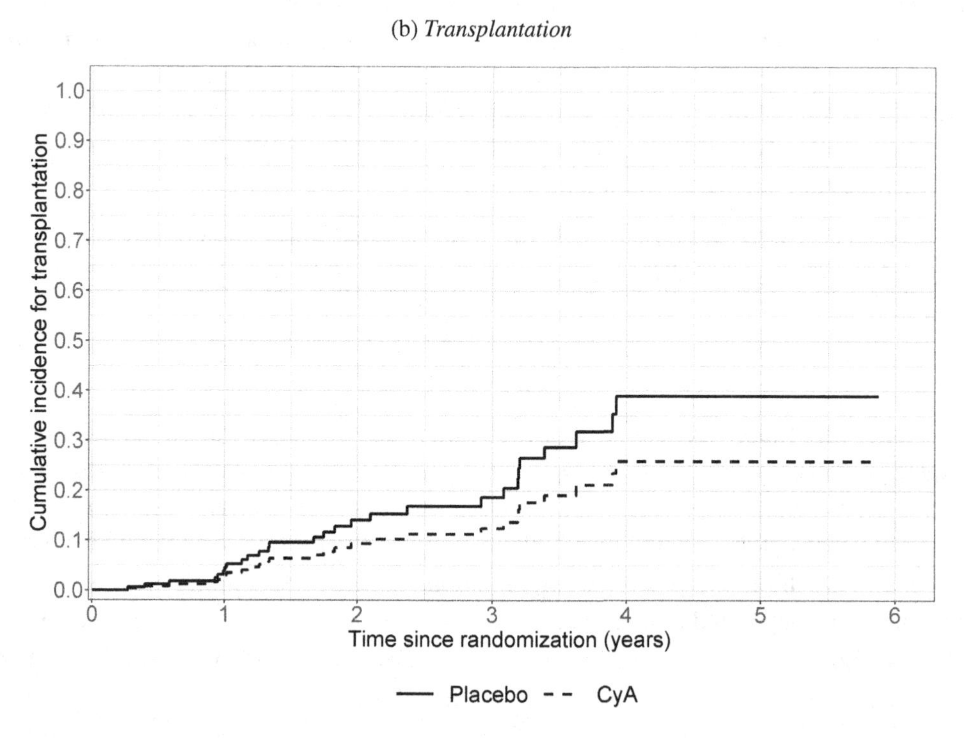

Figure 4.17 *PBC3 trial in liver cirrhosis: Predicted cumulative incidence for a 40-year-old woman with albumin equal to 38 g/L and bilirubin equal to 45 μmol/L based on Fine-Gray models.*

$0.967 - 1.377 = -0.410$, and $0.043 - 0.080 = -0.037$, respectively for transplantation for the three choices made, and $0.061 - 0.090 = -0.029$, $0.302 - 0.364 = -0.062$, and $0.256 - 0.373 = -0.117$, respectively for death without transplantation – all numbers in years). By fitting the direct linear model, we get a single treatment effect of $\widehat{\beta}_{11} = -0.067$ years for transplantation and $\widehat{\beta}_{21} = -0.068$ years for death without transplantation. In the final model in Table 4.6 (right column) we only adjust for the two biochemical variables albumin and $\log_2$(bilirubin) and we can compare with the results for the 3-year restricted mean life time in Table 4.4. This is because $\varepsilon_0(\tau) + \varepsilon_1(\tau) + \varepsilon_2(\tau) = \tau$ (Section 1.2.2) and, therefore, coefficients $\beta_{0j}, \beta_{1j}, \beta_{2j}$ in linear models for the three ε-parameters will satisfy $\beta_{1j} + \beta_{2j} = -\beta_{0j}$. Adding up the β-parameters for the two end-points from the last model in Table 4.6 we get, $-0.068 - 0.079 = -0.147$ for treatment, $-0.002 - 0.028 = -0.030$ for albumin, and $0.091 + 0.124 = 0.215$ for $\log_2$(bilirubin), to be compared with the coefficients 0.148, 0.030, and -0.215 for the models for $\varepsilon_0(\tau)$.

The overall average time lost may be estimated using the g-formula (Section 1.2.5). The last models in Table 4.6 were re-fitted on 1,000 bootstrap samples and this provided average time lost due to transplantation of 0.143 years (bootstrap SD = 0.040) for placebo and 0.073 years (bootstrap SD = 0.030) for CyA. This gives an average treatment effect of -0.070 years (0.050) – close to the estimated treatment effect in the linear model of Table 4.6, however with a somewhat smaller SD. The corresponding numbers for death without transplantation were, respectively, 0.288 years (0.057) for placebo and 0.208 years (0.048) for CyA, yielding an average treatment effect of -0.080 years (0.072), both numbers close to the values from the table.

Table 4.6 *PBC3 trial in liver cirrhosis: Estimated coefficients (and SD) from direct linear models for time lost (in years) due to death without transplantation or to transplantation before $\tau = 3$ years (Bili: Bilirubin).*

(a) *Death without transplantation*

Covariate		$\widehat{\beta}$	SD	$\widehat{\beta}$	SD	$\widehat{\beta}$	SD
Intercept		0.244					
Treatment	CyA vs. placebo	0.000	0.088	-0.082	0.078	-0.079	0.078
Albumin	per 1 g/L			-0.019	0.007	-0.028	0.007
$\log_2$(Bili)	per doubling			0.159	0.037	0.124	0.037
Sex	male vs. female			0.129	0.125		
Age	per year			0.013	0.004		

(b) *Transplantation*

Covariate		$\widehat{\beta}$	SD	$\widehat{\beta}$	SD	$\widehat{\beta}$	SD
Intercept		0.138					
Treatment	CyA vs. placebo	-0.049	0.089	-0.067	0.085	-0.068	0.083
Albumin	per 1 g/L			-0.007	0.008	-0.002	0.008
$\log_2$(Bili)	per doubling			0.072	0.044	0.091	0.042
Sex	male vs. female			0.132	0.139		
Age	per year			-0.010	0.004		

Fine-Gray model

The Fine-Gray model is a direct model for the cumulative incidence that expresses the association with covariates as sub-distribution hazard ratios. Therefore, interpretation of the resulting parameters is challenging. If using the model, this should be done for all causes in the competing risks model because the association between a given covariate and a given cumulative incidence may be a result of the way in which the covariate affects other causes. Based on Fine-Gray models for all causes, the estimated overall failure risk for given subjects may exceed 1.

4.2.3 *Recurrent events*

For recurrent events one may, in principle, focus on the same marginal parameters ($Q_h(t)$ and $\varepsilon_h(t)$) as exemplified in the previous sections. However, since it is the same type of event that may occur repeatedly, there is another marginal parameter that is of interest to study. This is the *mean function* or *expected number* of recurrent events over time, i.e.,

$$\mu(t) = E(N(t))$$

where $N(t)$ is the process counting the number of recurrent events in the interval from 0 to t.

No terminal event

We will focus on the situation depicted in Figure 1.5 and first look at the case where the mortality rate is negligible, i.e., state D on that figure is not relevant. In that case it turns out that the estimating equations that are set up for $\mu(t)$ are solved by the Nelson-Aalen estimator

$$\widehat{\mu}(t) = \sum_{\text{event times } X \leq t} \frac{dN(X)}{Y(X)} \tag{4.11}$$

(Lawless and Nadeau, 1995). To compute confidence limits around this estimator, robust estimators of the SD should be used and the confidence interval will typically be based on symmetric confidence limits for $\log(\mu(t))$ – similarly to confidence limits around the Nelson-Aalen estimator for the cumulative hazard (Section 2.1.1). A regression model

$$\mu(t \mid Z) = \mu_0(t) \exp(\text{LP}) \tag{4.12}$$

may also be analyzed quite simply since it may be shown (Lawless and Nadeau, 1995; Lin et al., 2000) that solving what are formally *score equations* based on a Cox partial log likelihood

$$l(\beta) = \sum_{\text{event times, } X} \log\Big(\frac{\exp(\text{LPevent})}{\sum_{j \text{ at risk at time } X} \exp(\text{LP}_j)}\Big)$$

leads to valid estimators for β (more details to be given in Section 5.5.4). Robust standard deviations must be used. A Breslow-type estimator for the baseline mean function $\mu_0(t)$ also exists. The model in Equation (4.12) is often denoted the *LWYY model* after the

Table 4.7 *Recurrent episodes in affective disorders: Estimated ratios between mean numbers of psychiatric episodes between patients with bipolar vs. unipolar diagnosis (c.i.: confidence interval).*

Model	Mortality treated as	$\exp(\widehat{\beta})$	95% c.i.
LWYY model	Censoring	1.52	(1.07, 2.17)
Ghosh-Lin model	Competing risk	1.95	(1.48, 2.56)

authors of Lin et al. (2000). Just as it was the case for the Cox model (Section 2.2.2), attention should be paid to the goodness-of-fit of the multiplicative model in Equation (4.12). Methods for doing this are discussed in Section 5.7.4.

We will exemplify this using data from Example 1.1.5 on recurrent episodes in affective disorders. By focusing on times from one re-admission to the next, disregarding the fact that there are in-hospital periods during which the event does not occur (see, e.g., Andersen et al., 2019), we are in the situation of Figure 1.5. The parameter $\mu(t)$, the expected number of re-admissions in $[0,t]$, refers to a population where the duration of these periods has a certain distribution, and one should realize that, in a population with another distribution of these durations, the parameter would have been different. Most importantly, the parameter $\mu(t)$ also refers to a population where patients cannot die. This is a completely unreasonable assumption, and we include this example, mainly to demonstrate the bias that arises when we, incorrectly, treat patients who die as censorings in Equation (4.11). We shall see that this bias is similar to that seen for competing risks in Section 4.1.2 when, incorrectly, estimating the cumulative incidence using '1−Kaplan-Meier' and we will below return to the correct analysis, properly taking mortality into account. Figure 4.18 shows the estimated values of $\mu(t)$ for patients whose initial diagnosis was either unipolar or bipolar, obtained using Equation (4.11). Note that (except for the fact that mortality is treated incorrectly) the vertical axis has the attractive interpretation as the average numbers of re-admissions over time since diagnosis and note that bipolar patients, on average, have more re-admissions than unipolar patients. This discrepancy can be quantified using the multiplicative regression model from Equation (4.12) and, as seen in Table 4.7, the ratio (assumed constant) between the two mean curves is estimated to be 1.52.

Terminal event

To perform a satisfactory analysis of the data on recurrent episodes in affective disorders, we need to estimate $\mu(t)$ in the presence of a non-negligible mortality rate. Let $S(t)$ be the (marginal) survival function, i.e., $S(t) = P(T_D > t)$ where T_D is time to entry into state D of Figure 1.5 without consideration of re-admissions before time t. The mean number of recurrent events in the interval from 0 to t may be estimated using the following modification of the estimator in Equation (4.11)

$$\widehat{\mu}(t) = \sum_{\text{event times } X \leq t} \widehat{S}(X-)\frac{dN(X)}{Y(X)} \tag{4.13}$$

(Cook and Lawless, 1997; Ghosh and Lin, 2000). We will denote Equation (4.13) the *Cook-Lawless estimator*. Here, $\widehat{S}(\cdot)$ is the Kaplan-Meier estimator for S and the minus sign in

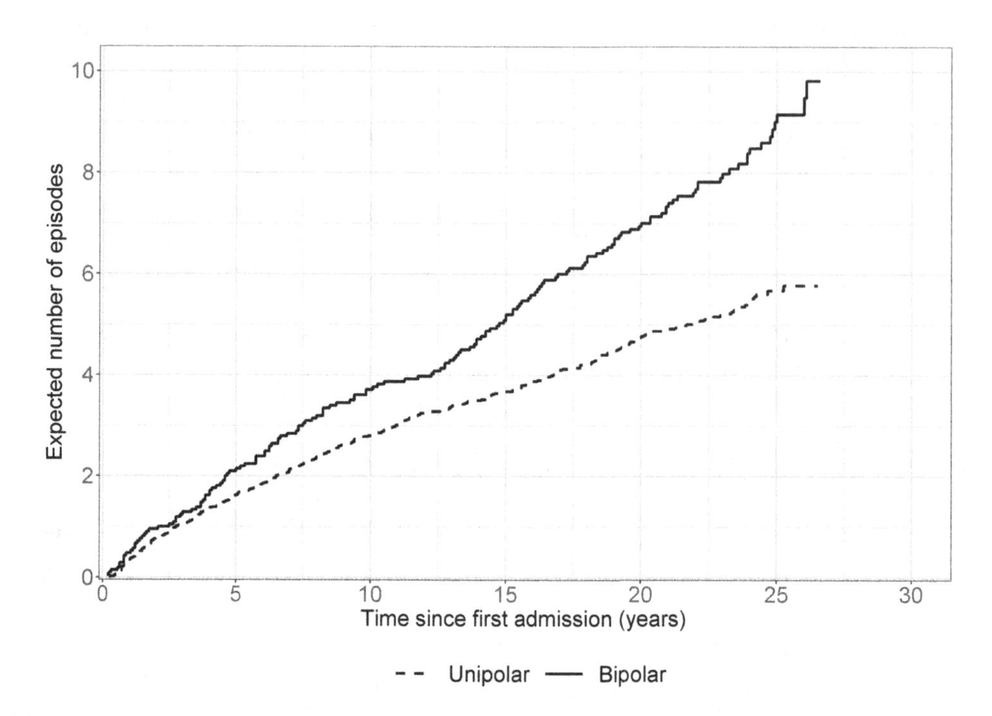

Figure 4.18 *Recurrent episodes in affective disorders: Estimated average numbers of psychiatric episodes after initial diagnosis for patients with unipolar or bipolar disorder. NB: mortality is treated as censoring.*

$\widehat{S}(X-)$ means that a death event at time X is not included in the Kaplan-Meier estimator at that time. An SD for this estimator is also available whereby confidence limits may be computed. Since the Kaplan-Meier estimator is ≤ 1, it is seen by comparing Equations (4.11) and (4.13) that treating mortality as censoring leads to an *upwards biased* estimator for $\mu(t)$. The intuition behind this bias is the same as that discussed when comparing the correct Aalen-Johansen estimator and the biased '1−Kaplan-Meier' estimator for the cumulative incidence with competing risks – namely that by treating dead patients as censored we pretend that, had they not been 'censored', then they would still be at risk for the recurrent event. The bias is clearly seen in Figure 4.19 when comparing with Figure 4.18 (and even clearer on the cover figure where the birds sit on the correctly estimated curve for unipolar patients).

The discrepancy between the two curves in Figure 4.19 may be quantified using a multiplicative regression model

$$\mu(t \mid Z) = \mu_0(t)\exp(\text{LP})$$

just like for the situation with no mortality, Equation (4.12). However, the estimating equations now need to be modified to properly account for the presence of a non-negligible mortality rate (Ghosh and Lin, 2002; more details to be given in Section 5.5.4). We will refer to this model as the *Ghosh-Lin model.* Table 4.7 also shows the estimated mean ratio from this model which is seen to be 1.95. For both estimates in Table 4.7, it has been assumed that censoring does not depend on covariates. The Ghosh-Lin estimate is seen to be larger than that based on the incorrect assumption of no mortality. The explanation

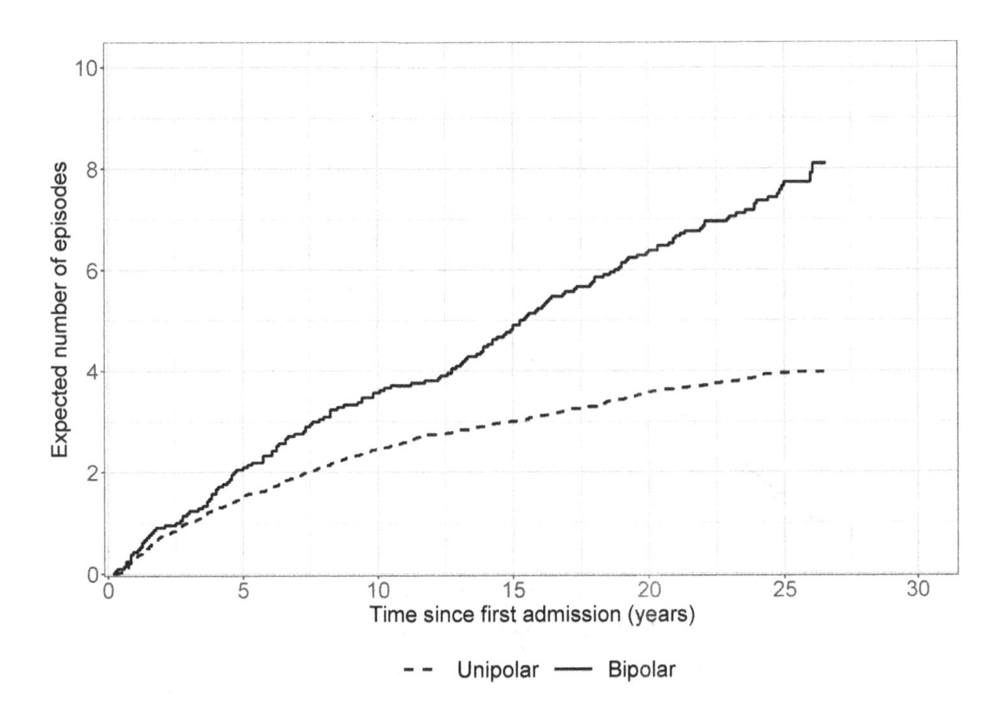

Figure 4.19 *Recurrent episodes in affective disorders: Estimated average numbers of psychiatric admissions after initial diagnosis for patients with unipolar or bipolar disorder. NB: Mortality is treated as a competing risk using the Cook-Lawless estimator.*

is that the bias affects the curves for the two groups differently because unipolar patients have a higher mortality rate than bipolar patients (estimated hazard ratio between bipolar and unipolar patients in a Cox model for the marginal mortality rate is 0.410 with 95% confidence limits from 0.204 to 0.825).

A critique that can be raised against the use of the marginal mean $\mu(t)$ in the presence of a competing risk is that a treatment may appear beneficial if it quickly kills the patient and, thereby, prevents further recurrent events from happening. Therefore, the occurrence of the competing event ('death') must somehow be considered jointly with the recurrent events process $N(t)$, at the least by also quoting results from an analysis of the mortality rate. One approach in this direction is the Mao-Lin (2016) model for the composite end-point consisting of recurrent events and death – to be discussed in Section 5.5.4.

LEADER cardiovascular trial in type 2 diabetes

Similar analyses on the data from the LEADER trial (Example 1.1.6) were conducted by Furberg et al. (2022). Figure 4.20 shows the non-parametric estimates of the mean functions for recurrent MI in the two treatment groups with or without proper adjustment for the competing mortality risk. The bias imposed by not taking death into account as a competing risk is rather small in this example as a consequence of the relatively low mortality rate (Table 1.3). Note that the Nelson-Aalen estimates are identical to the curves shown in Figure 2.15 and correctly estimate the *cumulative intensities*. The estimated mean ratio

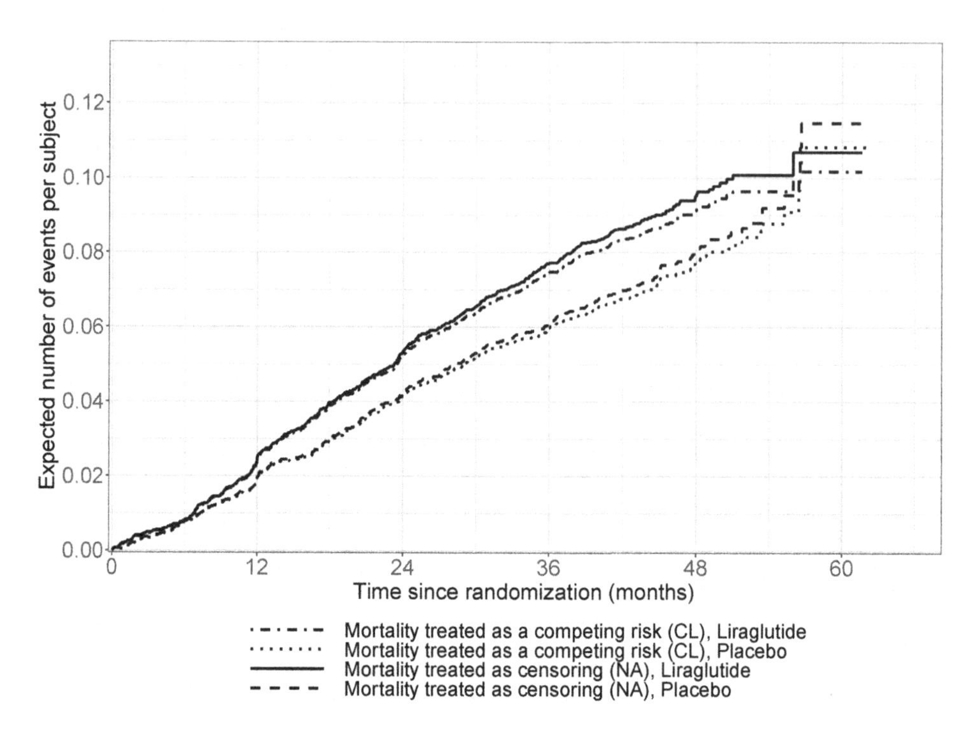

Figure 4.20 *LEADER cardiovascular trial in type 2 diabetes: Estimated average numbers of myocardial infarctions. NB: One curve for each treatment group where mortality is treated as censoring and one for each group where mortality is treated as a competing risk (CL: Cook-Lawless estimates, NA: Nelson-Aalen estimates).*

between liraglutide and placebo without taking mortality into account (LWYY model) is $\exp(-0.164) = 0.849$ (95% confidence limits from 0.714 to 1.009), while that obtained in the Ghosh-Lin model is $\exp(-0.159) = 0.853$ (0.718, 1.013). The latter is slightly closer to 1 because the mortality rate in the placebo group is slightly higher (Cox model for all-cause mortality gives a log(hazard ratio) for placebo vs. liraglutide of 0.166 (SD = 0.070)). We notice the need for both studying the recurrent events and mortality.

Mean function and terminal event

In the presence of a terminal event in a recurrent events multi-state process, the Nelson-Aalen estimator is upwards biased for the mean function. This is because, by censoring for the terminal event, one pretends to be in a population without that event. To correctly take the competing risk of the terminal event into account, the Cook-Lawless estimator must be used. In analogy with our recommendation to always study all competing events in a competing risks model (and not just the 'cause of interest'), we emphasize that the occurrence of the terminal event must be studied together with the recurrent events.

Direct models

A direct model may be set up for the way in which a marginal parameter depends on covariates. This requires (1) specification of a *link function* that gives the scale on which parameters are to be interpreted, and (2) setting up a set of *generalized estimating equations* (GEEs), the solutions to which are the desired parameter estimates. Direct modeling has some advantages compared to plug-in and micro-simulation. (1) it provides a set of regression coefficients that directly explain the association on the scale of the chosen link function, and (2) it targets directly the marginal parameter of interest and does not rely on a correct specification of all intensity models. However, a direct marginal model (1) does not provide information on the dynamics of the multi-state process, and it is not possible to simulate paths of the process based on a marginal model, and (2) requires a correct specification of the censoring distribution.

4.3 Marginal hazard models

In Sections 1.2.2 and 1.4.1, the marginal parameter *distribution of time of first entry into state h* (say, T_h) was introduced for situations where all subjects occupy the same state (0) at time 0. Examples include time to occurrence no. $h = 1, 2, \dots$ of a recurrent event (Figure 1.5) or time to relapse or GvHD in the model for the disease course after bone marrow transplantation (Figure 1.6). As mentioned there, the random variable T_h may, formally, be infinite because the event in question will not necessarily happen for all subjects (T_h is *improper*). If several such times (e.g., times for different event numbers, h in a recurrent events study) are studied simultaneously, then an assumption that these times are *independent* within subjects (i) will typically not be reasonable, see, e.g., the discussion in Section 3.9.

As discussed in that section, there is a quite different situation that gives rise to dependent event history data, namely when subjects come in clusters, such as members of the same family, or patients attending the same medical center. Here, the random variable T_h would be time to the event of interest for subject h in a given cluster (i).

In Section 3.9, the potential association within subjects/clusters was taken into account by using a *frailty* model, whereby, estimated regression parameters have a *within-cluster* interpretation. However, inference for the *marginal* time to event distributions without a specification of the intra-cluster/patient association is a useful alternative. This goal may be achieved using an approach based on generalized estimating equations (GEEs) as discussed, e.g., by Wei et al. (1989) and by Lin (1994). In Section 5.6, the mathematical background for the *marginal Cox model* will be described, and in Section 7.2, we will summarize the discussion of analysis of dependent event history data. In the present section, we will introduce the idea of marginal hazard models and discuss the extent to which it is applicable to the examples mentioned and to similar examples. We will first consider clustered data and next turn to the situation of marginal distributions of times into entry of different states in a multi-state model.

4.3.1 Clustered data

In situations where subjects come in clusters, it is relevant to account for the cluster structure in the analysis of the event history data. This can be done by setting up standard Cox models for the event intensity for each subject separately, i.e., by specifying the marginal hazard for subject h as

$$\alpha_h(t) \approx P(T_h \leq t+dt \mid T_h > t)/dt. \tag{4.14}$$

This hazard is *marginal* in the sense that the life course of other cluster members is not taken into account (even though this may be informative for subject h due to the suspected within-cluster correlation). As an example, one could specify the following model for subject h in cluster i with covariates $(Z_{ih1}, \ldots, Z_{ihp})$

$$\alpha_{ih}(t \mid Z) = \alpha_0(t)\exp(\text{LP}_{ih}),$$

which is just a standard Cox model (in which stratification is also possible, i.e., different baseline hazards in certain sub-groups). Estimation of the β coefficients and the baseline hazard(s) follow exactly the same lines as described previously (Section 2.2.1 and Section 3.3), and the estimates are exactly the same as they would have been under independence. The standard deviations of the estimates will be different because the cluster structure is taken into account when computing the *robust* standard deviations instead of the *model-based* standard deviations used in previous examples of the Cox model. The robust standard deviations will often be larger than the model-based since the latter will over-estimate the amount of precision by over-estimating the number of independent units in the study. This is typically the case when there is a positive within-cluster correlation. However, in situations with a negative within-cluster correlation or when the covariate varies within, rather than among clusters, they may also be smaller. We will consider the bone marrow transplantation data (Example 1.1.7) and take the cluster structure implied by patients attending different medical centers into account. We will study models for three different outcomes: relapse, relapse-free survival (i.e., the composite end-point of either relapse or death in remission – leaving state 0 in Figure 4.13), or overall survival (time until entry into either state 2 or 3 in that figure). Table 4.8 shows results from models including the three covariates graft type, disease, and age. For relapse, the estimated coefficients and the model-based standard deviations are close to those found in Table 3.12 where adjustment for the time-dependent covariate graft versus host disease was also conducted – without that adjustment the two sets of estimates and standard deviations would have been identical. The robust standard deviations tend to be larger than the model-based since a positive within-center association is suspected. This is, in particular, the case for the covariate age that has a larger variation among centers than within (F-ratio in a one-way ANOVA is 5.64). This tendency is confirmed by the overall Wald significance tests for the three coefficients: 21.60 based on the model-based results and 16.36 for the robust for the outcome relapse. The results for relapse-free survival and overall survival are similar since most events for the composite end-point are deaths (Table 1.4). Robust standard deviations tend to be larger, and the three degree of freedom Wald tests for all coefficients are more significant when based on the model-based results (81.52 vs. 52.59 for relapse-free survival and 76.06 vs. 48.78 for overall survival).

Table 4.8 *Bone marrow transplantation in acute leukemia: Estimated coefficients, model-based SD, and robust SD from marginal hazard models for relapse, relapse-free survival, and overall survival taking clustering by medical center into account (BM: Bone marrow, PB: Peripheral blood, AML: Acute myelogenous leukemia, ALL: Acute lymphoblastic leukemia).*

			SD		
		$\widehat{\beta}$	Model-based	Robust	Ratio
Relapse					
Graft type	BM only vs. BM/PB	-0.108	0.134	0.138	1.025
Disease	ALL vs. AML	0.549	0.129	0.174	1.345
Age	per 10 years	-0.045	0.044	0.075	1.686
Relapse-free survival					
Graft type	BM only vs. BM/PB	-0.161	0.077	0.077	0.997
Disease	ALL vs. AML	0.455	0.078	0.078	1.004
Age	per 10 years	0.169	0.026	0.033	1.286
Overall survival					
Graft type	BM only vs. BM/PB	-0.160	0.079	0.081	1.022
Disease	ALL vs. AML	0.405	0.080	0.078	0.975
Age	per 10 years	0.173	0.026	0.033	1.267

4.3.2 *Recurrent events*

For recurrent events, times to first, second, third, etc. event occurrence within each patient are correlated. Analyses of these times may, in the case of no competing risks in the form of a terminal event that prevents further occurrences of the recurrent event (Figure 1.5 without state D), be analyzed as just described for the bone marrow transplantation data. Thus, Cox models

$$\alpha_{hi}(t \mid Z) = \alpha_{h0}(t)\exp(\text{LP}_i), \quad h = 1, 2, \ldots, K$$

for the hazard of the time to event occurrence no. h for a patient, i with covariates $(Z_{i1}, \ldots, Z_{ip})$ may be set up. Here, the maximum number of events, K, to be studied needs specification which must be done on a case-by-case basis. Sub-models with common covariate effects for several or all h are possible (i.e., β_j instead of β_{hj}). The model is known as the *WLW model* for recurrent events (Wei, Lin and Weisfeld, 1989). An example could be the data on recurrent episodes of affective disorders, ignoring the fact that patients may die during follow-up (Example 1.1.5). However, in previous analyses of these data (e.g., Section 4.2.3) we have seen that neglecting mortality, i.e., treating it as censoring, may lead to considerably biased estimates in some situations and we will not pursue this idea any further. An analysis of times to first, second, ... event occurrence needs to properly address the competing risk of death. The problem is that the marginal hazard for time T_h, Equation (4.14)

$$\alpha_h(t) \approx P(T_h \leq t + dt \mid T_h > t)/dt,$$

is now marginal in the sense that consideration of neither times of previous events (i.e., no. $1, 2, \ldots h-1$), nor time to death, T_D is given. The latter means that, either $\alpha_h(t)$ should be interpreted in a hypothetical population without mortality, or the conditioning event $T_h > t$

Table 4.9 *Recurrent episodes in affective disorders: Numbers of recurrences and deaths used in marginal Cox models (WLW-models).*

Episode no.	Recurrence	Death	Total
$h = 1$	99	16	115
$h = 2$	82	28	110
$h = 3$	62	45	107
$h = K = 4$	47	55	102

should be taken to mean that the subject either is alive at time t but has not yet experienced event no. h, or the subject has already died at time t without having had h recurrences. We have already dismissed in Section 1.3 the first possibility as being unrealistic, and the second possibility means that $\alpha_h(t)$ is a *sub-distribution* hazard rather than an ordinary hazard (Section 4.2.2). In the latter case, one may turn to marginal Fine-Gray models for the cumulative incidences for event occurrence no. $h = 1, \dots .K$ (Zhou et al., 2010).

To have well-defined marginal hazards, the definition of 'event occurrence no. h' could be modified to being the *composite end-point* 'occurrence of event no. h *or* death', much like earlier definitions of recurrence-free survival in the bone marrow transplantation study, Example 1.1.7, or failure of medical treatment (transplantation-free survival) in the PBC3 trial, Example 1.1.1. This possibility was discussed by Li and Lagakos (1997) together with a suggestion to model the *cause-specific hazards* for recurrence no. $h = 1, \dots, K$, taking mortality into account as a competing risk. In the latter case, hazards are no longer marginal in the sense of Equation (4.14).

We will illustrate marginal hazard models using the data on recurrence and death in patients with affective disorders (Example 1.1.5). As in Section 2.5 we will restrict attention to the first $K = 4$ recurrences for which the numbers of events (recurrences and/or deaths) are shown in Table 4.9. Table 4.10 shows the results from analyses including only the covariate bipolar vs. unipolar disorder. For the composite end-point, the hazard ratios tend to decrease with episode number (h) while there is rather an opposite trend for the models for the cause-specific hazards for recurrence no. $h = 1, 2, 3, K = 4$. The likely explanation is that, as seen in Table 4.9, the fraction of deaths for the composite end-point increases with h and, as we have seen earlier (Section 4.2), mortality is higher for unipolar than for bipolar patients. The estimates for the separate coefficients β_h may be compared using a three degree of freedom Wald test which for the composite end-point is 6.08 ($P = 0.11$) and for the cause-specific hazards is 8.27 ($P = 0.04$). Even though the latter is borderline statistically significant, Table 4.10 also shows, for both analyses, the estimated log(hazard ratio) in the model where β is the same for all h.

4.3.3 *Illness-death model*

Consider an illness-death model for a simplified situation for the bone marrow transplantation data, i.e., without consideration of GvHD. This is Figure 1.3 with state 1 corresponding to relapse and state 2 to death. Suppose one wishes to make inference for both 'time to death' and 'time to relapse'. Could the marginal Cox model be applicable for this purpose?

Table 4.10 *Recurrent episodes in affective disorders: Estimated coefficients (and robust SD) for bipolar versus unipolar disorder from marginal Cox models (WLW-models) for the composite end-point of recurrence or death and for the cause-specific hazards of recurrence.*

	Composite end-point		Cause-specific hazard	
Episode no.	$\widehat{\beta}$	SD	$\widehat{\beta}$	SD
$h = 1$	0.380	0.209	0.495	0.202
$h = 2$	0.291	0.255	0.640	0.242
$h = 3$	0.003	0.246	0.534	0.269
$h = K = 4$	0.107	0.237	0.879	0.283
Joint	0.193	0.204	0.615	0.211

The answer seems to be 'no' because the marginal hazard for relapse given in Equation (4.14) is not well defined in the relevant population where death also operates. There have been attempts in the literature to do this anyway (taking into account the 'informative censoring by death') under the heading of semi-competing risks (e.g., Fine et al. 2001) , but we will not follow that idea here and refer to further discussion in Section 4.4.4. Instead, we will proceed as in the recurrent events case and re-define the problem to jointly study times to death and times to the composite end-point of either relapse or death (relapse-free survival). These times are correlated within each patient, since all deaths in remission count as events of both types but their marginal hazards are well defined. The numbers of events are 737 overall deaths and $764 (= 259 + 737 - 232$, cf. Table 1.4) occurrences of relapse or death in remission. Table 4.11 shows results from models including the covariates graft type, disease, and age. The models are fitted using a stratified Cox model, stratified for the two types of events with type-specific covariates and using robust SD. Note that, for both end-points, the estimates are the same as those found in Table 4.8. This is because they solve the same estimating equations. The robust SD are also close but not identical because another clustering is now taken into account (patient rather than center as in Table 4.8). The two sets of coefficients are strongly correlated: The estimated correlations are 0.98, 0.96, and 0.97, respectively, for graft type, disease, and age. These correlations are accounted for in Wald tests for equality of the two sets of coefficients. These are 0.004 ($P = 0.95$), 5.17 ($P = 0.023$), and 0.32 ($P = 0.57$), respectively, for the three covariates. Under the hypothesis of equal coefficients for graft type and age, the estimates are -0.161 (0.077) and 0.171 (0.026). Note that the SDs are not much reduced as a consequence of the high correlations.

If one, further, wishes to include 'time to GvHD' in the analysis, then this needs to be defined as GvHD-free survival, i.e., events for this outcome are either GvHD or death, whatever comes first. For this outcome there are $1,324 (= 976 + 737 - 389$, cf. Table 1.4) events of which 976 are GvHD occurrences. The results from an analysis of all three outcomes are found in Table 4.11 (where those for overall and relapse-free survival are the same as in the previous model). It is seen that the estimated coefficients for GvHD-free survival differ somewhat from those for the other two end-points since the majority of these events are not deaths.

Table 4.11 *Bone marrow transplantation in acute leukemia: Estimated coefficients (and robust SD) from marginal hazard models for relapse-free survival, overall survival, and GvHD-free survival (BM: Bone marrow, PB: Peripheral blood, AML: Acute myelogenous leukemia, ALL: Acute lymphoblastic leukemia).*

		$\widehat{\beta}$	SD
Relapse-free survival			
Graft type	BM only vs. BM/PB	-0.161	0.077
Disease	ALL vs. AML	0.455	0.078
Age	per 10 years	0.169	0.026
Overall survival			
Graft type	BM only vs. BM/PB	-0.160	0.079
Disease	ALL vs. AML	0.405	0.079
Age	per 10 years	0.173	0.027
GvHD-free survival			
Graft type	BM only vs. BM/PB	-0.260	0.059
Disease	ALL vs. AML	0.293	0.060
Age	per 10 years	0.117	0.019

Marginal hazard models

A marginal hazard model describes the marginal distribution of the time to a certain event. For clustered data, this is carried out without consideration of the event times for other cluster members and without having to specify the within-cluster association, and, in this situation, marginal hazard models are useful. The same may be the case in a recurrent events situation without a terminal event, in which case the marginal time to first, second, third, etc. event may be analyzed without having to specify their dependence (the WLW model).
However, in situations with competing risks (both for recurrent events and other multi-state models) the concept of a marginal hazard is less obvious.

Model-based and robust SD

Intensity-based models build on the likelihood approach that directly provides *model-based* SD. Direct marginal models aim at establishing the association between a marginal parameter and covariates and builds on setting up generalized estimating equations (GEEs). In this case, robust values for SDs are based on the GEE using the *sandwich* formula. These SD are robust to certain model deviations; however, the link between the marginal parameter and the covariates should be correctly specified (the GEE should be *unbiased*).

4.4 Independent censoring – revisited

4.4.1 Investigating the censoring distribution

In the beginning of Section 4.2, we mentioned that fitting direct models for a marginal parameter would typically require estimation of the distribution $G(t) = P(C > t)$ of *censoring times*. To do this, we imagine that in a study, any given subject has a potential time where observation of that subject would be terminated if he or she had not yet died (or, more precisely, not yet reached an absorbing state in the multi-state process) prior to that time. This potential censoring time will, at the latest, be the time from entry into the study and until the study is terminated, but subjects may drop out of the study prior to that time – see the examples in Section 1.1. For some subjects, C is observed but for others, observation of C is precluded if the survival time T is less than C. This suggests that $G(t)$ may be estimated by a Kaplan-Meier estimator where 'censoring' is the event and 'death' (absorbing states) acts as censoring. If $G(t)$ depends on covariates, then the distribution may be estimated via, e.g., a Cox model for the censoring hazard. To obtain valid estimation of $G(t)$, an *independent censoring* condition (similar to Equation (1.6)) must be fulfilled. Since, in essence, Equation (1.6) implies that C and T are conditionally independent given covariates and since conditional independence is a condition that is symmetric in C and T, it follows that the same condition will also ensure valid estimation of $G(t)$. As explained in Section 1.3, this condition cannot be directly evaluated based on the available data – except from the fact that the way in which the distributions of T and C depend on covariates may be investigated using regression models.

Even though the censoring distribution is typically not of primary scientific interest, it may be informative to study $G(t)$ to get an overview of the follow-up time distribution (e.g., van Houwelingen and Putter, 2012, ch. 1; Andersen et al., 2021). In what follows, we will do so in a number of examples.

PBC3 trial in liver cirrhosis

In the PBC3 trial (Example 1.1.1), most censoring was *administrative* and only four patients were lost to follow-up prior to end-of-study. This means that the distribution of censoring times mostly reflects the distribution of calendar entry times. Figure 4.21 shows the resulting $\widehat{G}(t)$ together with the Kaplan-Meier estimator $\widehat{S}(t)$ for the overall survival function (probability of no treatment failure before time t) for both treatment groups combined. It is seen that the censoring distribution is rather uniform (survival function close to linear) over the 6 years where the trial was conducted; however, with relatively fewer short censoring times reflecting that few patients were recruited towards the end of the trial.

The function $\widehat{G}(t)$ gives the probability of being uncensored at time t and $\widehat{S}(t)$ the probability of having had no treatment failure by that time. Thus, the product gives the probability of still being in the study at t and it holds that

$$\widehat{G}(t-)\widehat{S}(t-) = \frac{Y(t)}{n}, \tag{4.15}$$

the fraction of subjects still in the study just before t (where, as previously, the $t-$ means that a possible jump in the estimator at t is not yet included).

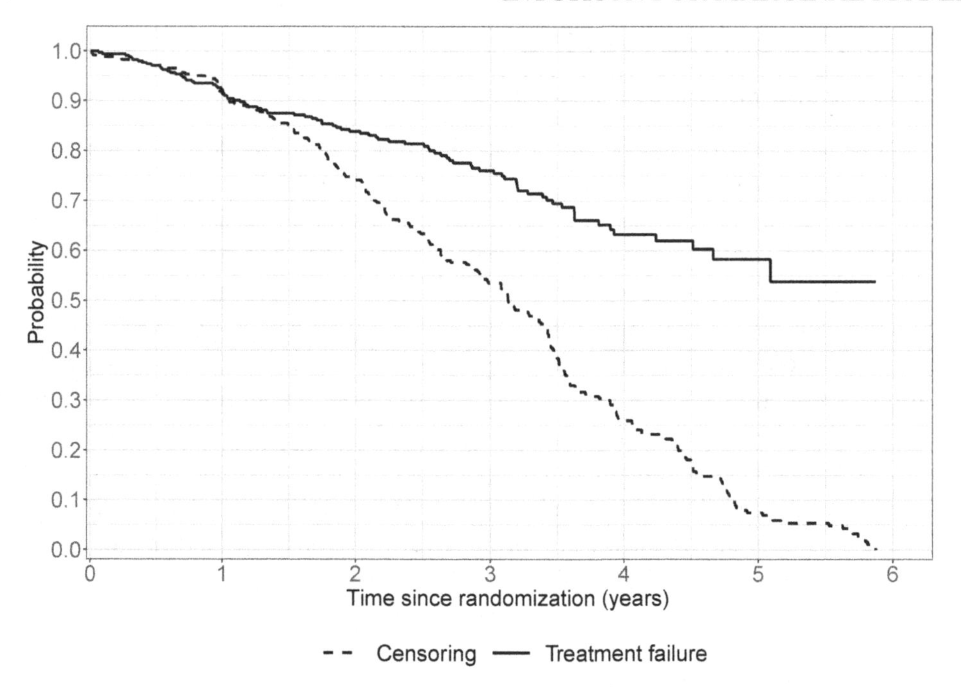

Figure 4.21 *PBC3 trial in liver cirrhosis: Kaplan-Meier estimates for censoring and treatment failure.*

To study if censoring depends on covariates, simple Cox models were studied including, respectively, treatment, albumin or $\log_2(\text{bilirubin})$. The estimates (SD) were, respectively, $\widehat{\beta} = 0.084\,(0.126)$, $\widehat{\beta} = 0.0010\,(0.013)$, and $\widehat{\beta} = -0.0025\,(0.0018)$ with associated (Wald) P-values 0.50, 0.94, and 0.16 showing that the censoring times (entry times) have a distribution that is independent of the prognostic variables.

PROVA trial in liver cirrhosis

In the PROVA trial, 75 patients out of 286 died and 20 dropped out prior to the end of the trial, leaving 191 patients alive at the end of the study (Example 1.1.4, Table 1.2). Figure 4.22 shows the estimated distribution of censoring times which is rather uniform between 1 and 4 years with few censorings before or after that interval. Wald P-values for the association between the censoring hazard and covariates based on a series of simple Cox models are shown in Table 4.12 where it is seen that censoring depends strongly on age but not much on the other covariates. Previous analyses of these data (e.g., Section 3.6.3) showed that adjustment for age had little impact on the results of hazard regression models.

Recurrent episodes in affective disorder

In Example 1.1.5 on recurrence in affective disorders, out of 119 patients included, 78 had died, leaving 41 censored observations out of whom 38 were in the study by its end in 1985. Figure 4.23 shows $\widehat{G}(t)$ for these data, and it is seen that most censorings happen between 22 and 26 years in accordance with entry between 1959 and 1963 and study termination in 1985. For these data, there is no association between censoring and the initial diagnosis

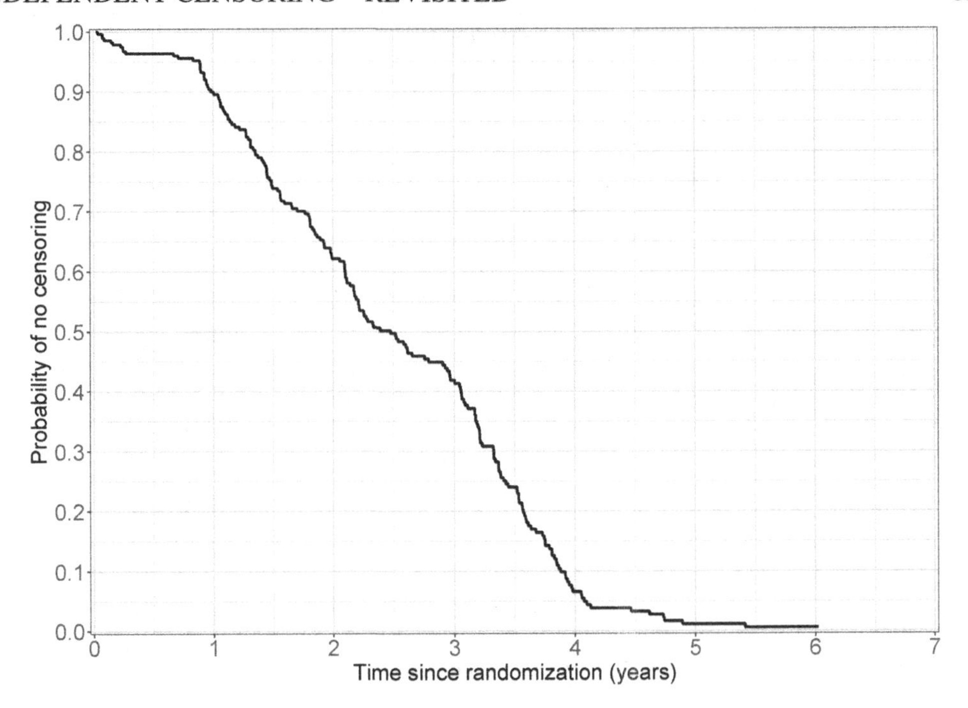

Figure 4.22 *PROVA trial in liver cirrhosis: Kaplan-Meier estimate for censoring.*

Table 4.12 *PROVA trial in liver cirrhosis: Tests for association between censoring and covariates.*

Covariate	*P*-value
Treatment	0.98
Size of varices	0.10
Sex	0.59
Coagulation factors	0.28
$\log_2$(Bilirubin)	0.18
Age	0.003

(bipolar versus unipolar, $P = 0.20$), but a very strong association with calendar time at initial diagnosis ($P < 0.001$). It was seen in Table 3.6 that the variable was also associated with the recurrence intensity, but also that adjustment did not much affect the estimate for the initial diagnosis.

4.4.2 *Censoring and covariates – a review*

We have studied two classes of models for multi-state survival data: Intensity-based models and marginal models and, for both, the issue of right-censoring had to be dealt with. As indicated in Section 4.2 (and as will be further discussed in Sections 5.5.1 and 6.1.8), a direct model for a marginal parameter typically requires that a model for the censoring distribution is correctly specified. The most simple situation arises when censoring can be considered independent of covariates, and it was seen in the previous section how censoring models can be investigated.

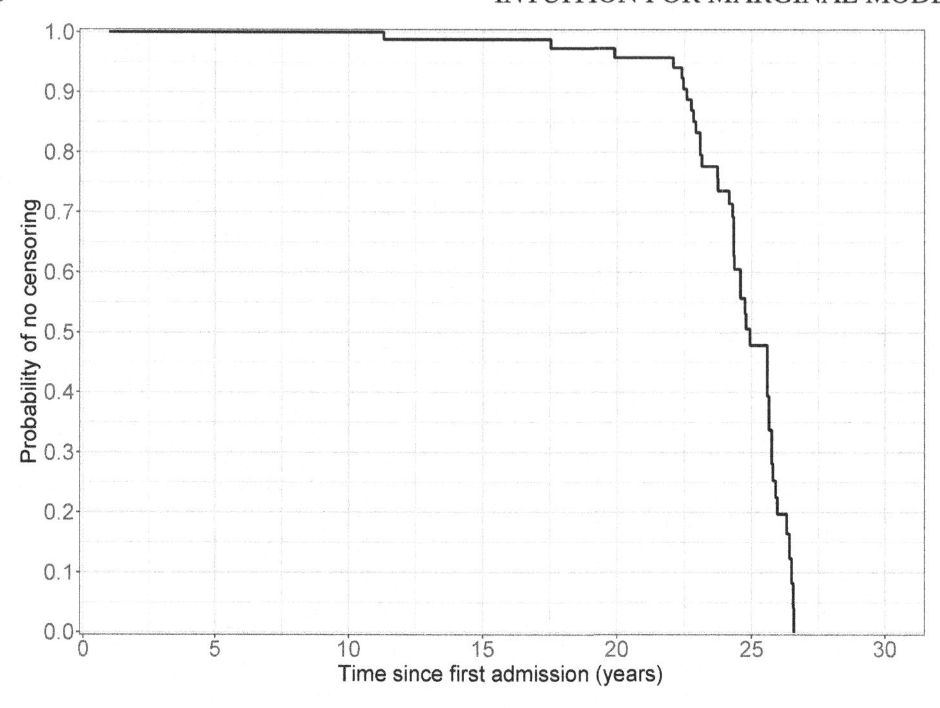

Figure 4.23 *Recurrent episodes in affective disorders: Kaplan-Meier estimate for censoring.*

For models based on intensities, the situation is different. If censoring is *independent*, i.e., censoring is conditionally independent of the multi-state process given covariates, then the methods discussed in Chapters 2 and 3 apply. However, this strictly requires that the covariates which create the conditional independence are accounted for in the hazard model. We hinted at this in Sections 1.3 and 2.2. The mathematical explanation follows from Section 3.1 where we emphasized that observation of the censoring times gives rise to likelihood contributions which factorize from the failure contributions under the conditional independence assumption. This means that the contributions arising from censoring can be disregarded when analyzing the failure intensities, and one needs not worry about the way in which censoring is affected by covariates ('the censoring model may be misspecified') except from the fact that the covariates that create the conditional independence are accounted for in the intensity model.

So, the conclusion is that intensity-based models and, thereby, inference for marginal parameters based on plug-in, are less sensitive to the way in which covariates may affect censoring. However, to achieve this flexibility, one has to assume that censoring is independent given covariates and covariates affecting the censoring distribution should be accounted for in the hazard model provided that they are associated with the hazard. Investigations along these line were exemplified in the previous section.

4.4.3 *Independent competing risks – a misnomer (*)*

In Section 4.4.1 (see also Section 1.3), the point of view was, as follows. In the complete population, there is a time-to-event distribution that we wish to estimate. The observations are incomplete because of censoring, and all subjects have a potential time to censoring that may or may not be observed. Thus, both time to event and time to censoring are well-defined (*proper*) random variables that have some joint distribution. To make inference on either of the marginal distributions, i.e., either the time to event in the absence of censoring (the distribution of scientific interest), or (as we did Section 4.4.1) the marginal time to censoring, an assumption of *independent censoring* is needed. We may be willing to make this assumption after careful discussion of the mechanisms that lead to censoring.

A question is then: Could a similar approach be used for competing risks? We will argue that the answer is negative, first and foremost because the question is ill-posed – we do not believe that there is a well-defined 'time to failure from cause 1 in the absence of cause 2' or vice versa. As explained in Section 1.3, in the complete population without censoring, both causes are operating, and the occurrence of failure from one cause prevents failure from the other cause from ever happening. As a consequence, and as explained earlier (Section 1.2.2, see also Section 5.1.2), the times to entry into states 1 or 2 in the competing risks model (Figure 1.2) are *improper* random variables because the corresponding event (failure from a specific cause) may never happen.

Nonetheless, there is a literature on classical competing risks building on random variables, each representing time to failure from a specific cause (e.g., Crowder, 2001, ch. 3). We will discuss this in the case of $k = 2$ competing causes, in which case there are two random variables $\widetilde{T}_1, \widetilde{T}_2$ representing time to failure from cause 1 or 2, respectively. These variables are known as the *latent failure times* and they have a joint survival distribution

$$\widetilde{S}(t_1,t_2) = P(\widetilde{T}_1 > t_1, \widetilde{T}_2 > t_2),$$

with marginal distributions $\widetilde{S}_h(t_h) = P(\widetilde{T}_h > t_h), h = 1,2$ ($\widetilde{S}_1(t_1) = \widetilde{S}(t_1,0)$ and $\widetilde{S}_2(t_2) = \widetilde{S}(0,t_2)$) and associated marginal (or net) hazards. Complete observations would be the smaller of the latent failure times, $\min(\widetilde{T}_1, \widetilde{T}_2)$ and the corresponding cause (the index of the minimum). Incomplete observation may be a consequence of right-censoring at C, in which case it is only known that both $\widetilde{T}_1$ and $\widetilde{T}_2$ are greater than C, and the cause of failure is unknown.

A major problem with this approach is that, as discussed, e.g., by Tsiatis (1975), with no further assumptions, the joint distribution $\widetilde{S}(t_1,t_2)$ is *unidentifiable* from the available observations – also in the case of no censoring. What is identifiable are the cause-specific hazard functions $\alpha_h(t)$ and functions thereof, such as the overall survival function $S(t)$ and the cumulative incidences $F_h(t)$ (e.g., Prentice et al., 1978), i.e., exactly the quantities that we have been focusing on. In terms of the joint survival function, $S(t) = \widetilde{S}(t,t)$ and the cause-specific hazards are obtained as

$$\alpha_h(t) = -\frac{\partial}{\partial t_h} \log \widetilde{S}(t_1,t_2)_{|t_1=t_2=t}.$$

An assumption that would make $\widetilde{S}(t_1,t_2)$ identifiable is independence of $\widetilde{T}_1, \widetilde{T}_2$ – a situation known as *independent competing risks* and, under that assumption, the hazard of $\widetilde{T}_h$ (the net hazard) equals $\alpha_h(t)$. However, as just mentioned, independence cannot be identified by observing only the minimum of the latent failure times and the associated cause. Another assumption that would make $\widetilde{S}(t_1,t_2)$ identifiable was discussed by Zheng and Klein (1995) who showed that if the dependence structure is specified by some specific shared frailty model (see Section 3.9), then $\widetilde{S}(t_1,t_2)$ can be estimated. The problem is that there is no support in the data for any such dependence structure.

These problems were nicely illustrated by Kalbfleisch and Prentice (1980, ch. 7) who gave an example of two different joint survival functions, one corresponding to independence, the other not necessarily, with identical cause-specific hazards. More specifically the following two joint survival functions were studied

$$\widetilde{S}(t_1,t_2) = \exp(1-\alpha_1 t_1 - \alpha_2 t_2 - \exp(\alpha_{12}(\alpha_1 t_1 + \alpha_2 t_2)))$$

(with $\alpha_1, \alpha_2 > 0$ and $\alpha_{12} > -1$) and

$$\widetilde{S}^*(t_1,t_2) = \exp\left(1-\alpha_1 t_1 - \alpha_2 t_2 - \frac{\alpha_1 e^{\alpha_{12}(\alpha_1+\alpha_2)t_1} + \alpha_2 e^{\alpha_{12}(\alpha_1+\alpha_2)t_2}}{\alpha_1+\alpha_2}\right).$$

For $\widetilde{S}$, $\widetilde{T}_1$ and $\widetilde{T}_2$ are independent if $\alpha_{12} = 0$ (in which case $\widetilde{S}(t_1,t_2)$ is the product $\exp(-\alpha_1 t_1)\exp(-\alpha_2 t_2)$) and the parameter α_{12}, therefore, quantifies deviations from independence. On the other hand, $\widetilde{S}^*(t_1,t_2)$ corresponds to independence no matter the value of α_{12} (it is always a product of survival functions of t_1 and t_2); however, the cause-specific hazards for $\widetilde{S}$ and $\widetilde{S}^*$ are the same, namely

$$\alpha_h(t) = \alpha_h(1+\alpha_{12}\exp(\alpha_{12}(\alpha_1+\alpha_2)t)), \quad h=1,2.$$

So, even though $\alpha_1, \alpha_2, \alpha_{12}$ can all be estimated if this model for the cause-specific hazards were postulated, the estimated value of α_{12} cannot be taken to quantify discrepancies from independence, since it cannot be ascertained from the data whether $\widetilde{S}$ or $\widetilde{S}^*$ (or some other model with the same cause-specific hazards) is correctly specified.

Such considerations have had the consequence that the latent failure time approach to competing risks has been more or less abandoned in biostatistics (e.g., Prentice et al., 1978; Andersen et al., 2012), and emphasis has been on cause-specific hazards and cumulative incidences. Even when simulating competing risks data, it has been argued (Beyersmann et al., 2009) that one should not use latent failure times even though this provides data with the same distribution as when using cause-specific hazards. We will return to this question in Section 5.4.

In summary, both concepts of independent censoring and independent competing risks relate to independence between two random time-variables. However, it is only in the former case that these two random variables (time to event and time to censoring) are well-defined, and, for that reason, we find the concept of independent censoring important, whereas we find the concept of independent competing risks (and the associated latent failure time approach) less relevant.

4.4.4 Semi-competing risks (*)

As mentioned in Section 4.3, Fine at al. (2001) studied the illness-death model under the heading of semi-competing risks with the focus of studying the marginal distribution of time to illness (entry into state 1 in Figure 1.3). This activity is quite similar to the study of marginal distributions of times to failure from specific causes in the competing risks model as discussed in the previous section. The approach of Fine et al. (2001) was to assume a shared gamma frailty model for the joint distribution of time to illness and time to death, and, under such an assumption, the marginal distributions are identifiable from the available data where occurrence of death (entry into state 2) prevents the illness from ever happening. However, since this (or similar) association structures cannot be supported by the observed data, we will follow recommendations of Xu et al. (2010) and Andersen and Keiding (2012) and not study the illness-death model from the point of view of semi-competing risks.

Independent censoring/competing risks

Both of the concepts of independent censoring and independent competing risks relate to independence between two random time-variables. However, it is only in the former case that these two random variables (time to event and time to censoring) are well-defined, and, for that reason, we find the concept of independent censoring important, whereas we find the concept of independent competing risks (and the associated latent failure time approach) less relevant.

4.5 Exercises

Exercise 4.1 Consider the data from the Copenhagen Holter study and estimate the probabilities of stroke-free survival for subjects with or without ESVEA using the Kaplan-Meier estimator.

Exercise 4.2 Consider the Cox model for stroke-free survival in the Copenhagen Holter study including the covariates ESVEA, sex, age, and systolic blood pressure (Exercise 2.4).

1. Estimate the survival functions for a female subject aged 65 years and with systolic blood pressure equal to 150 *mm*Hg – either with or without ESVEA.
2. Estimate the survival functions for patients with or without ESVEA using the g-formula.

Exercise 4.3 Consider the data from the Copenhagen Holter study and fit a linear model for the 3-year restricted mean time to the composite end-point stroke or death including ESVEA, sex, age, and systolic blood pressure.

Exercise 4.4 Consider the Cox models for the cause-specific hazards for the outcomes stroke and death without stroke in the Copenhagen Holter study including ESVEA, sex, age, and systolic blood pressure (Exercise 2.7). Estimate (using plug-in) the cumulative incidences for both end-points for a female subject aged 65 years and with systolic blood pressure equal to 150 *mm*Hg – either with or without ESVEA.

Exercise 4.5

1. Repeat the previous question using instead Fine-Gray models.
2. Estimate the cumulative incidence functions for patients with or without ESVEA using the g-formula.

Exercise 4.6 Consider the data from the Copenhagen Holter study and fit linear models for the expected time lost (numbers of years) before 3 years due to either stroke or death without stroke including ESVEA, sex, age, and systolic blood pressure.

Exercise 4.7 Consider an illness-death model for the Copenhagen Holter study with states '0: Alive without AF or stroke', '1: Alive with AF and no stroke', '2: Dead or stroke', see Figures 1.3 and 1.7.

1. Estimate the prevalence of AF.
2. Estimate the expected lengths of stay in states 0 or 1 up to 3 years.
3. Evaluate the SD of the expected lengths of stay using the bootstrap.

Exercise 4.8 Consider the data on mortality in relation to childhood vaccinations in Guinea-Bissau, Example 1.1.2.

1. Fit a marginal hazard model for the mortality rate, adjusting for cluster '(village)' and including binary variables for BCG and DTP vaccination and adjusting for age at recruitment (i.e., using time since recruitment as time-variable).

2. Compare the results with those from the gamma frailty model (Exercise 3.12).

Exercise 4.9 Consider the data on recurrent episodes in affective disorder, Example 1.1.5.

1. Estimate the mean number of episodes, $\mu(t)$, in $[0,t]$ for unipolar and bipolar patients, taking the mortality into account.

2. Estimate, incorrectly, the same mean curves by treating death as censoring and compare with the correct curves from the first question, thereby, re-constructing the cover figure from this book (unipolar patients).

Exercise 4.10 Consider the data from the Copenhagen Holter study.

1. Estimate the distribution, $G(t)$ of censoring.

2. Examine to what extent this distribution depends on the variables ESVEA, sex, age, and systolic blood pressure.

Chapter 5

Marginal models

This chapter explains some of the mathematical foundation for the methods illustrated via the practical examples in Chapter 4. As we did in that chapter, we will separately study methods based on plug-in of results from intensity models and, here, it turns out to be crucial whether the multi-state process is *Markovian* (Section 5.1) or not (Section 5.2). Along the way, we will also introduce methods that were not exemplified in Chapter 4, including the techniques of *landmarking* and *micro-simulation* (Sections 5.3 and 5.4), both of which also build on plug-in of hazard models. The second part of this chapter, Sections 5.5-5.7, describes the background for direct models for marginal parameters based on generalized estimating equations (GEEs), including a general method based on *cumulative residuals* for assessment of goodness-of-fit. Finally, Section 5.8 provides practical examples of new methods presented.

5.1 Plug-in for Markov processes (*)

Let $V(t)$ be a multi-state process with state space $\mathscr{S} = \{0, 1, \dots, k\}$ and assume that $V(t)$ is *Markovian*, i.e., the transition intensities satisfy the Markov property

$$\alpha_{hj}(t)dt \approx P(V(t+dt) = j \mid V(t) = h, \mathscr{H}_{t-}) = P(V(t+dt) = j \mid V(t) = h, \mathbf{Z}) \qquad (5.1)$$

for all states $h, j \in \mathscr{S}, j \neq h$. Here, the $\mathbf{Z}$ are *time-fixed covariates* included in $\mathscr{H}_t$, i.e., the only way in which the intensities depend on the past is via the current state $h = V(t)$ and via $\mathbf{Z}$. Define the cumulative intensities $A_{hj}(t) = \int_0^t \alpha_{hj}(u)du$ and let

$$A_{hh}(t) = -\sum_{j \in \mathscr{S}, j \neq h} A_{hj}(t).$$

We can now collect all $A_{hj}(t), h, j \in \mathscr{S}$ in a $(k+1) \times (k+1)$-matrix $\mathbf{A}(t)$. The *product-integral* of $\mathbf{A}(\cdot)$ over the interval $(s,t]$ is defined as the $(k+1) \times (k+1)$-matrix

$$\begin{aligned} \mathbf{P}(s,t) &= \prod_{(s,t]} (\mathbf{I} + d\mathbf{A}(u)) \\ &= \lim_{\max |u_i - u_{i-1}| \to 0} \prod (\mathbf{I} + \mathbf{A}(u_i) - \mathbf{A}(u_{i-1})) \end{aligned} \qquad (5.2)$$

for any partition $s = u_0 < u_1 < \cdots < u_N = t$ of $(s,t]$ (Gill and Johansen, 1990). Here, $\mathbf{I}$ is the $(k+1) \times (k+1)$ identity matrix. We defined the A_{hj} as integrated intensities, but (5.2) is also well-defined if the A_{hj} have jumps, and in the case where the A_{hj} correspond to purely discrete measures, the product-integral (5.2) is just a finite matrix product over the jump times in $(s,t]$ (reflecting the *Chapman-Kolmogorov equations*). In the special case where all intensities are time-constant on $(s,t]$, the product-integral (5.2) is the *matrix exponential*

$$\mathbf{P}(s,t) = \exp(\boldsymbol{\alpha}\cdot(t-s)) = \mathbf{I} + \sum_{i=1}^{\infty}\frac{1}{i!}(\boldsymbol{\alpha}\cdot(t-s))^i.$$

It now holds (Gill and Johansen, 1990), that $\mathbf{P}(s,t)$ is the *transition probability matrix* for the Markov process $V(\cdot)$, i.e., element h,j is $P_{hj}(s,t) = P(V(t) = j|V(s) = h)$. If $\mathbf{A}$ is absolutely continuous, then, for given intensity matrix $\boldsymbol{\alpha}(t)$, the matrix $\mathbf{P}$ given by (5.2) solves the *Kolmogorov forward differential equations*

$$\frac{\partial}{\partial t}\mathbf{P}(s,t) = \mathbf{P}(s,t)\boldsymbol{\alpha}(t), \text{ with } \mathbf{P}(s,s) = \mathbf{I}. \tag{5.3}$$

This suggests plug-in estimators for $\mathbf{P}(s,t)$ based on models fitted for the intensities. A non-parametric estimator for $\mathbf{P}(s,t)$ for an assumed homogeneous group is obtained by plugging-in the Nelson-Aalen estimator $\widehat{\mathbf{A}}$, and the resulting estimator

$$\widehat{\mathbf{P}}(s,t) = \prod_{(s,t]}(\mathbf{I} + d\widehat{\mathbf{A}}(u)) \tag{5.4}$$

is the *Aalen-Johansen* estimator (Aalen and Johansen, 1978; Andersen et al., 1993; ch. IV). If the transition intensities for a homogeneous group are assumed piece-wise constant on $(s,t]$, e.g., $\boldsymbol{\alpha} = \boldsymbol{\alpha}_1$ on $(s,u]$ and $\boldsymbol{\alpha} = \boldsymbol{\alpha}_2$ on $(u,t]$, then $\boldsymbol{\alpha}_1$ and $\boldsymbol{\alpha}_2$ are estimated by separate occurrence/exposure rates in the two sub-intervals, and the plug-in estimator, using the Chapman-Kolmogorov equations, is

$$\widehat{\mathbf{P}}(s,t) = \widehat{\mathbf{P}}(s,u)\widehat{\mathbf{P}}(u,t) = \exp(\widehat{\boldsymbol{\alpha}}_1\cdot(u-s))\exp(\widehat{\boldsymbol{\alpha}}_2\cdot(t-u)), \tag{5.5}$$

with similar expressions if there are more than two sub-intervals of $(s,t]$ on which $\boldsymbol{\alpha}$ is constant. Both this expression and (5.4) also apply if the model for the intensities is a hazard regression model with time-fixed covariates, e.g., a Cox model or a Poisson model. The situation with time-dependent covariates is more challenging and will be discussed in Section 5.2.4. Aalen et al. (2001) studied plug-in estimation based on additive hazards models.

The *state occupation probabilities* are

$$Q_h(t) = P(V(t) = h) = \sum_j Q_j(0)P_{jh}(0,t), \quad j \in \mathcal{S}.$$

In the situation where all subjects are in the same state (0) at time $t = 0$, i.e., $Q_0(0) = 1$, these are $Q_h(t) = P_{0h}(0,t)$ and the product-limit estimator may be used for this *marginal parameter*. We will pay special attention to this situation in what follows. Based on the state occupation probabilities, another marginal parameter, the *expected length of stay* in state h before time τ, is directly obtained as

$$\varepsilon_h(\tau) = \int_0^{\tau} Q_h(t)dt.$$

Since these marginal parameters (and $\mathbf{P}(s,t)$) are *differentiable functionals* of the intensities, large-sample properties of the resulting plug-in estimators may be derived from those of the intensity estimators using *functional delta-methods*. The details are beyond the scope of this presentation and may be found in Andersen et al. (1993; ch. II and IV).

We will now look at some of the multi-state models introduced in Section 1.1.

5.1.1 Two-state model (*)

In Figure 1.11, there are two states and a single transition intensity, $\alpha_{01}(t) = \alpha(t)$ from 0 to 1, the hazard function for the distribution of the survival time T. The two-state process is born Markov and the resulting $\mathbf{A}$ matrix is

$$\mathbf{A}(t) = \begin{pmatrix} -A(t) & A(t) \\ 0 & 0 \end{pmatrix}.$$

For this model, the transition probability matrix is

$$\mathbf{P}(s,t) = \begin{pmatrix} P_{00}(s,t) & 1 - P_{00}(s,t) \\ 0 & 1 \end{pmatrix}$$

and the differential Equation (5.3) is

$$\frac{\partial}{\partial t} P_{00}(s,t) = -P_{00}(s,t)\alpha(t)$$

with the solution $P_{00}(s,t) = \exp(-\int_s^t \alpha(u)du)$. In Section 4.1, an intuitive argument for this expression was given and the survival function $S(t) = P_{00}(0,t)$ may be estimated by plug-in based on a model for $\alpha(t)$, e.g., a piece-wise constant hazard model. If $\mathbf{A}$ corresponds to a discrete distribution, such as that estimated by the Nelson-Aalen estimator or by a Cox model using the Breslow estimator for the cumulative baseline hazard, then it is the general product-integral in which plug-in should be made and not the exponential expression. This leads in the case of the Nelson-Aalen estimator to the *Kaplan-Meier estimator* for $S(t)$

$$\widehat{S}(t) = \prod_{[0,t]} (1 - d\widehat{A}(u)) = \prod_{X_i \leq t} (1 - \frac{dN(X_i)}{Y(X_i)})$$

(Kaplan and Meier, 1958) and the conditional Kaplan-Meier estimator for $P_{00}(s,t) = S(t \mid s)$

$$\widehat{S}(t \mid s) = \prod_{s < X_i \leq t} (1 - \frac{dN(X_i)}{Y(X_i)}),$$

where $X_1, X_2, \ldots$ are the observation times, $X_i = T_i \wedge C_i$. The general variance formula in this case reduces to the *Greenwood formula*

$$SD(\widehat{S}(t)) = \widehat{S}(t) \sqrt{\sum_{X_i \leq t} \frac{dN(X_i)}{Y(X_i)(Y(X_i) - 1)}}$$

(Kaplan and Meier, 1958). As discussed in Section 4.1.1, confidence limits for $S(t)$ are typically based on symmetric confidence limits for $\log(A(t)) = \log(-\log(S(t)))$.

Note that, in the special case of no censoring, the Kaplan-Meier estimator is the relative frequency $\widehat{S}(t) = Y(t)/n$ of survivors at time t, and the Greenwood formula reduces to the binomial variance formula.

For the Cox model with time-fixed covariates, the cumulative hazard for given covariates $\boldsymbol{Z}$ is $A_0(t)\exp(\boldsymbol{\beta}^\mathsf{T}\boldsymbol{Z})$ and the plug-in estimator for the survival probability is

$$\widehat{S}(t \mid \boldsymbol{Z}) = \prod_{X_i \le t} \Big(1 - \frac{dN(X_i)\exp(\widehat{\boldsymbol{\beta}}^\mathsf{T}\boldsymbol{Z})}{\sum_{j \in R(X_i)} \exp(\widehat{\boldsymbol{\beta}}^\mathsf{T}\boldsymbol{Z}_j)}\Big),$$

where $R(t) = \{j : Y_j(t) = 1\}$ is the risk set at time t, cf. Equation (3.19). This estimator may become negative and an alternative and commonly used estimator is

$$\widehat{S}(t \mid \boldsymbol{Z}) = \exp(-\widehat{A}_0(t)\exp(\widehat{\boldsymbol{\beta}}^\mathsf{T}\boldsymbol{Z})),$$

where $\widehat{A}_0(t)$ is the Breslow estimator (3.18). This estimator can be criticized for using the continuous-time version of the product-integral on a time-discrete estimator; however, from a practical point of view, the differences between these estimators tend to be of minor importance, cf. the discussion in Section 4.1.1 about not estimating $S(t)$ by 'exp(−Nelson-Aalen)'.

From estimates of the survival function S, the *restricted mean life time*

$$\varepsilon_0(\tau) = E(T \wedge \tau) = \int_0^\tau S(t)dt$$

may be estimated by plug-in, also for given covariates based on a hazard regression model. In the special case of no censoring, $\widehat{\varepsilon}_0(\tau) = (1/n)\sum_i (T_i \wedge \tau)$ is a simple average.

5.1.2 Competing risks (*)

We will focus on the competing risks model in Figure 1.2 with one transient state (0) and two absorbing states (1, 2). The general competing risks model with $k > 2$ absorbing states is similar and, in any case, the competing risks process is born Markov. For the three-state model, the $\mathbf{A}$ matrix is:

$$\mathbf{A}(t) = \begin{pmatrix} -A_1(t) - A_2(t) & A_1(t) & A_2(t) \\ 0 & 0 & 0 \\ 0 & 0 & 0 \end{pmatrix}$$

where A_1, A_2 are the cumulative cause-specific hazards for the two competing events. With the $\mathbf{P}$ matrix

$$\mathbf{P}(s,t) = \begin{pmatrix} P_{00}(s,t) & P_{01}(s,t) & P_{02}(s,t) \\ 0 & 1 & 0 \\ 0 & 0 & 1 \end{pmatrix}$$

(where $P_{00} = 1 - P_{01} - P_{02}$ and $P_{0h}(s,t)$, $h = 1,2$ are the conditional cumulative incidences), the Kolmogorov forward equations become

$$\begin{aligned} \frac{\partial}{\partial t} P_{00}(s,t) &= -P_{00}(s,t)(\alpha_1(t) + \alpha_2(t)) \\ \frac{\partial}{\partial t} P_{0h}(s,t) &= P_{00}(s,t)\alpha_h(t),\, h = 1,2 \end{aligned}$$

with solutions

$$\begin{aligned} P_{00}(s,t) &= \exp\left(-\int_s^t (\alpha_1(u) + \alpha_2(u))du\right) \\ P_{0h}(s,t) &= \int_s^t P_{00}(s,u)\alpha_h(u)du,\, h = 1,2. \end{aligned}$$

In Section 4.1, intuitive arguments for these expressions were given. The resulting non-parametric plug-in estimators (for $s = 0$) are the *overall Kaplan-Meier estimator*

$$\widehat{S}(t) = \widehat{P}_{00}(0,t) = \widehat{Q}_0(t) = \prod_{X_i \le t} \left(1 - \frac{dN(X_i)}{Y(X_i)}\right)$$

(jumping at times of failure from either cause) and the *cumulative incidence estimator*

$$\widehat{Q}_h(t) = \widehat{F}_h(t) = \widehat{P}_{0h}(0,t) = \int_0^t \widehat{S}(u-)\frac{dN_h(u)}{Y(u)}, \quad h = 1,2, \tag{5.6}$$

where $N_h(\cdot)$ counts failures from cause $h = 1,2$ and $Y(t)$ is the number of subjects at risk in state 0 just before time t. The estimator (5.6) was discussed by, e.g., Aalen (1978), and is often denoted the *Aalen-Johansen* estimator even though this name is also used for the general estimator (5.4) (Aalen and Johansen, 1978). Note that, in (5.6), the Kaplan-Meier estimator is evaluated just before a cause-h failure time at u.

In the special case of no censoring, $\widehat{Q}_h(t) = N_h(t)/n$.

Expressions similar to $\widehat{S}$ and (5.6) apply when estimating the state occupation probabilities $Q_h(t)$, $h = 0,1,2$ based on, e.g., Cox models, for the two cause-specific hazards: $\alpha_h(t \mid \mathbf{Z}) = \alpha_{h0}(t)\exp(\boldsymbol{\beta}_h^{\mathsf{T}}\mathbf{Z})$. In all cases, variance estimators are available (e.g., Andersen et al., 1991; 1993, ch. VII). Plug-in based on piece-wise exponential models may also be performed.

If T, as in the two-state model, is the life time (time spent in state 0), then the restricted mean life time is

$$\varepsilon_0(\tau) = E(T \wedge \tau) = \int_0^\tau S(t)dt$$

and plug-in estimation is straightforward. In the competing risks model, one can also study the random variables

$$T_h = \inf\{t > 0 : V(t) = h\}, \quad h = 1,2,$$

i.e., the times of entry into state $h = 1,2$. These are *improper* random variables because, e.g., $P(T_1 = \infty) = \lim_{t\to\infty} Q_2(t)$ is the positive probability that cause 1 never happens. The restricted random variables $T_h \wedge \tau$ are proper and

$$E(T_h \wedge \tau) = E\int_0^\tau I(T_h > t)dt = \tau - E\int_0^\tau I(T_h \le t)dt = \tau - \int_0^\tau Q_h(t)dt.$$

It follows that

$$\varepsilon_h(\tau) = \int_0^\tau Q_h(t)dt$$

can be interpreted as the *expected life time lost due to cause h* before time τ and plug-in estimation is straightforward (Andersen, 2013).

In the special case of no censoring, $\widehat{\varepsilon}_h(\tau) = (1/n)\sum_i(\tau - T_{hi} \wedge \tau)$ is a simple average.

5.1.3 Progressive illness-death model (*)

In Figure 1.3 there are three transition intensities, two transient states (0, 1) and one absorbing (2). This model is Markov whenever the α_{12} intensity only depends on t and not on duration in state 1. This leads to the **A** matrix

$$\mathbf{A}(t) = \begin{pmatrix} -A_{01}(t) - A_{02}(t) & A_{01}(t) & A_{02}(t) \\ 0 & -A_{12}(t) & A_{12}(t) \\ 0 & 0 & 0 \end{pmatrix}$$

and associated **P** matrix

$$\mathbf{P}(s,t) = \begin{pmatrix} P_{00}(s,t) & P_{01}(s,t) & P_{02}(s,t) \\ 0 & P_{11}(s,t) & P_{12}(s,t) \\ 0 & 0 & 1 \end{pmatrix}$$

with $P_{00} = 1 - P_{01} - P_{02}$ and $P_{11} = 1 - P_{12}$. The Kolmogorov forward equations become

$$\begin{aligned}
\frac{\partial}{\partial t}P_{00}(s,t) &= -P_{00}(s,t)(\alpha_{01}(t) + \alpha_{02}(t)) \\
\frac{\partial}{\partial t}P_{01}(s,t) &= P_{01}(s,t)\alpha_{01}(t) - P_{11}(s,t)\alpha_{12}(t) \\
\frac{\partial}{\partial t}P_{11}(s,t) &= -P_{11}(s,t)\alpha_{12}(t)
\end{aligned}$$

with solutions

$$\begin{aligned}
P_{00}(s,t) &= \exp(-\int_s^t (\alpha_{01}(u) + \alpha_{02}(u))du) \\
P_{11}(s,t) &= \exp(-\int_s^t \alpha_{12}(u)du) \\
P_{01}(s,t) &= \int_s^t P_{00}(s,u)\alpha_{01}(u)P_{11}(u,t)du.
\end{aligned}$$

In Section 4.1.3, intuitive arguments for the latter expression were given, and plug-in estimation is possible. Classical work on the illness-death model include Fix and Neyman (1951) and Sverdrup (1965).

Additional marginal parameters for the illness-death model (with or without recovery) include

$$\varepsilon_h(\tau) = \int_0^\tau Q_h(t)dt, h = 0, 1,$$

the *expected length of stay* in $[0, \tau]$, respectively, alive without or with the illness. The *prevalence* is

$$\frac{Q_1(t)}{Q_0(t)+Q_1(t)},$$

i.e., the probability of living with the illness at time t given alive at that time.

5.1.4 Recurrent events (*)

In Section 1.1, two diagrams for recurrent events were studied, corresponding to either intervals or no intervals between periods where subjects are at risk for the recurrent event (Figures 1.4 and 1.5, respectively). In the former case (the illness-death model with recovery), there are two transient states (0 and 1), one absorbing state (2), and the $\mathbf{A}(t)$ matrix is

$$\mathbf{A}(t) = \begin{pmatrix} -A_{01}(t)-A_{02}(t) & A_{01}(t) & A_{02}(t) \\ A_{10}(t) & -A_{10}(t)-A_{12}(t) & A_{12}(t) \\ 0 & 0 & 0 \end{pmatrix}.$$

The associated $\mathbf{P}$ matrix is

$$\mathbf{P}(s,t) = \begin{pmatrix} P_{00}(s,t) & P_{01}(s,t) & P_{02}(s,t) \\ P_{10}(s,t) & P_{11}(s,t) & P_{12}(s,t) \\ 0 & 0 & 1 \end{pmatrix}$$

with $P_{00} = 1-P_{01}-P_{02}$ and $P_{11} = 1-P_{10}-P_{12}$. In contrast to the progressive illness-death model, the Kolmogorov forward equations do not have an explicit solution but for given Nelson-Aalen estimates $\widehat{A}_{hj}$, $\mathbf{P}$ can be estimated from (5.4) using plug-in. For the recurrent events intensity $\alpha_{01}(t)$, this corresponds to an AG-type Markov model, see Section 2.5. Assuming, further, that $Q_0(0) = 1$, both the $\varepsilon_h(\tau)$ and the prevalence may be estimated as in the previous section. This was exemplified in Section 4.1.3 using the data on recurrence in affective disorders (Example 1.1.5).

For the case with no intervals between at-risk periods, a maximum number (say, K) of recurrent events to be considered must be decided upon in order to get a transition matrix with finite dimension. Letting, e.g., $K = 2$ this corresponds to having states 0 and 1 in Figure 1.5 transient and states 2 and D absorbing (i.e., transitions out of state 2 are not considered). Re-labelling state D as 3, this gives the $\mathbf{A}(t)$ matrix

$$\mathbf{A}(t) = \begin{pmatrix} -A_{01}(t)-A_{03}(t) & A_{01}(t) & 0 & A_{03}(t) \\ 0 & -A_{12}(t)-A_{13}(t) & A_{12}(t) & A_{13}(t) \\ 0 & 0 & 0 & 0 \\ 0 & 0 & 0 & 0 \end{pmatrix}$$

and associated $\mathbf{P}$ matrix

$$\mathbf{P}(s,t) = \begin{pmatrix} P_{00}(s,t) & P_{01}(s,t) & P_{02}(s,t) & P_{03}(s,t) \\ 0 & P_{11}(s,t) & P_{12}(s,t) & P_{13}(s,t) \\ 0 & 0 & 1 & 0 \\ 0 & 0 & 0 & 1 \end{pmatrix}$$

with $P_{00} = 1 - P_{01} - P_{02} - P_{03}$ and $P_{11} = 1 - P_{12} - P_{13}$. In this case, the model is *progressive* and explicit expressions for the elements of $\mathbf{P}$ exist (see next section) but plug-in using the general expression (5.4) is also possible. Models for the recurrence intensities $\alpha_{h,h+1}(t)$ could be of PWP Markov type (i.e., separate models for each h) or of AG Markov type (one common model for all h), see Section 2.3. For this model, another marginal parameter of potential interest – also estimable from $\widehat{\mathbf{P}}$ – is the probability $P(N(t) \geq h)$ of seeing at least h recurrences in $[0,t]$. For situations where mortality is negligible, simplified versions of the two models considered are available (e.g., Exercise 5.1).

5.1.5 Progressive multi-state models (*)

In some models, explicit expressions for the state occupation probabilities $Q_h(t)$ were available and we have seen that these expressions follow directly from the general product-integral representation of the transition probability matrix $\mathbf{P}$. The strength of the general approach is that it applies to any multi-state Markov process including, as we saw in the previous section, the model depicted in Figure 1.4 where subjects may return to a previously occupied state. In *progressive* (or *irreversible*) models such as that shown in Figure 1.6 where subjects do not return to a previously occupied state, some state occupation probabilities may be expressed explicitly in terms of the intensities. As an example, for the model in Figure 1.6, the probability $Q_2(t)$ of being alive after a relapse may be expressed, as follows. There are two paths leading to state 2: Either (a) directly from the initial state 0, or (b) from that state via state 1 (GvHD). The probability of the former path is given by

$$Q_2^{(a)}(t) = \int_0^t P_{00}(0,u)\alpha_{02}(u)P_{22}(u,t)du$$

because, at some time $u < t$, a $0 \to 2$-transition must have happened and, between times u and t, the subject must stay in state 2. For the other path, i.e., via state 1, the probability is

$$Q_2^{(b)}(t) = \int_0^t P_{00}(0,u)\alpha_{01}(u) \int_u^t P_{11}(u,x)\alpha_{12}(x)P_{22}(x,t)dxdu,$$

reflecting that, first a $0 \to 1$-transition must happen (at $u < t$), next the subject must stay in state 1 from time u to a time x between u and t, make a $1 \to 2$-transition at x and stay in state 2 between x and t. Similar expressions, though cumbersome, may be derived for $Q_h(t)$ parameters in other progressive processes. These are not crucial for Markov processes as discussed so far in this chapter because the general product-integral representation is available; however, similar arguments may be applied also to some semi-Markov processes where intensities not only depend on time t but also on the time spent in a given state. This will be discussed in Section 5.2.4.

5.2 Plug-in for non-Markov processes (*)

For Markov processes, the transition intensities satisfy the *Markov property* (5.1), and under this assumption the product-integral maps transition intensities onto transition probabilities and, thereby, state occupation probabilities. If the Markov property is not satisfied, then the transition intensities at time t depend on the past $\mathscr{H}_{t-}$ in a more complex way than through the current state and, possibly, time-fixed covariates, and the transition probability matrix $\mathbf{P}(s,t)$ cannot be computed using the product-integral. An explanation why product-integration of the Nelson-Aalen estimators does not estimate transition probabilities in non-Markov models (Titman, 2015) is that, at time u, $t > u > s$, the Nelson-Aalen estimator uses all subjects in a given state at that time – no matter the state occupied at s.

5.2.1 *State occupation probabilities (*)*

The product-integral can still be used for estimating state occupation probabilities as shown by Datta and Satten (2001), see also Overgaard (2019). The basis for this is a concept related to the intensity, namely the *partial transition rate* which is the right-hand side of Equation (5.1), say,

$$\alpha^*_{hj}(t) \approx P(V(t+dt) = j \mid V(t) = h, \mathbf{Z})/dt. \tag{5.7}$$

Note that, in (5.7), conditioning is only on the current state and baseline covariates and, under Markovianity, it equals the transition intensity. It was shown by Datta and Satten (2001) that, provided that censoring is independent of $V(t)$, the cumulative partial transition rate $A^*_{hj}(t) = \int_0^t \alpha^*_{hj}(u)du$ can be consistently estimated by the *Nelson-Aalen estimator*

$$\widehat{A}^*_{hj}(t) = \int_0^t \frac{dN_{hj}(u)}{Y_h(u)}$$

with N_{hj}, Y_h defined as previously. Extensions to the situation where censoring and $V(t)$ depend on (possibly, time-dependent) covariates were studied by Datta and Satten (2002) and Gunnes et al. (2007).

The partial transition rates may be of interest in their own right and asymptotic results follow from Glidden (2002). The partial rates are also important for marginal models for recurrent events as we shall see in Section 5.5.4 where we will also argue why the Nelson-Aalen estimator is consistent for $A^*_{hj}(t)$. However, their main interest lies in the fact that they provide a step towards estimating *state occupation probabilities*. Assume for simplicity that all subjects occupy the same state, 0 at time 0, i.e., $Q_0(0) = 1$. In that case, the top row of the $(k+1) \times (k+1)$ product-integral matrix

$$\prod_{(0,t]} (\mathbf{I} + d\mathbf{A}^*(u))$$

is the vector of state occupation probabilities $\mathbf{Q}(t) = (Q_0(t), Q_1(t), \ldots, Q_k(t))$, suggesting the plug-in estimator

$$\widehat{\mathbf{Q}}(t) = (1, 0, \ldots, 0) \prod_{(0,t]} (\mathbf{I} + d\widehat{\mathbf{A}}^*(u)) \tag{5.8}$$

which is the top row of the Aalen-Johansen estimator. Asymptotic results for (5.8) were also given by Glidden (2002), including both a complex variance estimator and a simulation-based way of assessing the uncertainty based on an idea of Lin et al. (1993) – an idea that we will return to in connection with goodness-of-fit examinations using cumulative residuals in Section 5.7.

The estimator (5.8) works for any state in a multi-state model, but for a *transient* state in a *progressive* model, an alternative is available. This estimator, originally proposed for the non-Markov irreversible illness-death model (Figure 1.3) by Pepe (1991) and Pepe et al.

(1991), builds on the difference between Kaplan-Meier estimators and is, as such, not a plug-in estimator based on intensities. We discuss it here for completeness and it works, as follows, for the illness-death model. If T_0 is the time spent in the initial state and T_2 is the time of death, both random variables observed, possibly with right-censoring, then the Kaplan-Meier estimator based on T_0 estimates $Q_0(t)$ while that based on T_2 estimates $1-Q_2(t)=Q_0(t)+Q_1(t)$, so, their difference estimates the probability $Q_1(t)$ of being in the transient state 1 at time t. The resulting estimator is known as the *Pepe estimator*, and Pepe (1991) also provided variance estimators. Alternatively, a non-parametric bootstrap may be applied to assess the variability of the estimator. This idea generalizes to any transient state in a progressive model. To estimate $Q_h(t)$ for such a state, one may use the difference between the Kaplan-Meier estimators of staying in the set of states, say $\mathscr{S}_h$ from which state h is reachable and that of staying in $\mathscr{S}_h \cup \{h\}$.

Based on an estimator for $Q_h(t)$, the expected length of stay in that state, $\varepsilon_h(\tau)=\int_0^\tau Q_h(t)dt$ may be estimated by plug-in.

5.2.2 *Transition probabilities (*)*

We mentioned in the introduction to Section 5.2 that an explanation why product-integration of the Nelson-Aalen estimators does not estimate transition probabilities in non-Markov models is that, at time $u, t>u>s$, the Nelson-Aalen estimator uses all subjects in a given state at that time – no matter the state occupied at s (Titman, 2015). Based on such an idea, Putter and Spitoni (2018) suggested to use the proposal of Datta and Satten (2001) combined with *sub-setting* for estimation of transition probabilities $P_{hj}(s,t)=P(V(t)=j \mid V(s)=h)$. To estimate this quantity for a fixed value of s and a fixed state $h \in \mathscr{S}$, attention was restricted to those processes $V_i(\cdot)$ in state h at time s, i.e., processes for which $Y_{hi}(s)=1$ and counting processes and at-risk processes were defined for this subset

$$dN_{j\ell}^{LM}(t)=\sum_i dN_{j\ell i}(t)Y_{hi}(s), \quad Y_j^{LM}(t)=\sum_i Y_{ji}(t)Y_{hi}(s), \quad t\geq s.$$

Here 'LM' stands for *landmarking*, a common name used for restricting attention to subjects who are still at risk at a landmark time point, here time s (Anderson et al., 1983; van Houwelingen, 2007). We will return to uses of the landmarking idea in Section 5.3 where models with time-dependent covariates are studied. The Nelson-Aalen estimators for the partial transition rates based on these sub-sets are

$$\widehat{A}_{j\ell}^{*LM}(t)=\int_s^t \frac{dN_{j\ell}^{LM}(u)}{Y_j^{LM}(u)}, \quad t\geq s.$$

These may be plugged-in to the product-integral to yield the *landmark Aalen-Johansen estimator*

$$\widehat{\mathbf{P}}^{LM}(s,t)=\mathbf{Q}^{LM}(s)\prod_{(s,t)}(\mathbf{I}+\widehat{\mathbf{A}}^{*LM}(u)), \tag{5.9}$$

where $\mathbf{Q}^{LM}(s)$ is the $(k+1)$ row vector with element h equal to 1 and other elements equal to 0. For fixed s, the asymptotic properties of (5.9) follow from the results of Glidden (2002).

The work by Titman (2015) on transition probabilities for non-Markov models should also be mentioned here, even though the methods are not based on plug-in of intensities. Following Uña-Alvarez and Meira-Machado (2015), Titman suggested a similar extension (i.e., based on sub-setting) of the Pepe estimator for a transient state j in a progressive model. To estimate $P_{hj}(s,t)$, one looks at the sub-set of processes $V_i(\cdot)$ in state h at time s and, for fixed s, this transition probability is estimated as the difference between Kaplan-Meier estimators of staying in sets of states $\mathscr{S}_{hj}$ and $\mathscr{S}_{hj} \cup \{j\}$, respectively, at time t where $\mathscr{S}_{hj}$ is the set of states reachable from h and from which j can be reached. A variance estimator was also presented.

For any state, j (absorbing or transient) in any multi-state model (progressive or not), Titman (2015) also suggested another estimator for $P_{hj}(s,t)$ based on sub-setting to processes in state h at time s, as follows. Define $\mathscr{R}_{hj}$ to be the set of states reachable from h but from which j cannot be reached. For the considered sub-set of processes, the following competing risks process for $u \geq s$ is defined when j is an *absorbing state*

$$V_s^*(u) = \begin{cases} 0 & \text{if} \quad V(u) \notin \mathscr{R}_{hj} \cup \{j\}, \\ 1 & \text{if} \quad V(u) \in \mathscr{R}_{hj}, \\ 2 & \text{if} \quad V(u) = j. \end{cases}$$

For the considered sub-set, this process is linked to $V(t)$ by the relation $P_{hj}(s,t) = P(V_s^*(t) = 2)$ and the desired transition probability can be estimated using the Aalen-Johansen estimator for the cause 2 cumulative incidence for $V_s^*(t)$. More specifically, if $N_{s\ell}^*(u)$ counts cause $\ell = 1,2$ events and $Y_s^*(u)$ is the number still at risk for cause 1 or 2 events at time $u-$ then the estimator is

$$\widehat{P}_{hj}^T(s,t) = \int_s^t \widehat{P}(V_s^*(u) = 0 \mid V_s^*(s) = 0) \frac{dN_{s2}^*(u)}{Y_s^*(u)},$$

where $\widehat{P}(V_s^*(u) = 0 \mid V_s^*(s) = 0)$ is estimated using the Kaplan-Meier estimator

$$\prod_{(s,u]} \Big(1 - \frac{dN_{s1}^*(v) + dN_{s2}^*(v)}{Y_s^*(v)}\Big).$$

If j is a *transient state*, then the following survival process for $u \geq s$ is defined for the considered sub-set of processes

$$V_s^*(u) = \begin{cases} 0 & \text{if} \quad V(u) \notin \mathscr{R}_{hj}, \\ 1 & \text{if} \quad V(u) \in \mathscr{R}_{hj}. \end{cases}$$

For this sub-set, the process $V_s^*(t)$ is related to $V(t)$ via $P_{hj}(s,t) = P(V_s^*(t) = 0)P(V(t) = j \mid V_s^*(t) = 0)$, where the first factor can be estimated by the Kaplan-Meier estimator for $V_s^*(t)$. Titman (2015) proposed to estimate the second factor by the relative frequency of processes in state j at time t among those for which $V_s^*(t) = 0$, i.e., by

$$\frac{\sum_i I(V_i(t) = j, V_i(s) = h, V_i(t) \notin \mathscr{R}_{hj})}{\sum_i I(V_i(s) = h, V_i(t) \notin \mathscr{R}_{hj})}.$$

Titman's construction extended that of Allignol et al. (2014) for the illness-death model. For this model, an alternative estimator was previously proposed by Meira-Machado et al. (2006); however, the latter proposal has the drawback that to obtain consistency, the support for the survival time distribution must be contained within that of the censoring distribution, and we will not discuss these estimators further. Both Titman (2015) and Putter and Spitoni (2018) presented simulation studies showing that, for Markov models the Aalen-Johansen estimator outperforms the more general estimators discussed in this section. For non-Markov models, the Aalen-Johansen estimator was biased, whereas both the landmark Aalen-Johansen estimator and the general estimator proposed by Titman had a satisfactory performance. Malzahn et al. (2021) extended the landmark Aalen-Johansen technique to hybrid situations where, only for some transitions, the Markov property fails, whereas, for others, the Markov assumption is compatible with the data.

Non-parametric *tests* for the Markov assumption have been studied for the irreversible illness-death model by Rodriguez-Girondo and de Uña-Alvarez (2012) building on estimates of Kendall's τ. For general multi-state models, Titman and Putter (2022) derived logrank-type tests (Section 3.2.2) for Markovianity comparing landmark Nelson-Aalen estimators $\widehat{A}^{*LM}_{j\ell}(t)$ to standard Nelson-Aalen estimators $\widehat{A}_{j\ell}(t)$.

5.2.3 *Recurrent events (*)*

For the situation with *recurrent events*, Figure 1.5, marginal parameters such as $Q_h(t)$ and $\varepsilon_h(t)$ may also be relevant as discussed in Section 5.1.4. In this situation, the *expected number of events*

$$\mu(t) = E(N(t)),$$

where $N(t)$ is the recurrent events counting process, is another marginal parameter to target, and it has an attractive interpretation. For this parameter, a plug-in estimator also exists. This is because

$$\mu(t) = \sum_{h=1}^{\infty} h \cdot P(N(t) = h) = \sum_{h=1}^{\infty} h \cdot Q_h(t),$$

and since the $Q_h(t)$ may be estimated by (5.8), we can estimate $\mu(t)$ by plug-in. A difficulty is, though, that one has to decide on the number of terms to include in the sum – a choice that need not be clear-cut since, for a large number, h of events, there may not be sufficient data to properly estimate $Q_h(t)$.

We shall later (Section 5.5.4) see how direct models for $\mu(t)$ may be set up using generalized estimating equations. This also leads to an alternative plug-in estimator suggested by Cook et al. (2009) that we will discuss there.

5.2.4 *Semi-Markov processes (*)*

A special class of non-Markov models is semi-Markov models where, at time t, transition intensities $\alpha_{hj}(\cdot)$ depend not only on t but also on the time spent in state h, i.e., the duration $d = t - T_h$ where T_h is the (last) time of entry into state h before time t. We have studied such models previously in connection with gap time models for recurrent events (Section 2.3) and when discussing adapted time-dependent covariates in Section 3.7, e.g., for the

PROVA trial (Example 1.1.4). For these data, we presented Cox models with either t or d as the baseline time-variable and with the effect of the other time-variable expressed as explicit functions using time-dependent covariates. Such models may form the basis for plug-in estimation of transition and state occupation probabilities as discussed by Andersen et al. (2022). We will illustrate this via estimation of the probability $Q_2(t)$ of being alive in the relapse state in the four-state model for the bone marrow transplantation data (Example 1.1.7, Figure 1.6). Assuming the multi-state process to be Markovian, we showed in Section 5.1.3 that $Q_2(t)$ is the sum of

$$Q_2^{(a)}(t) = \int_0^t P_{00}(0,u)\alpha_{02}(u)P_{22}(u,t)du$$

and

$$Q_2^{(b)}(t) = \int_0^t P_{00}(0,u)\alpha_{01}(u)\int_u^t P_{11}(u,x)\alpha_{12}(x)P_{22}(x,t)dxdu.$$

Suppose now that the transition intensities out of states 1 and 2 depend on both t and d and that we have modeled $\alpha_{12}(t,d)$, $\alpha_{13}(t,d)$ and $\alpha_{23}(t,d)$. In such a situation, for $t > s$, the probability of staying in state 1 between times s and t given entry into the state at time $T_1 \leq s$ is given by

$$\begin{aligned} P_{11}(s,t;T_1) &= P(V(t)=1 \mid V(s)=1, V(T_1-)=0, V(T_1)=1) \\ &= \exp(-\int_s^t (\alpha_{12}(u,u-T_1)+\alpha_{13}(u,u-T_1))du), \end{aligned}$$

because the waiting time distribution in state 1 given entry at T_1 has hazard function $\alpha_{12}(u,u-T_1)+\alpha_{13}(u,u-T_1)$ at time u. Similarly, the probability af staying in state 2 between times s and t given entry into the state at time $T_2 \leq s$ is given by

$$\begin{aligned} P_{22}(s,t;T_2) &= P(V(t)=2 \mid V(s)=2, V(T_2-)=0 \text{ or } 1, V(T_2)=2) \\ &= \exp(-\int_s^t \alpha_{23}(u,u-T_2)du). \end{aligned}$$

The probability of being in state 2 at time t is now the sum of

$$Q_2^{(a*)}(t) = \int_0^t P_{00}(0,u)\alpha_{02}(u)P_{22}(u,t;u)du$$

and

$$Q_2^{(b*)}(t) = \int_0^t P_{00}(0,u)\alpha_{01}(u)\int_u^t P_{11}(u,x;u)\alpha_{12}(x,x-u)P_{22}(x,t;x)dxdu.$$

Note that P_{22} in this expression could also depend on the time u of $0 \to 1$ transition (though, in that case the process would not be termed semi-Markov). This idea generalizes to other progressive semi-Markov processes and to multi-state processes where the dependence on the past is modeled using adapted time-dependent covariates; however, both the resulting expressions and the associated variance calculations tend to get rather complex, as demonstrated by Shu et al. (2007) for the irreversible illness-death model.

Markov and non-Markov processes

For Markov processes, the intensities at time t only depend on the past history via the state occupied at t and, possibly, via time-fixed covariates. They have attractive mathematical properties, most importantly that transition probabilities may be obtained using plug-in via the *product-integral*.
Both the two-state model and the competing risks model are born Markov; however, for more complicated multi-state models the Markov assumption is restrictive and may not be fulfilled in practical examples. Analysis of non-Markov processes is less straightforward, though state occupation probabilities (and expected length of stay in a state) may be obtained using product-integration.

5.3 Landmarking

Plug-in works for hazard models with time-fixed covariates and for some models with adapted time-dependent covariates as exemplified in Sections 3.7, 5.1, and 5.2.4. For hazard models with non-adapted time-dependent covariates $Z(t)$, it is typically not possible to express parameters such as transition or state occupation probabilities using only the transition intensities. This is because the future course of the process $V(t)$ will also depend on the way in which the time-dependent covariates develop. In such a situation, in order to estimate these probabilities, a *joint model* for $V(t)$ and $Z(t)$ is needed and we will briefly discuss such joint models in Section 7.4. One way of approximating these probabilities using plug-in estimators is based on *landmarking*. In Sections 5.3.1-5.3.3, we will discuss this concept in the framework of the two-state model for survival data (Figure 1.1) with an illustration using the example on bone marrow transplantation in acute leukemia (Example 1.1.7). In Section 5.3.4, we will briefly mention the extensions needed for more complex multi-state models, and Section 5.3.5 provides some of the mathematical details.

5.3.1 *Conditional survival probabilities*

We look at the two-state model for survival data and have in mind a Cox regression model for the hazard function $\alpha(t \mid Z(\cdot))$ including, possibly non-adapted, time-dependent covariates $Z(t)$. We aim at estimating conditional survival probabilities, such as,

$$P_{00}(s,t) = P(T > t \mid T > s, (Z(u), u \leq s)), \quad t > s,$$

i.e., given survival till a *landmark* time s (where T is the survival time) and given the course of $Z(\cdot)$ up to the landmark time. The model that will be used for approximating such conditional survival probabilities is the following Cox model

$$\alpha_s(t \mid (Z(u), u \leq s)) = \alpha_{0s}(t) \exp(\mathrm{LP}_s), \quad t \geq s \tag{5.10}$$

with a landmark-specific baseline hazard $\alpha_{0s}(t)$ and with linear predictor

$$\mathrm{LP}_s = \beta_{1s} Z_1(s) + \cdots + \beta_{ps} Z_p(s)$$

including the current covariate values at the landmark time s and with landmark-specific regression coefficients. Recall from Section 3.7.2 that the value of the time-dependent covariate at time s may be the lagged value, $Z_j(s - \Delta)$ or similar. It is seen that the covariates in the model (5.10) are kept fixed for $t > s$ at their value at time s and the model is fitted (using delayed entry) for all subjects still at risk at time s, i.e., subjects i, for whom $Y_i(s) = Y_{0i}(s) = 1$. From this model, the desired conditional survival probabilities can be estimated using plug-in as explained in Sections 4.1.1 and 5.1.1. Typically, only short-term predictions will be aimed at since the covariate values at time s are used when predicting at that time, and the further development of the time-dependent covariates for $t > s$ is not accounted for. As a consequence, when fitting the model (5.10), one often censors everyone at some *horizon time* $t_{hor}(s) > s$ and, thereby, restricts attention to predictions for a period of length at most $t_{hor}(s) - s$.

Following this basic idea, van Houwelingen (2007) and van Houwelingen and Putter (2012, ch. 7) suggested to study a series of landmark models at times $0 \leq s_1 < s_2 < \cdots < s_L$, each fitted, as just described, to subjects still at risk at the various landmark times. The model at a given landmark s_j has baseline hazard $\alpha_{0s_j}(t)$ and linear predictor $\text{LP}_{s_j} = \beta_{1s_j}Z_1(s_j) + \cdots + \beta_{ps_j}Z_p(s_j)$. If the horizon time when analyzing data at landmark s_j is taken to be no larger than the subsequent s_{j+1}, then the analyses at different landmarks will be independent in the sense that a given failure time will appear in at most one analysis. However, this is no requirement for the method, though it should be kept in mind that if intervals from landmark to horizon do overlap for two successive landmarks, then, for each time point t belonging to both intervals, the two landmark models will provide alternative descriptions for the hazard at t. Therefore, models based on several landmarks should be regarded as merely descriptions to be used for predictions and not as proper probability models for the data generation process.

5.3.2 *Landmark super models*

The separate landmark models introduced in the previous section will typically contain a large number of parameters ($p \cdot L$ regression coefficients and L baseline hazards), and it is of interest to reduce this number. Furthermore, if successive landmarks are close, then one would not expect the regression coefficients to substantially change from one landmark to the next. Rather, some smoothness across landmarks is expected for the parameters. This leads to the development of *landmark super models*, as follows. First, regression coefficients at different landmarks are connected by letting $\beta_{ks} = \beta_k + \sum_{\ell=1}^{m_k} \gamma_{k\ell} f_{k\ell}(s)$, $k = 1, \ldots, p$ for m_k suitably chosen functions $f_{k\ell}(s)$ that may vary among covariates (k). For ease of notation we will drop this generality and let each β_{ks}, $k = 1, \ldots, p$, be

$$\beta_{ks} = \beta_k + \gamma_{k1} f_1(s) + \cdots + \gamma_{km} f_m(s), \tag{5.11}$$

with the same smoothing functions for all covariates. If, in Equation (5.11), we let $f_1(s_1) = \cdots = f_m(s_1) = 0$, then β_k will be the effect of covariate k at the first landmark, s_1. A typical choice could be $m = 2$ and

$$f_1(s) = (s - s_1)/(s_L - s_1), \quad f_2(s) = ((s - s_1)/(s_L - s_1))^2.$$

To fit these models, a data duplication trick, following the lines of Section 3.8 can be applied. Thus, L copies of the data set are needed, where copy number j includes all subjects still at risk at landmark s_j, i.e., subjects i with $Y_i(s_j) = 1$. The Cox model is *stratified* on

Table 5.1 *Bone marrow transplantation in acute leukemia: Distribution of the time-dependent covariates ANC500 and GvHD at entry and at five landmarks (ANC: Absolute neutrophil count, GvHD: Graft versus host disease).*

	Landmark s_j (months)					
	0	0.5	1.0	1.5	2.0	2.5
At risk $Y(s_j)$	2099	1988	1949	1905	1876	1829
$\text{ANC500}(s_j) = 1$	0	906	1912	1899	1874	1828
$\text{GvHD}(s_j) = 1$	0	180	391	481	499	495

j (to yield separate baseline hazards $\alpha_{0s_j}(t)$), and the model in stratum j should include the interactions $Z \cdot f_\ell(s_j), \ell = 1, \dots, m$. For inference, robust standard deviations should be used.

Having separate baseline hazards for each landmark will provide different models for the hazard at some time points if the time horizons $t_{hor}(s_j)$ are chosen in such a way that prediction intervals overlap. Therefore, the baseline hazards could also be taken to vary from one landmark to the next in a smooth way by letting

$$\alpha_{0s}(t) = \alpha_0(t) \exp(\eta_1 g_1(s)) + \cdots + \eta_{m'} g_{m'}(s)), \tag{5.12}$$

with all $g_\ell(s_1) = 0$, such that $\alpha_0(t)$ refers to the first landmark, s_1. Often, one would choose the same smoothing functions for regression coefficients and baseline hazards, i.e., $m = m'$ and $g_\ell = f_\ell$. This model can also be fitted using the duplicated data set; however, stratification on j should no longer be imposed because only a single baseline hazard is needed. The model with baseline hazard given by Equation (5.12) provides a description for all values of s and, thereby, using this model conditional predictions may be obtained for all s, not only for the landmarks chosen when fitting the model.

Note the similarity with the tests for proportional hazards using time-dependent covariates as discussed in Section 3.7.7. It is seen that the idea of studying departures from proportionality using suitably defined functions $f_j(t)$ may also be used to obtain flexible models with non-proportional hazards – both for models with time-fixed and time-dependent covariates.

5.3.3 Bone marrow transplantation in acute leukemia

In Section 3.7.8, we studied models for the data on bone marrow transplantation (BMT) in acute leukemia (Example 1.1.7) including the two time-dependent covariates: Occurrence of graft versus host disease (GvHD) and reaching an Absolute Neutrophil Count above 500 cells per μL (ANC500). We will now estimate conditional probabilities of relapse-free survival, i.e., time to relapse or death in remission whatever comes first, based on past information on the two time-dependent covariates using landmarking. Both covariates take the value 0 at time $t = 0$ of BMT and, typically, a change of value from 0 to 1 takes place (if at all) relatively shortly after BMT. Table 5.1 shows the distribution of the two covariates at chosen landmarks 0.5, 1.0, 1.5, 2.0, and 2.5 months.

We first fit models like (5.10) at these landmarks using the horizon $t_{hor} = s_j + 6$ months, see Table 5.2. It is seen that both covariates have an effect on the hazard function – for GvHD,

Table 5.2 *Bone marrow transplantation in acute leukemia: Estimated effects (and robust SD) of time-dependent covariates ANC500 and GvHD at five landmarks using a 6-month horizon from landmarks (ANC: Absolute neutrophil count, GvHD: Graft versus host disease).*

Landmark		ANC500		GvHD	
j	s_j (months)	$\widehat{\beta}_{s_j}$	SD	$\widehat{\beta}_{s_j}$	SD
1	0.5	-0.335	0.107	0.703	0.149
2	1.0	-0.609	0.320	0.679	0.115
3	1.5	-1.686	0.610	0.863	0.113
4	2.0	-2.964	0.404	0.802	0.115
5	2.5	-3.354	0.178	0.831	0.121

Table 5.3 *Bone marrow transplantation in acute leukemia: Estimated (Est) smooth effects of time-dependent covariates ANC500 and GvHD (with robust SD) based on landmark super models using a 6-month horizon from landmarks (ANC: Absolute neutrophil count, GvHD: Graft versus host disease).*

		Stratified		Smoothed	
Covariate	Parameter	Est	SD	Est	SD
ANC500(s)	β_1	-0.322	0.106	-0.298	0.094
ANC500(s)$f_1(s)$	γ_{11}	-1.191	1.670	-1.393	1.630
ANC500(s)$f_2(s)$	γ_{12}	-2.257	1.791	-2.038	1.765
GvHD(s)	β_2	0.663	0.143	0.674	0.140
GvHD(s)$f_1(s)$	γ_{21}	0.391	0.430	0.333	0.417
GvHD(s)$f_2(s)$	γ_{22}	-0.226	0.342	-0.175	0.333
$g_1(s)$	η_1			1.414	1.606
$g_2(s)$	η_2			1.940	1.751

the effect is rather constant over time, and presence of ANC500 seems to be increasingly protective over time. Based on this model we predict the 6-month conditional relapse-free survival probabilities given still at risk at the respective landmarks, i.e., using $t_{hor}(s_j) = s_j + 6$ months. Figure 5.1 shows these predictions for subjects with either ANC500$(s_j) =$ GvHD$(s_j) = 0$ or ANC500$(s_j) =$ GvHD$(s_j) = 1$. It is seen that the effect of, in particular ANC500, over time has a quite marked influence on the curves.

Next, we fit a landmark super model with coefficients given by (5.11) choosing $f_1(s) = (s - s_1)/(s_L - s_1) = (s - 0.5)/2$ and $f_2(s) = f_1(s)^2$. Estimated coefficients are shown in Table 5.3 (stratified), and Figure 5.2 shows the associated conditional relapse-free survival curves. These are seen to be roughly consistent with those based on the simple landmark model. Table 5.3 (smoothed) also shows coefficients in a super model with a single baseline hazard $\alpha_0(t)$ (i.e., at $s_1 = 0.5$ months) and later landmark-specific baseline hazards (i.e., at $s_j > 0.5$ months) specified by (5.12) and, thereby, varying smoothly among landmarks. The smoothing functions were chosen as $g_\ell(s) = f_\ell(s), \ell = 1, 2$. Figure 5.3 shows the corresponding estimated conditional relapse-free survival probabilities – quite similar to those in the previous figures.

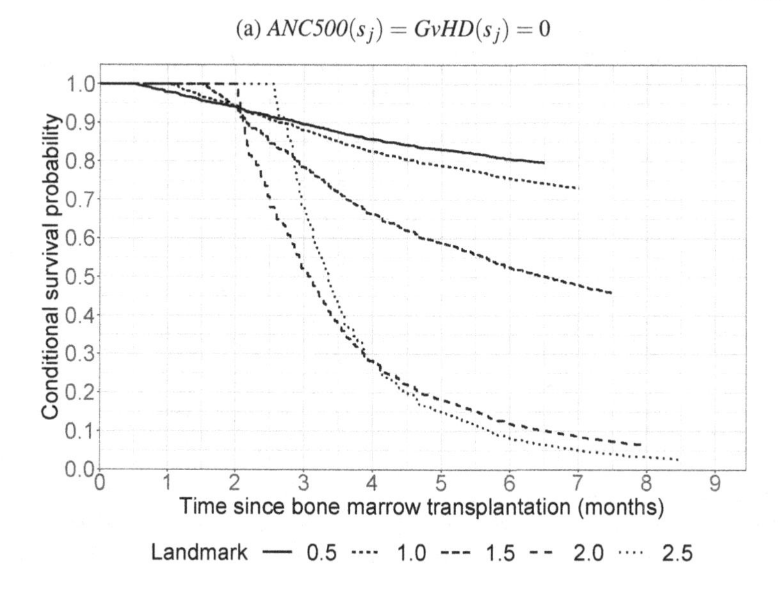

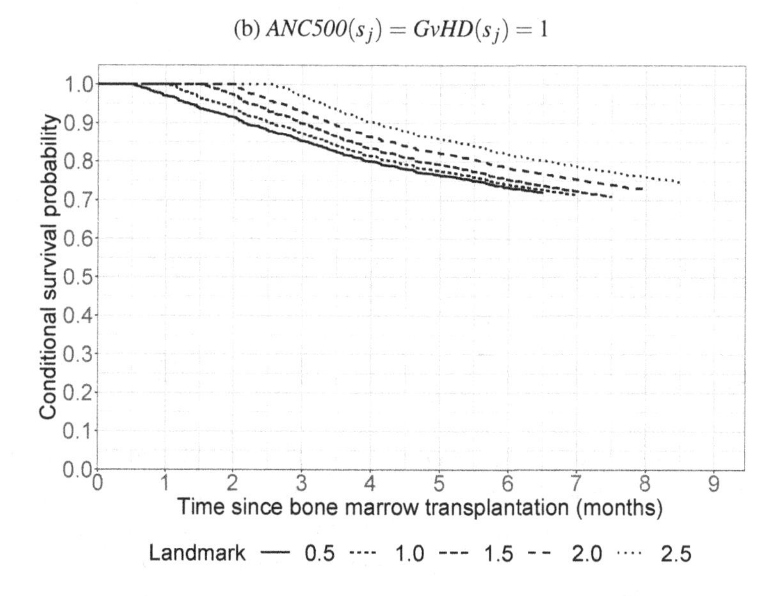

Figure 5.1 *Bone marrow transplantation in acute leukemia: Estimated 0- to 6-month conditional relapse-free survival probabilities given survival till landmarks* s_j = *0.5, 1.0, 1.5, 2.0, and 2.5 months.*

5.3.4 *Multi-state landmark models*

Landmarking for general multi-state models, i.e., not necessarily the two-state model, follows closely the techniques outlined in Sections 5.3.1-5.3.2. This is because, using the techniques discussed in these sections, landmark models may be set up for one transition intensity at a time and, once the landmark models are established, transition probabilities may be predicted by plug-in. When setting up the landmark models, all sorts of flexibility is available in the sense that different covariates and different smoothing functions $(f_\ell(s), g_\ell(s))$

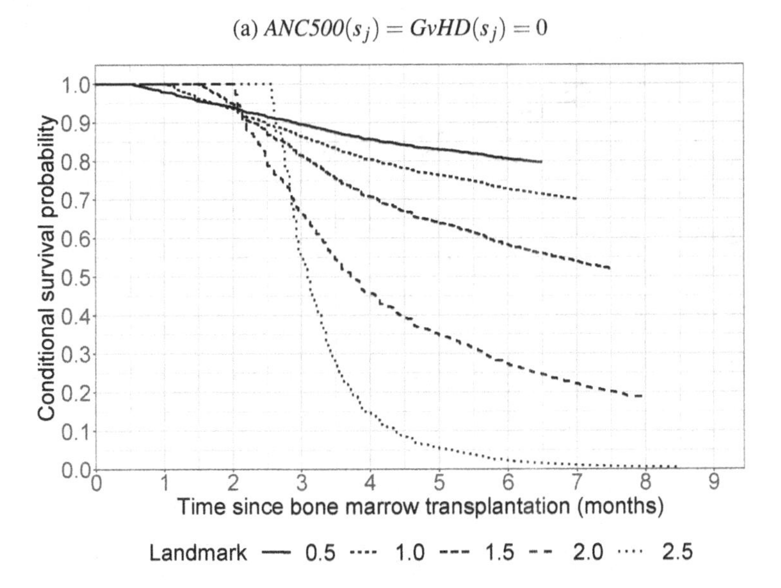

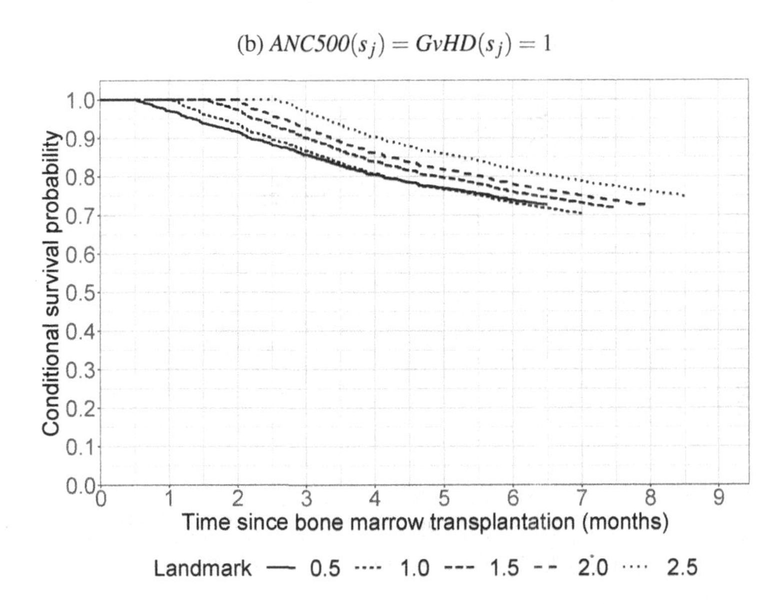

Figure 5.2 *Bone marrow transplantation in acute leukemia: Estimated 0- to 6-month conditional relapse-free survival probabilities given survival till landmarks* $s_j = 0.5, 1.0, 1.5, 2.0,$ *and 2.5 months. Estimates are based on a landmark super model with coefficients varying smoothly among landmarks and landmark-specific baseline hazards.*

may be chosen for the different transitions, though the same landmarks are typically used for all transition hazards.

5.3.5 Estimating equations (*)

In this section, we will explain how the estimates in models (5.10)-(5.12) are obtained. Model (5.10) for landmarks $s_1, \ldots, s_L$ consists of L standard Cox regression models, one

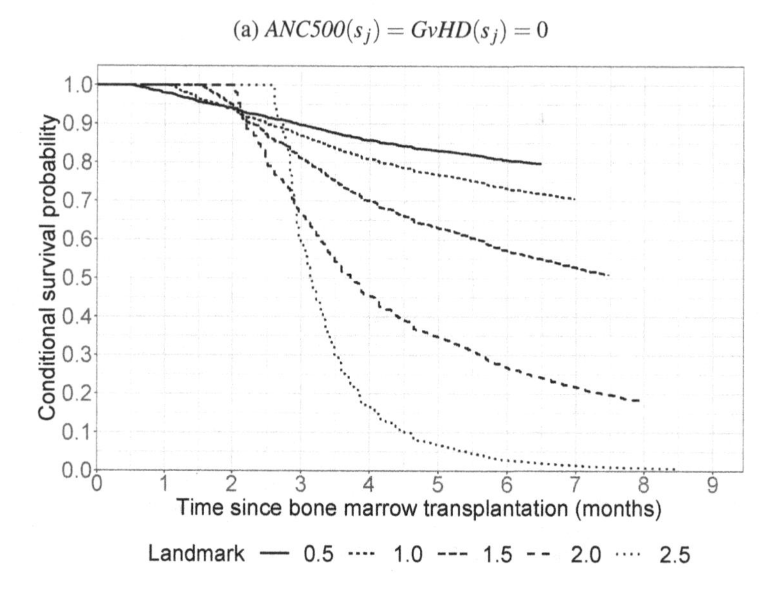

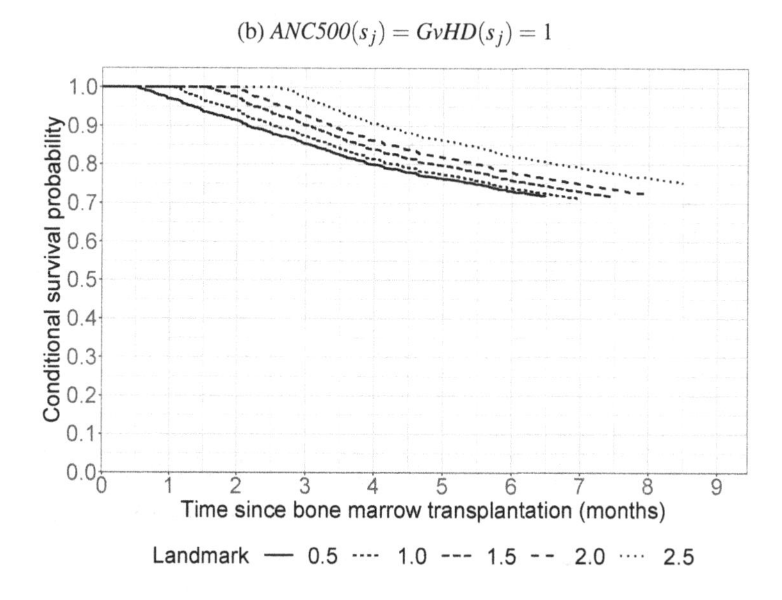

Figure 5.3 *Bone marrow transplantation study: Estimated 0- to 6-month conditional relapse-free survival probabilities given survival till landmarks $s_j = 0.5, 1.0, 1.5, 2.0$, and 2.5 months. Estimates are based on a landmark super model with coefficients and baseline hazards varying smoothly among landmarks.*

for each of the data sub-sets that together constitute the stacked data set. The estimating equations $\boldsymbol{\beta}_s$ are obtained from the Cox log-partial likelihood

$$\sum_{i=1}^{n} \int_{s}^{t_{hor}} Y_i(s) \log\Big(\frac{\exp(\mathrm{LP}_{s,i})}{\sum_k Y_k(t)\exp(\mathrm{LP}_{s,k})}\Big) dN_i(t), \quad s = s_1, \ldots, s_L,$$

by taking derivatives with respect to the parameters in the linear predictor $\mathrm{LP}_s = \beta_{1s}Z_1(s) + \cdots + \beta_{ps}Z_p(s)$ and equating to zero. The resulting Breslow estimator for the cumulative baseline hazard $A_{0s}(t), t < t_{hor}$ is

$$\widehat{A}_{0s}(t) = \int_s^t \frac{\sum_i Y_i(s) dN_i(u)}{\sum_i Y_i(u) \exp(\widehat{\mathrm{LP}}_{s,i})}.$$

This model contains L sets of parameters, each consisting of p regression parameters and a baseline hazard, estimated separately for each landmark $s_j, j = 1, \ldots, L$. If the horizon time, $t_{hor}(s_j)$, used when analyzing data at landmark s_j is no larger than the subsequent s_{j+1}, then the analyses at different landmarks will be independent and inference for parameters can in principle be performed using model-based variances; however, typically robust variances are used for all landmarking models.

The model (5.11) is also a stratified Cox model where all strata contribute to the estimating equations for regression parameters. The estimating equations are obtained from the log(pseudo-likelihood)

$$\sum_{j=1}^{L} \sum_{i=1}^{n} \int_{s_j}^{t_{hor}(s_j)} Y_i(s_j) \log\Big(\frac{\exp(\mathrm{LP}_{s_j,i})}{\sum_k Y_k(t) \exp(\mathrm{LP}_{s_j,k})}\Big) dN_i(t) \tag{5.13}$$

where the linear predictor is

$$\mathrm{LP}_{s_j} = \sum_{k=1}^{p} \beta_{ks_j} Z_k(s_j) = \sum_{k=1}^{p} (\beta_k + \sum_{\ell=1}^{m} \gamma_{k\ell} f_\ell(s_j) Z_k(s_j)),$$

see Exercise 5.5. The Breslow estimator is

$$\widehat{A}_{0s_j}(t) = \int_{s_j}^t \frac{\sum_i Y_i(s_j) dN_i(u)}{\sum_i Y_i(u) \exp(\widehat{\mathrm{LP}}_{s_j,i})}.$$

Also in model (5.12), all strata contribute to the estimation of the coefficients in the linear predictor

$$\begin{aligned}
\mathrm{LP}_{s_j} &= \sum_{\ell=1}^{m'} \eta_\ell g_\ell(s_j) + \sum_{k=1}^{p} \beta_{ks_j} Z_k(s_j) \\
&= \sum_{\ell=1}^{m'} \eta_\ell g_\ell(s_j) + \sum_{k=1}^{p} (\beta_k + \sum_{\ell=1}^{m} \gamma_{k\ell} f_\ell(s_j) Z_k(s_j)),
\end{aligned}$$

and the pseudo-log-likelihood has the same form as (5.13). Model (5.12) has only one cumulative baseline hazard which may be estimated by

$$\widehat{A}_0(t) = \int_0^t \frac{\sum_{j=1}^{L} \sum_i Y_i(s_j) dN_i(u)}{\sum_{j=1}^{L} \sum_i Y_i(u) \exp(\widehat{\mathrm{LP}}_{s_j,i})}.$$

Here, event times belonging to several prediction intervals give several contributions to the estimator. In all cases, inference for regression parameters (including γs and ηs) is based on robust variances.

5.4 Micro-simulation

In Section 4.1 and in previous sections of this chapter, we have utilized mathematical relationships between transition intensities and marginal parameters to obtain estimates for the latter via *plug-in*. This worked well for the more simple models like those depicted in Figures 1.1-1.3 and, more generally, for *Markov models*. However, when the multi-state models become more involved (e.g., Section 5.2), this approach becomes more cumbersome. In this section, we will briefly discuss a *brute force* approach to estimating marginal parameters based on transition intensities, namely *micro-simulation* (or *discrete event simulation*) (e.g., Mitton, 2000; Rutter et al., 2011).

5.4.1 Simulating multi-state processes

The idea is as follows: Based on (estimated) transition intensities, generate many paths for the multi-state process and estimate the marginal parameter of interest by averaging over the paths generated. This is doable because, at any time t when the process occupies state $h = V(t)$, the transition intensities govern what happens in the next little time interval from t to $t+dt$. Thus, in this time interval, the process moves to state $j \neq h$ with conditional probability $\alpha_{hj}(t)dt$ given the past (Equation (1.1)) and with conditional probability $1 - \sum_{j\neq h} \alpha_{hj}(t)dt$, the process stays in state h. Note that some of these intensities will be zero if a state j cannot be reached directly from state h. In this time interval, it is thus possible to simulate data $(dN_{hj}(t), j \neq h)$ from the process by a multinomial experiment

$$(dN_{hj}(t), j \neq h, 1 - \sum_{j\neq h} dN_{hj}(t)) \sim \text{mult}(1; \alpha_{hj}(t)dt, j \neq h, 1 - \sum_{j\neq h} \alpha_{hj}(t)dt).$$

The relevant multinomial distribution has index parameter 1 because at most one of the counting processes $N_{hj}(\cdot)$ jumps at t, and the probability parameters are given by the transition intensities (see also Section 3.1). The intensities may be given conditionally on time-fixed covariates Z and also conditionally on *adapted* time-dependent covariates $Z(t)$ (Section 3.7). In these cases, the time-fixed covariate pattern for which processes are generated must first be decided upon. Non-adapted time-dependent covariates $Z(t)$ involve extra randomness, and micro-simulations can in this case only be carried out if a model for this extra randomness is also set up. This will require joint modeling of $V(t)$ and $Z(t)$ (Section 7.4) and will not be further considered here.

It may be computationally involved to generate paths for $V(t)$ in steps of size dt as just described and, fortunately, the simulations may be carried out in more efficient ways. This is because, at any time t, we have locally a competing risks situation, and methods for generating competing risks data (e.g., Beyersmann et al., 2009) may be utilized. Following the recommended approach from that paper, the simulation algorithm goes, as follows:

1. Generate a time T_1 of transition out of the initial state 0 based on the survival function

$$S_0(t) = \exp(-\sum_{h\neq 0} \int_0^t \alpha_{0h}(u)du) = P_{00}(0,t).$$

   Given a transition at T_1, the process moves to state $h_1 \neq 0$ with probability

$$\frac{\alpha_{0h_1}(T_1)}{\sum_{h\neq 0} \alpha_{0h}(T_1)}.$$

2. If the state, say $\tilde{h}$, reached when drawing a state from this multinomial distribution is absorbing, then stop.

3. If state $\tilde{h}$ is transient, then generate a time $T_2 > T_1$ of transition out of that state based on the (conditional) survival function

$$S_1(t \mid T_1) = \exp(-\sum_{h \neq \tilde{h}} \int_{T_1}^{t} \alpha_{\tilde{h}h}(u \mid T_1)du) = P_{\tilde{h}\tilde{h}}(T_1, t).$$

Given a transition at T_2, the process moves to state $h_2 \neq \tilde{h}$ with probability

$$\frac{\alpha_{\tilde{h}h_2}(T_2)}{\sum_{h \neq \tilde{h}} \alpha_{\tilde{h}h}(T_2)}.$$

4. Go to step 2.

Note that, in step 3, as the algorithm progresses, the past will contain more and more information in the form of previous times $(T_1, T_2, \dots)$ and types of transition, and the transition intensities may depend on this information, e.g., in the form of adapted time-dependent covariates. The simulation process stops when reaching an absorbing state. For processes without an absorbing state (e.g., Figure 1.4 without state 2), one has to decide upon a time horizon (say, τ) within which the processes are generated. In this case, attention must be restricted to marginal parameters not relating to times $> \tau$.

The algorithm works well with a parametric specification of the intensities, e.g., as piecewise constant functions of time (Iacobelli and Carstensen, 2013). However, for non- or semi-parametric models, such as the Cox model, in the steps where the next state is determined given a transition at time T_ℓ, i.e., when computing the probabilities $\alpha_{h_{\ell-1}h_\ell}(T_\ell)/(\sum_{h \neq h_{\ell-1}} \alpha_{h_{\ell-1}h}(T_\ell))$ and drawing from the associated multinomial distribution, one would need the jumps for the cumulative hazards $\widehat{A}_{h_{\ell-1}h}(T_\ell)$ at that time and, typically, at most one of these will be > 0. For such situations, an alternative version of the algorithm is needed and, for that purpose, the method not recommended by Beyersmann et al. (2009) is applicable. This method builds on the latent failure time approach to competing risks discussed in Section 4.4. However, the method does provide data with the correct distribution. The algorithm then goes, as follows:

1. Generate independent latent potential times $\widetilde{T}_{11}, \dots, \widetilde{T}_{1k}$ of transition from the initial state 0 to each of the other states based on the 'survival functions'

$$S_{0h}(t) = \exp(-\int_0^t \alpha_{0h}(u)du), h = 1, \dots, k.$$

Let T_1 be the minimum of $\widetilde{T}_{11}, \dots, \widetilde{T}_{1k}$; if this minimum corresponds to $T_1 = \widetilde{T}_{1h_1}$, then the process moves to state h_1 at time T_1.

2. If the state, $\tilde{h}$, thus reached is absorbing, then stop.

3. If state $\tilde{h}$ is transient, then generate independent latent potential times $\widetilde{T}_{2h}$, $h \neq \tilde{h}$ all $> T_1$ of transition from that state based on the (conditional) 'survival functions'

$$S_{\tilde{h}h}(t \mid T_1) = \exp(-\int_{T_1}^{t} \alpha_{\tilde{h}h}(u \mid T_1)du), h \neq \tilde{h}.$$

Let T_2 be the minimum of $\widetilde{T}_{2h}, h \neq \tilde{h}$; if this minimum corresponds to $T_2 = \widetilde{T}_{2h_2}$, then the process moves to state h_2 at time T_2.

4. Go to step 2.

Based on N processes $(V_\ell^*(t), \ell = 1, 2, \ldots, N)$ generated in either of these ways, the desired marginal parameter may be estimated by a simple average, e.g.,

$$\widehat{Q}_h(t) = \frac{1}{N} \sum_\ell I(V_\ell^*(t) = h)$$

for the state h occupation probability $Q_h(t)$ at time t.

If the intensities used when generating the $V_\ell^*(t)$ were known, then the variability of the estimator could be obtained from the empirical variation across the processes generated. Thus, the SD of $\widehat{Q}_h(t)$ could be obtained as the square root of $1/(N(N-1)) \sum_\ell \big(I(V_\ell^*(t) = h) - \widehat{Q}_h(t)\big)^2$. This can be made small by simulating many processes (large N). However, such a measure of variability would not take into account that the intensities used when generating the $V_\ell^*(t)$ contain parameters $\boldsymbol{\theta}$ which are, themselves, estimated with a certain uncertainty. To account for this second level of uncertainty, the following double generation of processes can be applied (e.g., O'Hagan et al., 2007):

1. Draw parameter values $\boldsymbol{\theta}_1, \ldots, \boldsymbol{\theta}_B$ from the estimated distribution of the parameter estimate $\widehat{\boldsymbol{\theta}}$, typically a multivariate normal distribution.

2. For each $\boldsymbol{\theta}_b, b = 1, \ldots, B$, simulate N processes $V_{b\ell}^*(t), \ell = 1, \ldots, N$ based on intensities having parameter value $\boldsymbol{\theta}_b$, and calculate the corresponding estimate of the marginal parameter, e.g.,
$$\bar{Q}_{hb}(t) = \frac{1}{N} \sum_\ell I(V_{b\ell}^*(t) = h).$$
The associated (within-simulation runs) variance
$$\mathrm{SD}_{hb}^2(t) = \frac{1}{N(N-1)} \sum_\ell \big(I(V_{b\ell}^*(t) = h) - \bar{Q}_{hb}(t)\big)^2$$
can be made small by choosing N large.

3. The variation among estimates for different b is quantified by the (between-simulation runs) variance
$$\mathrm{SD}_h^2(t) = \frac{1}{(B-1)} \sum_b \big(\bar{Q}_{hb}(t) - \bar{\bar{Q}}_h(t)\big)^2 \tag{5.14}$$
where
$$\bar{\bar{Q}}_h(t) = \frac{1}{B} \sum_b \bar{Q}_{hb}(t).$$

The variability of the estimator $\widehat{Q}_h(t)$ obtained by simulating using the estimate $\widehat{\boldsymbol{\theta}}$ based on the observed data can now, for large N, be quantified by (5.14). An alternative way of evaluating this variability would be to use the bootstrap.

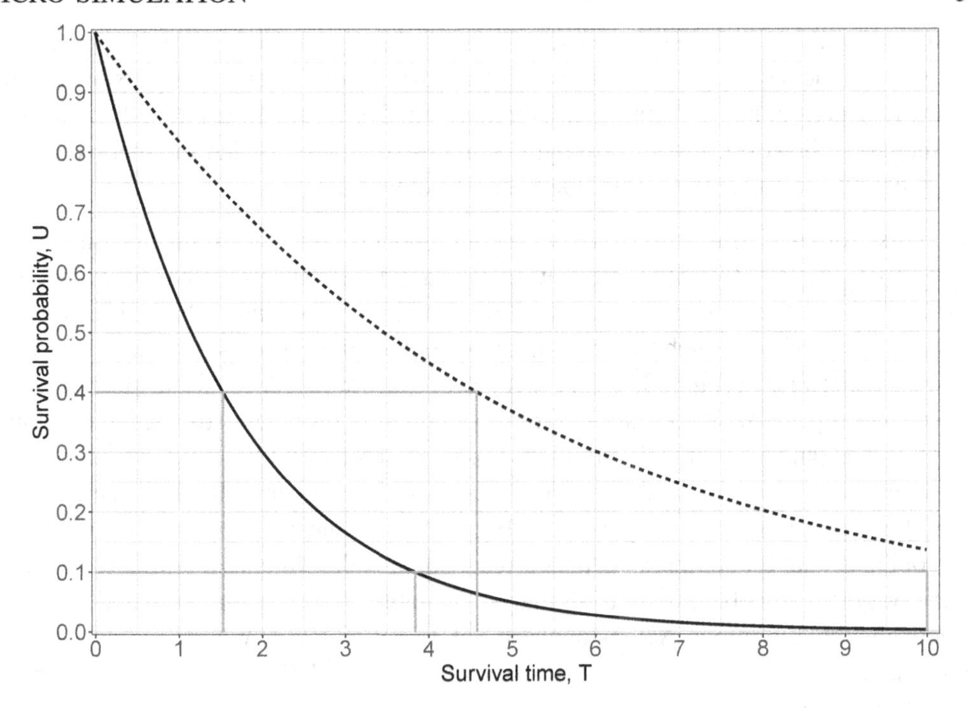

Figure 5.4 *Illustration of sampling from a survival function: For the solid curve, values of U of, respectively 0.4 or 0.1, provide survival times of 1.53 and 3.84, for the dashed curve $U = 0.4$ gives $T = 4.59$ and $U = 0.1$ a survival time censored at $\tau = 10$.*

5.4.2 Simulating from an improper distribution

When drawing a time T from a survival distribution $S(t)$, one would typically draw a random number U from a uniform $[0,1]$ distribution and find the T for which $S(T) = U$. This works well for a proper distribution, i.e., when $S(t) \to 0$ as $t \to \infty$. However, an estimated survival function $\widehat{S}(t)$ does often not reach 0 even for large values of t, in which case this method does not directly work (if a small value of U is drawn), see Figure 5.4 for further explanation.

Two approaches may be used here. First, ways of extrapolating $\widehat{S}(t)$ for large values of t could be applied and, here, use of a parametric model would be possible. This, however, is not entirely satisfactory since the reason the estimated survival function does not reach 0 is the lack of data to support a model for the right-hand tail of the distribution. This means that any parametric extrapolation will lack support from the data. A model with a piece-wise constant hazard may be the least unsatisfactory choice though, as just stated, there will be little data support for the estimate of the hazard in the last interval (ending in $+\infty$). Second, if data provide no support for the intensities beyond some threshold (say, τ), then attention could be restricted to marginal parameters not involving times $> \tau$. This is the approach that we will illustrate in the next section.

5.4.3 PROVA trial in liver cirrhosis

We will illustrate micro-simulation using data from the PROVA trial in liver cirrhosis (Example 1.1.4). For this example, other methods for analyzing marginal parameters, such as the probability $Q_1(t)$ of being alive at time t after a bleeding episode, or the expected time $\varepsilon_1(\tau)$ spent in the bleeding state before time τ are available. Thereby, it is possible to compare the results from micro-simulation with those obtained from these alternative methods. We emphasize, however, that the strength of micro-simulation is that the method applies quite generally, i.e., also when other methods do not work.

The basis for the simulations is a set of models for the transition intensities $\alpha_{01}(t), \alpha_{02}(t), \alpha_{12}(t-T_1)$ with a piece-wise constant dependence of either time since randomization (t) or time since bleeding ($t-T_1$). The number of intervals, respectively 17, 16, and 10, were chosen such that each interval contained 2-3 events and, first, no covariates were taken into account. The first method of generating illness-death processes was applied since a piece-wise constant hazard model directly provides estimates of the transition intensities (and not just their cumulatives). A censored event time was generated if the time of exit from state 0 would otherwise exceed 4.13 years, or if the time spent in state 1 would otherwise exceed 3.73 years (see Figure 5.4). Table 5.4 shows the resulting estimates of $Q_1(t)$ and, for comparison, the corresponding Aalen-Johansen estimates. It is seen that the two estimators behave quite similarly. The table also gives the estimated SD using, respectively, the asymptotic Aalen-Johansen SD, Equation (5.14) with $N=B=1{,}000$, or the bootstrap. The SD across replications for $b=1,\ldots,B=1{,}000$ is also illustrated by the width of the histogram shown in Figure 5.5.

Table 5.4 *PROVA trial in liver cirrhosis: Estimates (and SD) of the probability, $Q_1(t)$, of being in the bleeding state 1 at time t, using either the Aalen-Johansen estimator or micro-simulation. The SD for the estimate obtained using micro-simulation was either based on $SD_1(t)$ given by Equation (5.14) with $N=B=1{,}000$ or on $1{,}000$ bootstrap replications.*

	Aalen-Johansen		Micro-simulation		
t	$\widehat{Q}_1(t)$	SD	$\widehat{Q}_1(t)$	$SD_1(t)$	Bootstrap
0.5	0.050	0.013	0.053	0.015	0.011
1.0	0.081	0.016	0.074	0.020	0.016
1.5	0.091	0.018	0.091	0.022	0.017
2.0	0.093	0.019	0.088	0.021	0.018
2.5	0.089	0.019	0.082	0.021	0.018
3.0	0.089	0.019	0.075	0.022	0.019
3.5	0.079	0.019	0.069	0.023	0.020
4.0	0.063	0.020	0.051	0.017	0.016

From 10,000 simulated illness-death processes, the expected time (years) $\varepsilon_1(\tau)$ spent in the bleeding state before time τ was also estimated. Figure 5.6 shows the estimate as a function of τ together with the corresponding estimate based on the integrated Aalen-Johansen estimator. The two estimators are seen to coincide well.

In a second set of simulations, the binary covariate Z: Sclerotherapy (yes=1, no=0) was added to the three transition intensity models assuming proportional hazards, thus requiring three more parameters to be estimated. From $N=1{,}000$ processes, each re-sampled $B=1{,}000$ times, the probability $Q_1(2)$ was estimated for $Z=0,1$. Histograms of the resulting estimates are shown in Figure 5.7. It is seen that treatment with sclerotherapy reduces

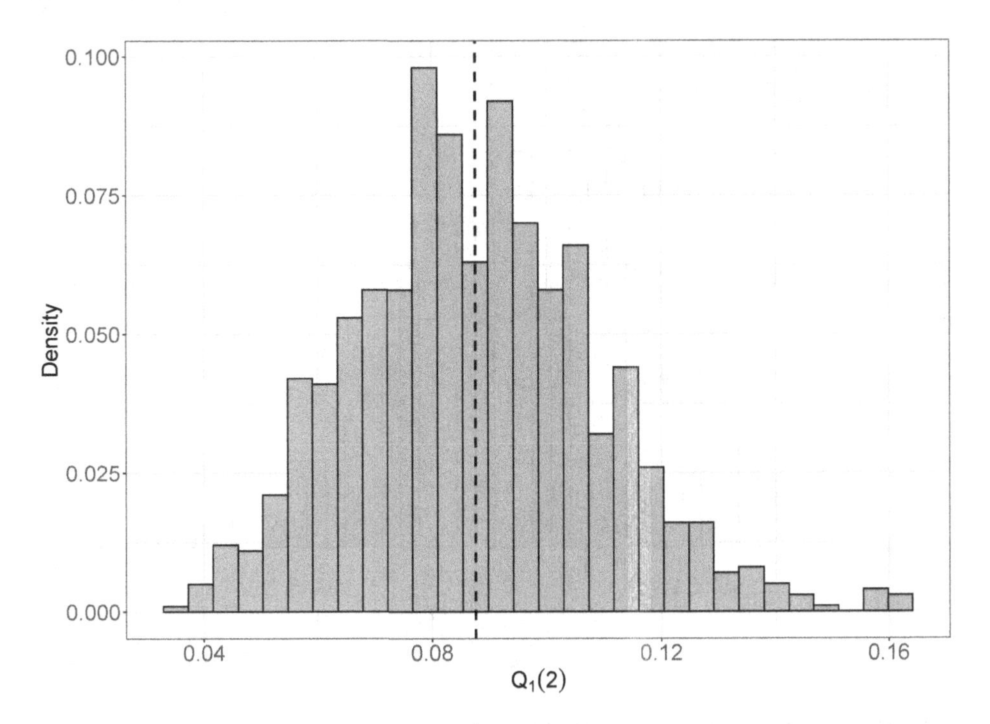

Figure 5.5 *PROVA trial in liver cirrhosis: Distribution of* $\widehat{Q}_1(2)$ *based on* $B = 1,000$ *estimates, each from* $N = 1,000$ *processes.*

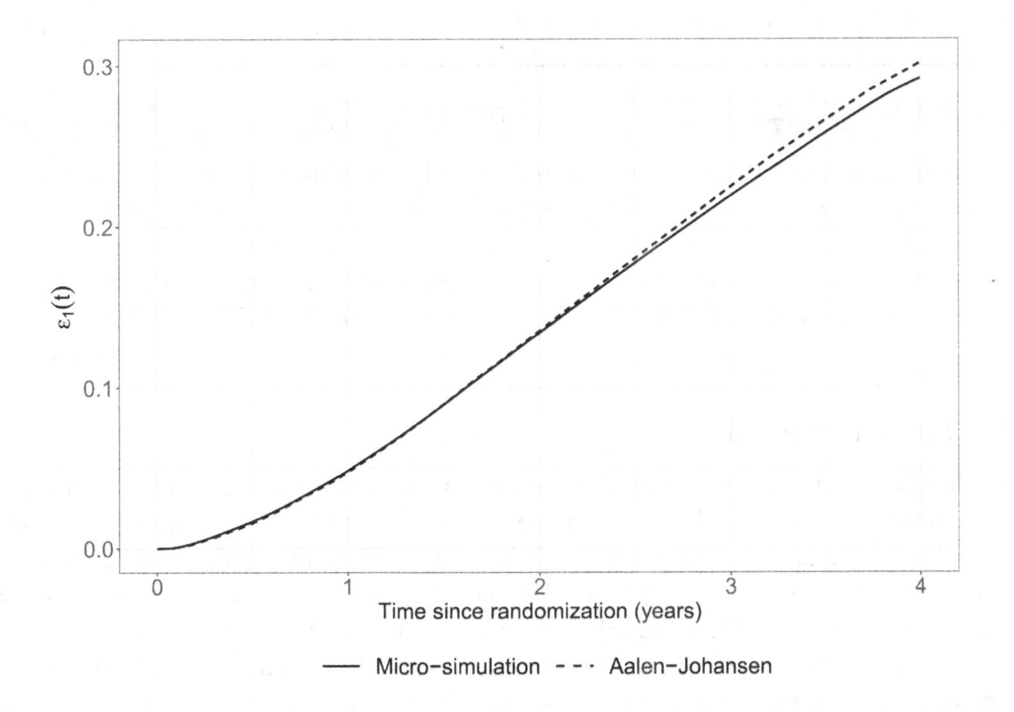

Figure 5.6 *PROVA trial in liver cirrhosis: Estimates of average time,* $\varepsilon_1(\tau)$ *spent in the bleeding state before time (year)* τ*; based on either the Aalen-Johansen estimator or on micro-simulation* ($N = 10,000$).

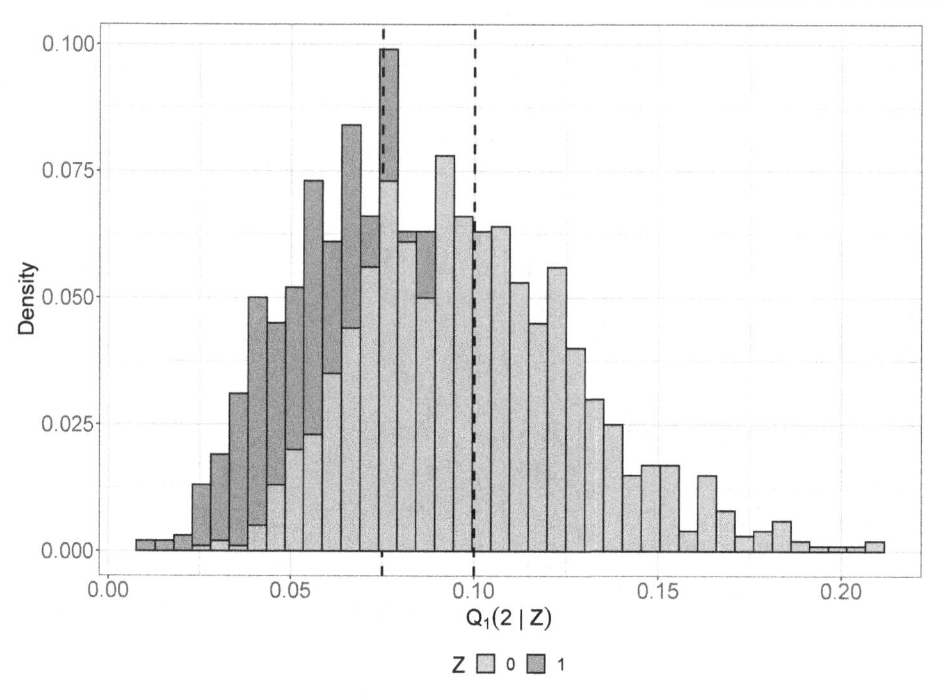

Figure 5.7 *PROVA trial in liver cirrhosis: Distribution of* $\widehat{Q}_1(2 \mid Z), Z = 0, 1,$ *(sclerotherapy: No, yes) based on* $B = 1,000$ *estimates, each from* $N = 1,000$ *processes.*

the probability of being in the bleeding state, likely owing to the fact that this treatment increases the death intensity without bleeding (Table 3.3).

Micro-simulation

Micro-simulation is a general *plug-in* technique for marginal parameters in a multi-state model when intensities have been specified. The strength of the method is its generality, and it is applicable in situations where plug-in using a mathematical expression is not feasible. This includes estimation based on a model where the intensities are functions of the past given by adapted time-dependent covariates.

5.5 Direct regression models

An alternative to basing estimation of marginal parameters for given covariates on models for all intensities is to directly set up a model for the way in which the marginal parameter depends on covariates. This requires specification of a *link function* that gives the scale on which parameters are to be interpreted (Section 1.2.5) and setting up a set of *generalized estimating equations* (GEEs), the solutions of which are the desired parameter estimates. Such an approach has some advantages compared to plug-in and micro-simulation. First, it provides a set of regression coefficients that directly explain the association on the scale of the chosen link function and, second, it targets directly the marginal parameter of interest and, thereby, it does not rely on a correct specification of all intensity models – a specification that may be difficult. A direct marginal model does not provide information on the dynamics of the multi-state process, and it is not possible to simulate paths of the process based on a marginal model. Also, direct modeling requires modeling of the censoring distribution.

In this section, the general ideas of GEE are first introduced (Section 5.5.1), and in the subsequent sections these ideas will be used to outline properties of estimators in direct regression models for marginal parameters in a number of multi-state models.

5.5.1 Generalized estimating equations (*)

Generalized estimating equation (GEE) is a technique for estimating parameters in a regression model for a marginal mean value parameter. Let, in a general setting, $T_1, \dots, T_n$ be *independent* random variables with conditional mean value given a p-vector of covariates $\mathbf{Z}$ (possibly including a constant term corresponding to the model intercept) specified as the *generalized linear model*

$$g(E(T \mid \mathbf{Z})) = \boldsymbol{\beta}_0^{\mathsf{T}} \mathbf{Z}, \tag{5.15}$$

i.e., the mean value transformed with the *link function* g is linear in the covariates $\mathbf{Z}$, and $\boldsymbol{\beta}_0$ is the true regression coefficient. To estimate this parameter, a set of *unbiased estimating equations* is set up

$$\mathbf{U}(\boldsymbol{\beta}) = \sum_i \mathbf{U}_i(\boldsymbol{\beta}) = \sum_i \mathbf{A}(\boldsymbol{\beta}, \mathbf{Z}_i)\big(T_i - g^{-1}(\boldsymbol{\beta}^{\mathsf{T}} \mathbf{Z}_i)\big) = \mathbf{0}. \tag{5.16}$$

Equations (5.16) are unbiased by (5.15) since, given covariates $\mathbf{Z}$, $E(\mathbf{U}(\boldsymbol{\beta}_0)) = \mathbf{0}$. The function $\mathbf{A}(\boldsymbol{\beta}, \mathbf{Z}_i)$ is typically the p-vector

$$\mathbf{A}(\boldsymbol{\beta}, \mathbf{Z}_i) = \Big(\frac{\partial}{\partial \beta_j} g^{-1}(\boldsymbol{\beta}^{\mathsf{T}} \mathbf{Z}_i), \quad j = 1, \dots, p\Big)$$

of partial derivatives of the mean function. The independent random variables T_i, $i = 1, \dots, n$ could be vector-valued with a dimension that may vary among i-values reflecting a clustered data structure, possibly with clusters of varying size, s_i. In that case, $\mathbf{A}(\boldsymbol{\beta}, \mathbf{Z}_i)$ would be a $(p \times s_i)$-matrix, possibly including an $(s_i \times s_i)$ working correlation matrix. However, we will for simplicity restrict attention to the scalar case where T_i is univariate.

The asymptotic properties of the solution $\widehat{\boldsymbol{\beta}}$ to (5.16) rely on a Taylor expansion of (5.16) around the true parameter value $\boldsymbol{\beta}_0$

$$\mathbf{U}(\boldsymbol{\beta}) \approx \mathbf{U}(\boldsymbol{\beta}_0) + \mathbf{DU}(\boldsymbol{\beta}_0)(\boldsymbol{\beta} - \boldsymbol{\beta}_0),$$

where $\mathbf{DU}(\boldsymbol{\beta})$ is the $(p \times p)$-matrix of partial derivatives of $\mathbf{U}$. Inserting $\widehat{\boldsymbol{\beta}}$, using $\mathbf{U}(\widehat{\boldsymbol{\beta}}) = \mathbf{0}$, and re-arranging we get

$$\sqrt{n}(\widehat{\boldsymbol{\beta}} - \boldsymbol{\beta}_0) \approx (-n^{-1}\mathbf{DU}(\widehat{\boldsymbol{\beta}}))^{-1} \frac{1}{\sqrt{n}} \mathbf{U}(\boldsymbol{\beta}_0).$$

Now, $\frac{1}{\sqrt{n}}\mathbf{U}(\boldsymbol{\beta}_0)$ is a sum of independent random variables to which a central limit theorem may be applied, i.e., conditions may be given under which it has a limiting $N(\mathbf{0}, \mathbf{V})$-distribution as $n \to \infty$ where the limiting covariance matrix may be estimated by

$$\widehat{\mathbf{V}} = \frac{1}{n} \sum_i \mathbf{U}_i(\widehat{\boldsymbol{\beta}}) \mathbf{U}_i(\widehat{\boldsymbol{\beta}})^{\mathsf{T}}.$$

If, further, $-n^{-1}\mathbf{DU}(\widehat{\boldsymbol{\beta}})$ converges in probability to a non-negative definite matrix, then it follows that $\sqrt{n}(\widehat{\boldsymbol{\beta}}-\boldsymbol{\beta}_0)$ also has a limiting zero-mean normal distribution with a covariance matrix that can be estimated by *the sandwich estimator*

$$\mathbf{DU}(\widehat{\boldsymbol{\beta}})^{-1}\widehat{\mathbf{V}}\mathbf{DU}(\widehat{\boldsymbol{\beta}})^{-1}. \tag{5.17}$$

The 'meat' of the sandwich, $\widehat{\mathbf{V}}$, is the covariance of the GEE and the 'bread' is the inverse matrix of the partial derivatives, $\mathbf{DU}(\boldsymbol{\beta})$, of the GEE. If the GEEs are obtained as *score equations* by equating log-likelihood derivatives to zero, then the variance estimator simplifies because minus the second log-likelihood derivative equals the inverse variance of the score (e.g., Andersen et al., 1993, ch. VI).

In our applications of GEE, we will often face the additional complication that 'our random variables T_i' are incompletely observed because of right-censoring in which case *inverse probability of censoring weighted* (IPCW) GEE

$$\mathbf{U}(\boldsymbol{\beta})=\sum_i \mathbf{U}_i(\boldsymbol{\beta})=\sum_i D_i\widehat{W}_i\mathbf{A}(\boldsymbol{\beta},\mathbf{Z}_i)\left(T_i-g^{-1}(\boldsymbol{\beta}^{\mathsf{T}}\mathbf{Z}_i)\right)=\mathbf{0} \tag{5.18}$$

must be used. In (5.18), $D_i=1$ if T_i is completely observed, $D_i=0$ if T_i is right-censored, and $\widehat{W}_i$ is a weight giving the inverse probability, possibly depending on covariates $\mathbf{Z}_i$, that the observation at T_i is uncensored. In this situation, the covariance of the GEE, i.e., the 'meat' of the sandwich, will involve an extra term owing to the need to estimate the censoring distribution. Assuming that censoring is independent and does not depend on $\mathbf{Z}$, this is typically done using the Kaplan-Meier estimator, say $\widehat{G}(t)$ with censoring being the event; otherwise, a regression model for censoring can be used, e.g., a Cox model leading to weights depending on estimates of $G(t\mid\mathbf{Z})$ (Section 4.4.1). Presence of *delayed entry* further complicates the situation and, in this case, the weights $\widehat{W}_i$ need to be modified to also reflect the (inverse) probability of no truncation. We will not go into details here but refer to Geskus (2016, ch. 2) for an example where such a modification was studied for the competing risks model.

5.5.2 *Two-state model (*)*

For the two-state model, as discussed in Section 4.2.1, a hazard regression model (e.g., multiplicative or additive) directly implies a model for the state occupation probabilities $Q_0(t)=S(t)$ and $Q_1(t)=F(t)$ (with, respectively, a cloglog or a $-$log link function). This approach implies a marginal model for all time points, t. If a model for a single (t_0) or a few time points and/or a model with other link functions is wanted, then direct regression can be achieved using *direct binomial regression* or using *pseudo-values*. We will return to this in Sections 5.5.5 and 6.1.1.

For the τ-restricted mean life time $\varepsilon_0(\tau)=E(T\wedge\tau)$ where T is the survival time, possibly observed with right-censoring, direct regression models were studied by Tian et al. (2014). Let the potential right-censoring time for subject i be C_i and let $D_i=I(T_i\wedge\tau\le C_i)$ be the indicator of observing the restricted life time for that subject. Let $X_i=T_i\wedge\tau\wedge C_i$ and assume the generalized linear model

$$\varepsilon_0(\tau\mid\mathbf{Z})=g^{-1}(\boldsymbol{\beta}^{\mathsf{T}}\mathbf{Z})$$

(where the vector $\mathbf{Z}$ now includes the constant covariate equal to 1). Typical link functions could be $g = \log$ or $g =$ identity. Now, $\boldsymbol{\beta}$ is estimated by solving the unbiased GEE

$$U(\boldsymbol{\beta}) = \sum_i \frac{D_i}{\widehat{G}(X_i)} \mathbf{Z}_i (X_i - g^{-1}(\boldsymbol{\beta}^\mathsf{T} \mathbf{Z}_i)) = \mathbf{0},$$

where $\widehat{G}$ is the Kaplan-Meier estimator for the distribution of C. Thus, subjects for whom $T_i \wedge \tau$ was observed are *up-weighted* to also represent subjects who were censored. Using counting process notation, the estimating equations become

$$U(\boldsymbol{\beta}) = \sum_i \int_0^\tau \frac{\mathbf{Z}_i}{\widehat{G}(t)} (t - g^{-1}(\boldsymbol{\beta}^\mathsf{T} \mathbf{Z}_i)) dN_i(t) = \mathbf{0}.$$

Tian et al. (2014) discussed asymptotic normality for $\widehat{\boldsymbol{\beta}}$, the solution to these GEEs, assuming, among other things that G is independent of $\mathbf{Z}$ with $G(\tau) > 0$, and derived an expression for the sandwich estimator of the variance of $\widehat{\boldsymbol{\beta}}$.

5.5.3 *Competing risks (*)*

In the competing risks model, the most important marginal parameter is the cumulative incidence, say for cause 1

$$F_1(t) = P(T \leq t, D = 1),$$

i.e., the state 1 occupation probability in the multi-state model of Figure 1.2. Here, T is the life time (time spent in state 0) and D the failure indicator, $D = V(\infty)$. Fine and Gray (1999) studied the following generalized linear model for this parameter

$$\log(-\log(1 - F_1(t \mid \mathbf{Z}))) = \log(\widetilde{A}_{01}(t)) + \boldsymbol{\beta}^\mathsf{T} \mathbf{Z}, \tag{5.19}$$

where the risk parameter is linked to the covariates in the same way as in the Cox model for survival data, i.e., using the cloglog link. (Fine and Gray allowed inclusion of deterministic time-dependent covariates, but we will skip this possibility in what follows.) Indeed, the *Fine-Gray model* (5.19) is a Cox model for the hazard function for the improper random variable

$$T_1 = \inf\{t : V(t) = 1\},$$

which is the time of entry into state 1. This hazard function is the cause 1 *sub-distribution hazard* given by

$$\begin{aligned} \widetilde{\alpha}_1(t) &= \lim_{dt \to 0} P(T_1 \leq t + dt \mid T_1 > t)/dt \qquad (5.20) \\ &= \lim_{dt \to 0} P(T \leq t + dt, D = 1 \mid T > t \text{ or } (T \leq t \text{ and } D \neq 1))/dt. \end{aligned}$$

It follows from this expression that the sub-distribution hazard has a rather unintuitive interpretation, being the cause-1 mortality rate among subjects who are either alive or have already failed by a competing cause. We will discuss other choices of link function for the cumulative incidence later (Section 5.5.5 and Chapter 6), and there we will see that this may lead to other difficulties for the resulting model. Nice features of the link function in the

Fine-Gray model include the fact that predicted failure probabilities stay within the admissible range between 0 and 1 and, as we shall see now, that it suggests estimating equations inspired by the score equations resulting from a Cox model.

In the special case of no censoring, the 'risk set'

$$\widetilde{R}_1(t) = \{i : T_i \geq t \text{ or } (T_i \leq t \text{ and } D_i \neq 1)\}$$

for the sub-distribution hazard is completely observed, and we let $\widetilde{Y}_i(t) = I(T_i \geq t \text{ or } (T_i \leq t \text{ and } D_i \neq 1)) = 1 - N_{1i}(t-)$ (where $N_{1i}(t) = I(T_i \leq t, D_i = 1)$ is the counting process for cause 1 failures) be the membership indicator for this risk set. In this case the 'Cox score' is

$$\mathbf{U}_1(\boldsymbol{\beta}) = \sum_i \int_0^\infty (\mathbf{Z}_i - \frac{\sum_j \widetilde{Y}_j(t)\mathbf{Z}_j \exp(\boldsymbol{\beta}^\mathsf{T}\mathbf{Z}_j)}{\sum_j \widetilde{Y}_j(t)\exp(\boldsymbol{\beta}^\mathsf{T}\mathbf{Z}_j)})dN_{1i}(t).$$

Fine and Gray (1999) used martingale results to ascertain that the resulting score equation is unbiased and to obtain asymptotic normality for its solution. Similar results were obtained in the case with right-censoring, conditionally independent of $V(t)$ for given covariates, and where the censoring times C_i are known for all i (e.g., administrative censoring). In that case the risk set is re-defined as

$$\widetilde{R}_1(t) = \{i : C_i \wedge T_i \geq t \text{ or } (T_i \leq t \text{ and } D \neq 1 \text{ and } C_i \geq t)\},$$

i.e., subjects who are either still alive and uncensored at t or have failed from a competing cause before t and, at the same time, have a censoring time, C_i, exceeding t. The membership indicator for this risk set is $Y_i^*(t) = \widetilde{Y}_i(t)I(C_i \geq t)$ and, thus, subjects who fail from a competing cause stay in the risk set, not indefinitely, but until their time of censoring. This leads to the 'Cox score'

$$\mathbf{U}_1^*(\boldsymbol{\beta}) = \sum_i \int_0^\infty (\mathbf{Z}_i - \frac{\sum_j Y_j^*(t)\mathbf{Z}_j \exp(\boldsymbol{\beta}^\mathsf{T}\mathbf{Z}_j)}{\sum_j Y_j^*(t)\exp(\boldsymbol{\beta}^\mathsf{T}\mathbf{Z}_j)})dN_{1i}(t).$$

In the general case of right-censoring (assumed to be conditionally independent of $V(t)$ for given covariates) but now with C_i unobserved for failing subjects, IPCW techniques were used, as follows. According to the situation where all censoring times were known, subjects j failing from a competing cause should stay in the risk set until their time, C_j of censoring. Now, C_j is not observed, so those subjects stay in the risk set with a weight that diminishes over time reflecting a decreasing probability of still being uncensored. This is obtained using the weights

$$W_j(t) = \frac{I(C_j \geq T_j \wedge t)\widehat{G}(t)}{\widehat{G}(T_j \wedge C_j \wedge t)} \tag{5.21}$$

where $\widehat{G}$ estimates the censoring distribution either non-parametrically using the Kaplan-Meier estimator or via a regression model. There are three kinds of subjects who were not observed to fail from cause 1 before t:

1. j is still alive and uncensored in which case $W_j(t) = 1$,

2. j was censored before t in which case $W_j(t) = 0$,

3. j failed from a competing cause before t in which case $W_j(t) = \widehat{G}(t)/\widehat{G}(T_j)$, the conditional probability of being still uncensored at t given uncensored at the failure time T_j.

The resulting GEE are $\mathbf{U}_1^W(\boldsymbol{\beta}) = \mathbf{0}$ where

$$\mathbf{U}_1^W(\boldsymbol{\beta}) = \sum_i \int_0^\infty (\mathbf{Z}_i - \frac{\sum_j W_j(t)\widetilde{Y}_j(t)\mathbf{Z}_j \exp(\boldsymbol{\beta}^\mathsf{T}\mathbf{Z}_j)}{\sum_j W_j(t)\widetilde{Y}_j(t)\exp(\boldsymbol{\beta}^\mathsf{T}\mathbf{Z}_j)})W_i(t)dN_{1i}(t), \tag{5.22}$$

and $\widetilde{Y}_i(t) = 1 - N_{1i}(t-)$, the indicator of no observed cause 1 failure before time t. Fine and Gray (1999) showed that these equations are approximately unbiased, that their solutions are asymptotically normal, and derived a consistent sandwich variance estimator. The estimator

$$\widehat{\widetilde{A}}_{01}(t) = \sum_i \int_0^t \frac{W_i(u)dN_i(u)}{\sum_j W_j(u)\widetilde{Y}_j(u)\exp(\widehat{\boldsymbol{\beta}}^\mathsf{T}\mathbf{Z}_j)}$$

for the cumulative baseline sub-distribution hazard $\widetilde{A}_{01}(t)$ was also presented with asymptotic results.

The Fine-Gray model can be used for a single cause or for all causes – one at a time – and, as we have seen, inference requires modeling of the censoring distribution. When all causes are modeled, there is no guarantee that, for any given covariate pattern, one minus the sum of the estimated cumulative incidences given that covariate pattern is a proper survival function (e.g., Austin et al., 2021). Furthermore, the partial likelihood approach is not fully efficient. Based on such concerns, Mao and Lin (2017) proposed an alternative non-parametric likelihood approach to joint modeling of all cumulative incidences. The Jacod formula (3.1) for the competing risks model was re-written in terms of the cumulative incidences and their derivatives – the sub-distribution densities

$$f_j(t) = \frac{d}{dt}F_j(t) = \alpha_j(t)S(t),$$

as follows. For two causes of failure, the contribution to the Jacod formula from an observation at time X is, with the notation previously used,

$$L = \alpha_1(X)^{I(D=1)}\alpha_2(X)^{I(D=2)}S(X).$$

This can be re-written as

$$\begin{aligned} L &= (\alpha_1(X)S(X))^{I(D=1)}(\alpha_2(X)S(X))^{I(D=2)}S(X)^{1-I(D=1)-I(D=2)} \\ &= f_1(X)^{I(D=1)}f_2(X)^{I(D=2)}(1-F_1(X)-F_2(X))^{1-I(D=1)-I(D=2)}, \end{aligned}$$

in which the cumulative incidences may be parametrized, e.g., as in the Fine-Gray model or using other link functions. Mao and Lin (2017) showed that, under suitable conditions, the resulting estimators are efficient and asymptotically normal. Similar to modeling via hazard functions, this approach does not require a model for censoring.

Cause-specific time lost

Conner and Trinquart (2021) used the approach of Tian et al. (2014) to study regression models for the τ-restricted cause-specific time lost in the competing risks model. Following Section 5.1.2, the parameter of interest is

$$\varepsilon_h(\tau) = \tau - E(T_h \wedge \tau) = \int_0^\tau Q_h(u)du$$

where T_h is the time of entry into state h, possibly observed with right-censoring. Let the potential right-censoring time for subject i be C_i and assume the generalized linear model

$$\varepsilon_h(\tau \mid \mathbf{Z}) = g^{-1}(\boldsymbol{\beta}_h^\mathsf{T}\mathbf{Z})$$

(where the vector $\mathbf{Z}$ includes the constant covariate equal to 1). Typical link functions could be $g = \log$ or $g =$ identity. Now, $\boldsymbol{\beta}_h$ is estimated by solving the unbiased GEE

$$U(\boldsymbol{\beta}_h) = \sum_i \int_0^\tau \frac{\mathbf{Z}_i}{\widehat{G}(t)}(\tau - t - g^{-1}(\boldsymbol{\beta}_h^\mathsf{T}\mathbf{Z}_i))dN_{hi}(t) = \mathbf{0}$$

where $\widehat{G}$ is the Kaplan-Meier estimator for the distribution of C and $N_{hi}(t)$ the counting process for h-events for subject i.

Conner and Trinquart (2021) discussed conditions for asymptotic normality of the resulting solution $\widehat{\boldsymbol{\beta}}_h$ to these GEEs and derived an expression for the sandwich estimator for the variance of $\widehat{\boldsymbol{\beta}}_h$.

5.5.4 Recurrent events (*)

For recurrent events we will focus on the mean function

$$\mu(t) = E(N(t)),$$

where $N(t)$ counts the number of events in $[0,t]$. We will distinguish between the two situations where either there are competing risks in the form of a terminal event, the occurrence of which prevents further recurrent events from happening, or there is no such terminal event.

No terminal event

We will begin by considering the latter situation. The parameter $\mu(t)$ is closely linked to a partial transition rate as introduced in (5.7), as follows. The partial transition rates for this model are (approximately for small $dt > 0$)

$$\alpha^*_{h,h+1}(t) \approx P(V(t+dt) = h+1 \mid V(t) = h)/dt,$$

and if these are assumed independent of h, then they equal

$$P(N(t+dt) = N(t)+1)/dt = E(dN(t))/dt = d\mu(t),$$

the derivative of the mean function. In Section 3.2.1 we derived the score equation (3.9) for the cumulative hazard and the idea is now to use this as the basis for an unbiased GEE for $\mu(t)$

$$\sum_i Y_i(t)\big(dN_i(t) - d\mu(t)\big) = 0, \tag{5.23}$$

where $Y_i(t) = 1$ if i is still uncensored at time t, i.e., $Y_i(t) = I(C_i > t)$ (note that, for recurrent events without competing risks, times C_i of censoring will always be observed). Equation (5.23) is solved by

$$d\mu(t) = \frac{\sum_i Y_i(t) dN_i(t)}{\sum_i Y_i(t)}$$

corresponding to estimating the mean function by the Nelson-Aalen estimator

$$\widehat{\mu}(t) = \int_0^t \frac{\sum_i dN_i(u)}{\sum_i Y_i(u)}. \tag{5.24}$$

(Note that we only have $dN_i(t) = 1$ if $Y_i(t) = 1$.) Equation (5.23) is unbiased if censoring is independent of the multi-state process in which case the estimator can be shown to be consistent (Lawless and Nadeau, 1995; Lin et al., 2000). For more general censoring, (5.23) may be replaced by the IPCW GEE

$$\sum_i \frac{Y_i(t)}{\widehat{G}_i(t)}\big(dN_i(t) - d\mu(t)\big) = 0$$

leading to the weighted Nelson-Aalen estimator

$$\widehat{\mu}(t) = \int_0^t \frac{\sum_i dN_i(u)/\widehat{G}_i(u)}{\sum_i Y_i(u)/\widehat{G}_i(u)},$$

where $\widehat{G}_i(t)$ estimates the probability $E(Y_i(t))$ that subject i is uncensored at time t, possibly via a regression model. For both estimators, a sandwich variance estimator is available, or bootstrap methods may be used.

A multiplicative regression model for the mean function, inspired by the Cox regression model, is

$$\mu(t \mid \mathbf{Z}) = \mu_0(t)\exp(\boldsymbol{\beta}^\mathsf{T}\mathbf{Z}), \tag{5.25}$$

see Lawless and Nadeau (1995) and Lin et al. (2000), often referred to as the *LWYY model.* Unbiased GEE may be established from a working intensity model with a Cox type intensity where the score equations are

$$\sum_i Y_i(t)\big(dN_i(t) - d\mu_0(t)\exp(\boldsymbol{\beta}^\mathsf{T}\mathbf{Z}_i)\big) = 0 \tag{5.26}$$

and

$$\sum_i Y_i(t)\mathbf{Z}_i\big(dN_i(t) - d\mu_0(t)\exp(\boldsymbol{\beta}^\mathsf{T}\mathbf{Z}_i)\big) = \mathbf{0}. \tag{5.27}$$

Equation (5.26) is, following the lines of Section 3.3, for fixed $\boldsymbol{\beta}$ solved by

$$\widehat{\mu}_0(t) = \int_0^t \frac{\sum_i dN_i(u)}{\sum_i Y_i(u)\exp(\boldsymbol{\beta}^\mathsf{T}\mathbf{Z}_i)} \tag{5.28}$$

and inserting this solution into (5.27) leads to the equation

$$\sum_i \left(\mathbf{Z}_i - \frac{\sum_j Y_j(t)\mathbf{Z}_j \exp(\boldsymbol{\beta}^\mathsf{T}\mathbf{Z}_j)}{\sum_j Y_j(t)\exp(\boldsymbol{\beta}^\mathsf{T}\mathbf{Z}_j)} \right) dN_i(t) = \mathbf{0}$$

which is identical to the Cox score Equation (3.17). To assess the uncertainty of the estimator, a sandwich estimator, as derived by Lin et al. (2000) must be used instead of the model-based SD obtained from the derivative of the score. Further, the baseline mean function $\mu_0(t)$ may be estimated by the Breslow-type estimator obtained by inserting $\widehat{\boldsymbol{\beta}}$ into (5.28).

Terminal event

The situation where there are events competing with the recurrent events process was studied by Cook and Lawless (1997) and by Ghosh and Lin (2000, 2002), see also Cook et al. (2009). In this situation, the partial transition rate

$$\alpha^*_{h,h+1}(t) \approx P(V(t+dt) = h+1 \mid V(t) = h)/dt$$

(which we assume to be independent of h) has a slightly different interpretation, namely $\alpha^*(t) \approx E(dN(t) \mid T_D > t)/dt$ where T_D is the time to the competing event, i.e., the time of entry into state D in Figure 1.5, typically the time to death. We define

$$A^*(t) = E(N(t) \mid T_D > t) = \int_0^t \alpha^*(u)du \tag{5.29}$$

and, as in the case of no competing risks, it may be estimated by the Nelson-Aalen estimator

$$\widehat{A}^*(t) = \int_0^t \frac{\sum_i dN_i(u)}{\sum_i Y_i(u)}.$$

The quantity $A^*(t)$ is not of much independent interest (it conditions on the future); however, since in the case of competing risks we have

$$E(N(t)) = \int_0^t S(u)dA^*(u),$$

this suggests the plug-in estimator, the *Cook-Lawless estimator*,

$$\widehat{\mu}(t) = \int_0^t \widehat{S}(u-)d\widehat{A}^*(u)$$

where $\widehat{S}(\cdot)$ is the Kaplan-Meier estimator for $S(t) = P(T_D > t)$. Asymptotic results for this estimator were presented by Ghosh and Lin (2000).

Regression analysis for $\mu(t)$ in the presence of a terminal event can proceed in two directions. Cook et al. (2009) discussed a plug-in estimator combining a regression model for $S(t)$ via a Cox model for the marginal death intensity and one for $A^*(t)$ using the estimating Equations (5.26)-(5.27) for $\mu(t \mid \mathbf{Z})$ without competing risks. As it was the case for the plug-in models discussed previously, this enables prediction of $E(N(t) \mid \mathbf{Z})$ but does not provide regression parameters that directly quantify the association. To obtain this, the direct model for $\mu(t \mid \mathbf{Z})$ discussed by Ghosh and Lin (2002) is applicable. This model also has the multiplicative structure (5.25), and direct IPCW GEE for this marginal parameter were set up, as follows. Ghosh and Lin (2002), following Fine and Gray (1999), first considered the case with purely administrative censoring, i.e., where the censoring times C_i are known for all subjects i and, next, for general censoring, IPCW GEE were studied. The resulting equations are, except for the factors $\widetilde{Y}_i(t) = 1 - N_i(t-)$ appearing in Equation (5.22) for the Fine-Gray model, identical to that equation, i.e.,

$$\mathbf{U}_1^W(\boldsymbol{\beta}) = \sum_i \int_0^\infty \left(\mathbf{Z}_i - \frac{\sum_j W_j(t)\mathbf{Z}_j \exp(\boldsymbol{\beta}^\mathsf{T}\mathbf{Z}_j)}{\sum_j W_j(t)\exp(\boldsymbol{\beta}^\mathsf{T}\mathbf{Z}_j)} \right) W_i(t) dN_{1i}(t), \tag{5.30}$$

where the weights $W_i(t)$ are given by (5.21). Ghosh and Lin (2002) presented asymptotic results for the solution $\widehat{\boldsymbol{\beta}}$, including a sandwich-type variance estimator, and for the Breslow-type estimator

$$\widehat{\mu}_0(t) = \sum_i \int_0^t \frac{W_i(u)dN_i(u)}{\sum_j W_j(u)\exp(\widehat{\boldsymbol{\beta}}^\mathsf{T}\mathbf{Z}_j)}$$

for the baseline mean function.

As discussed in Section 4.2.3, the occurrence of the competing event ('death') must be considered jointly with the recurrent events process $N(t)$ when a terminal event is present. To this end, Ghosh and Lin (2002) also studied an *inverse probability of survival weighted* (IPSW) estimator, as follows. In (5.30), the weights are re-defined as

$$W_i^D(t) = \frac{I(T_{D_i} \wedge C_i \geq t)}{\widehat{S}(t \mid Z_i)}$$

where the denominator estimates the conditional probability given covariates of survival past time t. Ghosh and Lin showed that the corresponding GEE are approximately unbiased and derived asymptotic properties of the resulting estimator $\widehat{\boldsymbol{\beta}}$. Though the details were not given, this also provides the joint asymptotic distribution of $\widehat{\boldsymbol{\beta}}$ and, say $\widehat{\boldsymbol{\beta}}_D$, the estimated regression coefficient in a Cox model for the survival time distribution. Note that, compared to the IPCW approach, the IPSW approach has the advantage of not having to estimate the censoring distribution, but instead the survival time distribution which is typically of greater scientific interest.

Mao-Lin model

Mao and Lin (2016) defined a *composite end-point* combining information on $N(t)$ and survival. They considered multi-type recurrent events processes $N_h(t), h = 1, \ldots, k$, $N_h(t)$ counting events of type h together with, say $N_0(t)$, the counting process for the terminal event and assumed that each event type and death can be equipped with a *severity weight* (or *utility*) c_h, $h = 0, 1, \ldots, k$. They defined the weighted process

$$\bar{N}(t) = \sum_{h=0}^{k} c_h N_h(t)$$

(which is a counting process if all $c_h = 1$) and considered a multiplicative model

$$E(\bar{N}(t)) = \mu_0(t)\exp(\boldsymbol{\beta}^\mathsf{T}\mathbf{Z}) \tag{5.31}$$

for its mean, sometimes referred to as the *Mao-Lin model.* Approximately unbiased GEE for $\boldsymbol{\beta}$ are exactly equal to (5.30) and also the estimator for $\mu_0(t)$ suggested by Mao and Lin (2016) equals that from the Ghosh-Lin model. Asymptotic results for these estimators were provided. Furberg et al. (2022) studied a situation with competing risks where some causes of death were included in a composite end-point, but others were considered as events competing with the composite end-point.

5.5.5 *State occupation probabilities (*)*

For a general multi-state process, $V(t)$, the state occupation probability $Q_h(t) = P(V(t) = h)$ is the expectation of the random variable $I(V(t) = h)$. In a situation with no censoring, regression models for $E(I(V(t_0) = h) \mid \mathbf{Z})$, for a fixed time point t_0, could be fitted using GEE with the binary outcome variable $I(V(t_0) = h)$. In the more realistic setting with censoring, direct binomial regression for the state occupation probability $Q_h(t_0)$ was studied by Scheike et al. (2008) for the special case of the competing risks model and, more generally, by Scheike and Zhang (2007). Azarang et al. (2017) used a similar approach for the transition probability in the progressive illness-death model (Figure 1.3).

The technique of Scheike and Zhang (2007) is closely related to what we have demonstrated in Sections 5.5.1-5.5.4. If the right-censoring time for subject i is C_i, then the indicator $I(V_i(t_0) = h)I(C_i > t_0)$ is always observed and can be used, suitably weighted, as response in the GEE

$$\boldsymbol{U}(\boldsymbol{\beta}) = \sum_i \boldsymbol{A}(\boldsymbol{\beta}, \mathbf{Z}_i)\Big(\frac{I(V_i(t_0) = h)I(C_i > t_0)}{\widehat{G}(t_0)} - g^{-1}(\boldsymbol{\beta}^\mathsf{T}\mathbf{Z}_i)\Big) \tag{5.32}$$

for the regression parameter $\boldsymbol{\beta} = (\beta_0, \beta_1, \ldots, \beta_p)^\mathsf{T}$. This parameter vector includes an intercept β_0 depending on the chosen value of t_0. The model has link function g, i.e., $g(E(I(V_i(t_0) = h) \mid \mathbf{Z}_i)) = \boldsymbol{\beta}^\mathsf{T}\mathbf{Z}_i$. In (5.32), $\boldsymbol{A}(\boldsymbol{\beta}, \mathbf{Z}_i)$ is usually the $(p+1)$-vector of partial derivatives

$$\mathbf{A}(\boldsymbol{\beta}, \mathbf{Z}_i) = \Big(\frac{\partial}{\partial \beta_j} g^{-1}(\boldsymbol{\beta}^\mathsf{T}\mathbf{Z}_i), \quad j = 0, 1, \ldots, p\Big)$$

of the mean function (see Section 5.5.1) and $\widehat{G}$ is the Kaplan-Meier estimator if C_i, $i = 1, \ldots, n$ are assumed i.i.d. with survival distribution $G(\cdot)$.

The asymptotic distribution of $\widehat{\boldsymbol{\beta}}$ was derived by Scheike and Zhang (2007) together with a variance estimator using the sandwich formula. However, bootstrap or an *i.i.d. decomposition* (to be further discussed in Section 5.7) are also possible when estimating the asymptotic variance. Extensions to a model for several time points simultaneously have also been considered (e.g., Grøn and Gerds, 2014). Blanche et al. (2023) compared, for the competing risks model, estimates based on (5.32) with those obtained by solving the GEE of the form (5.18) with $I(V_i(t_0) = h)$ as the response variable (see also Exercise 5.4).

For the analysis of the cumulative incidence in a competing risks model, the cloglog link function $\log(-\log(1-p))$ will provide regression coefficients with a similar interpretation as those in the Fine-Gray model (Section 5.5.3); however, using the direct binomial approach other link functions may also be studied. This is also possible using pseudo-observations to be discussed in Chapter 6. As discussed, e.g., by Gerds et al. (2012), a log-link gives parameters with a relative risk interpretation; however, this comes with the price that estimates may be unstable for time points close to 0 and that predicted risks may exceed 1.

5.6 Marginal hazard models (*)

In Section 4.3, we introduced analysis of the marginal parameter 'distribution of time, T_h of (first) entry into state h' in a multi-state model via models for the *marginal hazard*. Within any subject (i), different T_h, e.g., times to event no. $h = 1,2,...$ in a model for recurrent events, cannot reasonably be assumed independent, and both in that section and in Section 3.9, the situation was treated together with that of *clustered data* which also gives rise to dependent event history data. In the latter situation, time-to-event information for subjects from the same family, medical center or the like, is studied and independence *within* clusters is questionable, whereas independence *among* clusters may still be reasonable. In a *frailty* model (Section 3.9), regression parameters with a *within cluster* interpretation are estimated.

In this section, we will discuss inference for the *marginal* time to event distributions without a specification of the intra-cluster/subject association using *marginal Cox models* as discussed, e.g., by Wei et al. (1989) and by Lin (1994). We will, furthermore, in Sections 5.6.3-5.6.5 discuss to what extent this approach and the very concept of a marginal hazard are applicable in the different situations. In Section 7.2, we will summarize the discussion of analysis of dependent event history data.

5.6.1 *Cox score equations – revisited (*)*

Before discussing the marginal Cox model, recall the score equations for $\boldsymbol{\beta}$ for a Cox model

$$\lambda_i(t) = Y_i(t)\alpha_0(t)\exp(\boldsymbol{\beta}^\mathsf{T}\mathbf{Z}_i(t))$$

for the intensity process $\lambda_i(t)$ for the counting process $N_i(t) = I(X_i \leq t, D_i = 1)$ counting occurrences of the event of interest where, as usual, $Y_i(t) = I(X_i \geq t)$. These are $\boldsymbol{U}(\boldsymbol{\beta}) = \mathbf{0}$ where

$$\boldsymbol{U}(\boldsymbol{\beta}) = \sum_i \int_0^\infty (\mathbf{Z}_i(t) - \bar{\mathbf{Z}}(\boldsymbol{\beta},t))dN_i(t),$$

cf. (3.17). Here,

$$\bar{\mathbf{Z}}(\boldsymbol{\beta},t) = \frac{\mathbf{S}_1(\boldsymbol{\beta},t)}{S_0(\boldsymbol{\beta},t)}$$

and $S_0(\boldsymbol{\beta},t) = \sum_i Y_i(t)\exp(\boldsymbol{\beta}^\mathsf{T}\mathbf{Z}_i(t))$, $\mathbf{S}_1(\boldsymbol{\beta},t) = \sum_i Y_i(t)\mathbf{Z}_i(t)\exp(\boldsymbol{\beta}^\mathsf{T}\mathbf{Z}_i(t))$. Note that, since $\sum_i \int_0^t (\mathbf{Z}_i(t) - \bar{\mathbf{Z}}(\boldsymbol{\beta},t))Y_i(t)\exp(\boldsymbol{\beta}^\mathsf{T}\mathbf{Z}_i(t))\alpha_0(t)dt = \mathbf{0}$, the score may be re-written as

$$\boldsymbol{U}(\boldsymbol{\beta}) = \sum_i \int_0^\infty (\mathbf{Z}_i(t) - \bar{\mathbf{Z}}(\boldsymbol{\beta},t))dM_i(t) \tag{5.33}$$

where $M_i(t) = N_i(t) - \int_0^t Y_i(u)\exp(\boldsymbol{\beta}^\mathsf{T}\mathbf{Z}_i(u))\alpha_0(u)du$ is the counting process martingale (Equation (1.21)). This shows that the score evaluated based on data on the interval $[0,t]$ and evaluated at the true parameter value $\boldsymbol{\beta}_0$ is a *martingale*, and the martingale central limit theorem may be used to show asymptotic normality of the score. Thereby, asymptotic normality of the solution $\widehat{\boldsymbol{\beta}}$ follows as in Section 5.5.1 with the simplification that, as explained below (5.17), minus the derivative $\boldsymbol{DU}(\widehat{\boldsymbol{\beta}})$ of the score estimates the inverse variance of the score, such that the variance of $\widehat{\boldsymbol{\beta}}$ may be estimated by $\boldsymbol{DU}(\widehat{\boldsymbol{\beta}})^{-1}$ with

$$\boldsymbol{DU}(\boldsymbol{\beta}) = \sum_i \int_0^\infty \left(\frac{\boldsymbol{S}_2(\boldsymbol{\beta},t)}{S_0(\boldsymbol{\beta},t)} - \frac{\boldsymbol{S}_1(\boldsymbol{\beta},t)\boldsymbol{S}_1(\boldsymbol{\beta},t)^\mathsf{T}}{S_0(\boldsymbol{\beta},t)^2} \right) dN_i(t) \tag{5.34}$$

and $\boldsymbol{S}_2(\boldsymbol{\beta},t) = \sum_i Y_i(t)\mathbf{Z}_i(t)\mathbf{Z}_i(t)^\mathsf{T}\exp(\boldsymbol{\beta}^\mathsf{T}\mathbf{Z}_i(t))$. This is the *model-based* variance estimate. A *robust* estimator of the variance of $\widehat{\boldsymbol{\beta}}$ may also be derived as in Section 5.5.1 following Lin and Wei (1989) where the 'meat' of the sandwich in (5.17) is

$$\widehat{\boldsymbol{V}} = \sum_i \int_0^\infty (\mathbf{Z}_i(t) - \bar{\mathbf{Z}}(\widehat{\boldsymbol{\beta}},t))d\widehat{M}_i(t) \int_0^\infty (\mathbf{Z}_i(t) - \bar{\mathbf{Z}}(\widehat{\boldsymbol{\beta}},t))^\mathsf{T} d\widehat{M}_i(t). \tag{5.35}$$

Here, $\widehat{M}_i$ is obtained by plugging-in $\widehat{\boldsymbol{\beta}}$ and the Breslow estimator for $\alpha_0(t)dt$ into the expression for M_i. The resulting robust variance-covariance matrix is then, as in (5.17), $\boldsymbol{DU}(\widehat{\boldsymbol{\beta}})^{-1}\widehat{\boldsymbol{V}}\boldsymbol{DU}(\widehat{\boldsymbol{\beta}})^{-1}$.

5.6.2 *Multivariate Cox model (*)*

For the Cox model for the intensity of a single event, the robust variance derived in the previous section seems of minor importance since, in this case, it gives the variance of the estimator for a least false parameter for a misspecified Cox model, though it may be useful for hypothesis testing, see Lin and Wei (1989).

For a multivariate situation, the sandwich variance is needed because it is robust against misspecification of the within-cluster correlation structure. We will now discuss the multivariate situation in more detail. Suppose that there are *independent* units, $i = 1, \ldots, n$, within which there are K types of event of interest. We will denote these units as 'clusters' even though, in some of the situations to be studied, the units correspond to subjects. The associated counting processes are $N_{hi}(t) = I(X_{hi} \leq t, D_{hi} = 1)$, $h = 1, \ldots, K$, where $X_{hi} = T_{hi} \wedge C_{hi}$ and $D_{hi} = I(X_{hi} = T_{hi})$. Here, $T_{hi} \leq \infty$ is the uncensored time of the type h event in cluster i and C_{hi} the associated time of right-censoring. The marginal hazard for events of type h in a given cluster i is

$$\alpha_h(t) = \lim_{dt \to 0} P(T_{hi} \leq t + dt \mid T_{hi} > t, \mathbf{Z}_{hi}(t))/dt, \tag{5.36}$$

i.e., conditioning is only on $T_{hi} > t$ (and on covariates) and not on other information for cluster no. i.

The Cox model for the marginal intensity of type h events is

$$\lambda_{hi}(t) = Y_{hi}(t)\alpha_{h0}(t)\exp(\boldsymbol{\beta}_h^\mathsf{T}\mathbf{Z}_i(t)), \quad h = 1, \ldots, K$$

with $Y_{hi}(t) = I(X_{hi} \geq t)$. Following Lin (1994) we will use the feature of *type-specific covariates* (see Section 3.8) and re-write the model as

$$\lambda_{hi}(t) = Y_{hi}(t)\alpha_{h0}(t)\exp(\boldsymbol{\beta}^\mathsf{T}\mathbf{Z}_{hi}(t)), \quad h = 1, \ldots, K, \tag{5.37}$$

since this formulation, as discussed in Section 3.8, allows models with the same $\boldsymbol{\beta}$ for several types of events. The GEEs for $\boldsymbol{\beta}$ are now $\bar{\boldsymbol{U}}(\boldsymbol{\beta}) = \mathbf{0}$ where

$$\bar{\boldsymbol{U}}(\boldsymbol{\beta}) = \sum_i \sum_{h=1}^{K} \int_0^\infty (\mathbf{Z}_{hi}(t) - \bar{\mathbf{Z}}_h(\boldsymbol{\beta}, t)) dN_{hi}(t).$$

Lin (1994) also discussed a model with a common baseline hazard across event types and a more general, stratified model generalizing both of these models was discussed by Spikerman and Lin (1998) . The derivations for these models are similar to those for (5.37) and we will omit the corresponding details.

Lin (1994) discussed asymptotic normality of the solution $\widehat{\boldsymbol{\beta}}$ and presented the robust estimator of the variance-covariance matrix, which is $\boldsymbol{D}\bar{\boldsymbol{U}}(\widehat{\boldsymbol{\beta}})^{-1}\widehat{\bar{\boldsymbol{V}}}\boldsymbol{D}\bar{\boldsymbol{U}}(\widehat{\boldsymbol{\beta}})^{-1}$. Here, $\boldsymbol{D}\bar{\boldsymbol{U}}$ and $\widehat{\bar{\boldsymbol{V}}}$ are sums over types h of the corresponding type-specific quantities given by (5.34) and (5.35). Lin's asymptotic results hold true no matter the correlation among clusters i. However, the results build on the assumption that the vectors of event times $(T_{hi}, h = 1, \ldots, K)$ and right-censoring times $(C_{hi}, h = 1, \ldots, K)$ are conditionally independent given the covariates $(\mathbf{Z}_{hi}, h = 1, \ldots, K)$.

5.6.3 *Clustered data (*)*

For truly clustered data, e.g., a family with K members, an event of type h corresponds to the event of interest for family member no. h. In this situation, conditioning in (5.36) is only on subject h in family i being event-free at time t and not on the status of other family members. Individual censoring times C_{hi}, conditionally independent of T_{hi} given covariates could be a plausible assumption. For clustered data, one could generalize to a situation with competing risks and study Cox models for the cause-specific event hazard for each individual without having to specify the within-family correlation structure though the hazard function for subject h is no longer a marginal hazard in the sense of (5.36). Alternatively, in this situation one may follow the approach of Zhou et al. (2012) who extended the Fine-Gray model for the cause-specific cumulative incidence to clustered data.

5.6.4 *Recurrent events (*)*

For recurrent events, i corresponds to a subject, K is the maximum number of events of interest, and an event of type h is the hth occurrence of the event for the subject. For the situation with no competing risks (Figure 1.5 without the terminal state D), the WLW model (Wei et al., 1989 – see Section 4.3.2) would be applicable for the marginal distributions of time until first, second, third, etc. recurrent event. This is because both the marginal hazard for event recurrence no. h, i.e., without considering possible occurrence of events no. $1, 2, \ldots, h-1$, is well-defined, and also the existence of a censoring time C_i for each subject i, conditionally independent of the recurrent events process for given covariates is plausible (though, one has to get used to the fact that subject i is considered at risk for event no. h no matter if event no. $h-1$ has occurred).

When there are competing risks, the notion of a marginal hazard becomes less obvious and difficulties appear when applying the WLW model in this situation (which was done by Wei et al., 1989, in one of their examples). Here, one can argue that a marginal hazard for time to event no. h is not well-defined because it relates to a hypothetical population without mortality or, considering time to death, T_D, as a censoring time, one can argue that this cannot reasonably be considered independent of $T_1, T_2, \ldots$. If one, which makes more sense, treats death as a competing risk, then the model by Zhou et al. (2012) may be adapted to a (marginal) analysis of the cumulative incidences for event no. $h = 1, \ldots, K$.

One way of circumventing this problem would be to study the event times $(T_1 \wedge T_D, \ldots, T_K \wedge T_D)$, possibly jointly with T_D. These event times are all censored by a single C_i for subject i. This was one of the possibilities discussed by Li and Lagakos (1997), and an example of this approach was given (for the data on recurrent episodes in affective disorder) in Table 4.10. Alternatively, one could acknowledge the presence of competing risks by restricting attention to *cause-specific* hazards

$$\alpha_h(t) = \lim_{dt \to 0} P(T_{hi} \leq t + dt \mid T_{hi} > t, T_{Di} > t, \mathbf{Z}_{hi}(t))/dt,$$

i.e., also conditioning on being alive at t, but then the parameter is no longer a marginal hazard. This solution was also discussed by Li and Lagakos (1997) and exemplified in Table 4.10.

5.6.5 *Illness-death model (*)*

Difficulties, similar to those described for recurrent events, also appear when using the concept of a marginal hazard for the illness-death model (Figure 1.3) or for the model for the disease course after bone marrow transplantation (Figure 1.6). For the bone marrow transplantation study (Example 1.1.7), i would be the patient and $h = 1, 2, 3$ could correspond, respectively, to GvHD, relapse, and death for that patient. To pinpoint the problem, consider a simplified multi-state model for the bone marrow transplantation data, i.e., without consideration of GvHD. This is Figure 1.3 with state 1 corresponding to relapse and state 2 to death, in which case the time to relapse $T_1 = \inf_t(V(t) = 1)$ is an *improper* random variable with $P(T_1 = \infty)$ being the probability of making a direct $0 \to 2$ transition, i.e., experiencing death in remission. Here, the marginal hazard for T_1 given by (5.36) is mathematically well-defined. However, it is either the hazard function for the distribution of time to relapse in a population with no mortality, or it is the sub-distribution hazard for the cause 1 cumulative incidence, cf. Section 5.5.3. The former situation was touched upon by Lin (1994) in one of his examples where it was noted that if one attempts to make marginal Cox models for T_1 and time to death, $T_2 = \inf_t(V(t) = 2)$ by censoring for death when analyzing T_1, then the conditional independence assumption between event times (T_1, T_2) and the associated censoring times $(C \wedge T_2, C)$ given covariates will be violated. This situation, i.e., attempting to study the marginal distribution of T_1 in a population without death informatively censoring for T_2, was later referred to as *semi-competing risks* (e.g., Fine et al., 2001); however, as discussed in Section 4.4.4, in our opinion this is asking a wrong question for the illness-death model. In the latter situation, i.e., it is acknowledged that mortality is operating but no conditioning on T_2 is done when studying the hazard for T_1, one cannot make inference for this hazard without distinguishing between censoring and death. Models for the cumulative incidence of type 1 events, i.e., acknowledging the fact that an observed death in remission signals that $T_1 = \infty$, were studied by Bellach et al. (2019), thereby providing alternatives to the Fine-Gray model. This approach is related to the previously discussed model of Mao and Lin (2017) who studied joint models for all cumulative incidences, see Section 5.5.3.

A way of circumventing these problems, similar to what was discussed for recurrent events in the previous section, would be to follow Lin (1994) and re-define the problem for the illness-death model to study marginal Cox models, not for (T_1, T_2) but for (T_0, T_2) with $T_0 = T_1 \wedge T_2$ being the time spent in the initial state, 0, i.e., the times under study would be recurrence-free survival and overall survival times. For this pair of times, the marginal hazards are well-defined and the censoring times are (C, C) which may reasonably be assumed conditionally independent of the event times for given covariates. Thus, if for the bone marrow transplantation study, one wishes to analyze both time to relapse and time to GvHD, then a possibility would be to study marginal models for relapse-free survival (i.e., without censoring for GvHD) and for GvHD-free survival (i.e., without censoring for relapse), possibly jointly with time to death. Such analyses were exemplified for the data from Example 1.1.7 in Table 4.11.

5.7 Goodness-of-fit

In previous chapters and sections, a number of different models for multi-state survival data have been discussed, including models for intensities (rates) and direct models for marginal parameters (such as risks). All models impose a number of assumptions, such as proportional hazards, additivity of covariates (no interaction) and linearity of quantitative covariates in a linear predictor. We have in connection with examples shown how these assumptions may be checked, often by explicitly introducing parameters describing departures from these assumptions. Thus, interaction terms or quadratic terms have been added to a linear predictor (e.g., Section 2.2.1) as well as time-dependent covariates expressing interactions in a Cox model between covariates and time (e.g., Section 3.7.7).

In this section, some general techniques for assessment of goodness-of-fit will be reviewed, building on an idea put forward for the Cox model by Lin et al. (1993). The techniques are based on *cumulative residuals* and provide both a graphical model assessment and a numerical goodness-of-fit test – both using *re-sampling* from an approximate large-sample distribution and, as we shall see, they are applicable for both hazard models and marginal models. Section 5.7.1 presents the mathematical idea for the general method, with special attention to GEE and the Cox model. Examples and discussion of how graphs and tests are interpreted are given in Section 5.8.4.

5.7.1 *Cumulative residuals (*)*

Many of the estimators considered so far are obtained by solving equations of the form

$$U(\boldsymbol{\beta}) = \sum_i \boldsymbol{A}(\boldsymbol{\beta}, \boldsymbol{Z}_i) e_i$$

for a suitably defined set of *residuals* $e_i, i = 1, \ldots, n$. This is the case for the general GEE discussed in Section 5.5.1, but also the score equations based on Cox partial likelihood may be re-written into this form (Section 5.6.1). Solving the equations and, thereby, obtaining parameter estimates $\widehat{\boldsymbol{\beta}}$, a set of observed residuals $\widehat{e}_i$ are obtained which may be used for model checking by looking at processes of the form

$$W(z) = W_j(z) = \sum_i h(\mathbf{Z}_i) I(Z_{ij} \leq z) \widehat{e}_i \tag{5.38}$$

based on *cumulative sums of residuals*, where Z_{ij} is a single (quantitative) covariate.

5.7.2 Generalized estimating equations (*)

If the residual is

$$\widehat{e}_i = T_i - g^{-1}(\widehat{\boldsymbol{\beta}}, \mathbf{Z}_i)$$

for a suitably defined response T_i, assumed independent among subjects, and link function g then $W(z)$ in (5.38) may be re-written as

$$W(z) = \sum_i h(\mathbf{Z}_i) I(Z_{ij} \leq z) e_i + \sum_i h(\mathbf{Z}_i) I(Z_{ij} \leq z)(g^{-1}(\widehat{\boldsymbol{\beta}}, \mathbf{Z}_i) - g^{-1}(\boldsymbol{\beta}, \mathbf{Z}_i))$$

(Lin et al., 2002). Here, the second sum is Taylor expanded around the true parameter value $\boldsymbol{\beta}_0$

$$-\sum_i h(\mathbf{Z}_i) I(Z_{ij} \leq z) \frac{\partial}{\partial \boldsymbol{\beta}} g^{-1}(\boldsymbol{\beta}^*, \mathbf{Z}_i)(\widehat{\boldsymbol{\beta}} - \boldsymbol{\beta}_0)$$

and by another Taylor expansion, see Section 5.5.1,

$$\begin{aligned}\widehat{\boldsymbol{\beta}} - \boldsymbol{\beta}_0 &\approx (-\boldsymbol{DU}(\widehat{\boldsymbol{\beta}}))^{-1} \boldsymbol{U}(\boldsymbol{\beta}_0) \\ &= (-\boldsymbol{DU}(\widehat{\boldsymbol{\beta}}))^{-1} \sum_i \boldsymbol{A}(\boldsymbol{\beta}_0, \mathbf{Z}_i) e_i.\end{aligned}$$

Collecting terms, the goodness-of-fit process $W(z)$ is seen to have the same asymptotic distribution as the following sum $\sum_i f_z(V_i(\cdot)) e_i$ of i.i.d. terms, where

$$f_z(V_i(\cdot)) = h(\mathbf{Z}_i) I(Z_{ij} \leq z) \left(1 - \frac{\partial}{\partial \boldsymbol{\beta}} g^{-1}(\boldsymbol{\beta}_0, \mathbf{Z}_i)(-\boldsymbol{DU}(\boldsymbol{\beta}_0))^{-1} \sum_k \boldsymbol{A}(\boldsymbol{\beta}_0, \mathbf{Z}_k)\right).$$

The asymptotic distribution of $W(z)$ can now be approximated by generating i.i.d. standard normal variables $(U_1, \ldots, U_n)$ and calculating $\sum_i \widehat{f_z}(V_i(\cdot)) \widehat{e}_i U_i$ (sometimes referred to as the *conditional multiplier theorem* or the *wild bootstrap*, e.g., Martinussen and Scheike, 2006, ch. 2; Bluhmki et al., 2018) where, in $\widehat{f_z}(\cdot)$, $\boldsymbol{\beta}_0$ is replaced by $\widehat{\boldsymbol{\beta}}$. This *i.i.d. decomposition* gives rise to a number of plots of cumulative residuals and also to tests obtained by comparing the observed goodness-of-fit process to those obtained by repeated generation of i.i.d. standard normal variables. We will illustrate this in Section 5.8.4.

5.7.3 Cox model (*)

For the Cox regression model, the estimating equation is the Cox score Equation (5.33)

$$\boldsymbol{U}(\boldsymbol{\beta}) = \sum_i \int_0^\infty (\mathbf{Z}_i - \bar{\mathbf{Z}}(\boldsymbol{\beta}, t)) dM_i(t) = \mathbf{0},$$

where M_i is the *martingale residual* for the counting process N_i for subject i. This was shown in Section 5.6 where the definitions of $\bar{\mathbf{Z}}(\boldsymbol{\beta},t)$, $S_0(\boldsymbol{\beta},u)$, and $\boldsymbol{S}_1(\boldsymbol{\beta},u)$ are also found. The goodness-of-fit process is

$$W(t,z) = W_j(t,z) = \sum_i h(\mathbf{Z}_i) I(Z_{ij} \le z) \widehat{M}_i(t)$$

with

$$\widehat{M}_i(t) = N_i(t) - \int_0^t Y_i(u) \exp(\widehat{\boldsymbol{\beta}}^{\mathsf{T}} \mathbf{Z}_i) \frac{\sum_k dN_k(u)}{S_0(\widehat{\boldsymbol{\beta}},u)}.$$

Taylor expanding $\widehat{M}_i(t)$ as a function of $\boldsymbol{\beta}$ around $\boldsymbol{\beta}_0$ and using

$$\begin{aligned}
\widehat{\boldsymbol{\beta}} - \boldsymbol{\beta}_0 &\approx (-\boldsymbol{DU}(\boldsymbol{\beta}_0))^{-1} \boldsymbol{U}(\boldsymbol{\beta}_0) \\
&= (-\boldsymbol{DU}(\boldsymbol{\beta}_0))^{-1} \sum_i \int_0^\infty (\mathbf{Z}_i - \bar{\mathbf{Z}}(\boldsymbol{\beta}_0,t)) dM_i(t),
\end{aligned}$$

the process $W(t,z)$ is approximated by

$$\begin{aligned}
W(t,z) &\approx \sum_i h(\mathbf{Z}_i) I(Z_{ij} \le z) \left(N_i(t) - \int_0^t Y_i(u) \exp(\boldsymbol{\beta}_0^{\mathsf{T}} \mathbf{Z}_i) \frac{\sum_k dN_k(u)}{S_0(\boldsymbol{\beta}_0,u)} \right. \\
&- \int_0^t \frac{S_0(\boldsymbol{\beta}_0,u) \mathbf{Z}_i^{\mathsf{T}} \exp(\boldsymbol{\beta}_0^{\mathsf{T}} \mathbf{Z}_i) - \exp(\boldsymbol{\beta}_0^{\mathsf{T}} \mathbf{Z}_i) \boldsymbol{S}_1^{\mathsf{T}}(\boldsymbol{\beta}_0,u)}{S_0(\boldsymbol{\beta}_0,u)^2} d\sum_k N_k(u) \\
&\times \left. (-\boldsymbol{DU}(\boldsymbol{\beta}_0))^{-1} \sum_k \int_0^\infty (\mathbf{Z}_k - \bar{\mathbf{Z}}(\boldsymbol{\beta}_0,u)) dM_k(u) \right).
\end{aligned}$$

In this expression, the Doob-Meyer decomposition (1.21) of N_i is used to get that $W(t,z)$ has the same asymptotic distribution as

$$\sum_i \int_0^t (h(\mathbf{Z}_i) I(Z_{ij} \le z) - g_j(\boldsymbol{\beta}_0,u,z)) dM_i(u)$$

$$-\sum_i \int_0^t Y_i(u) \exp(\boldsymbol{\beta}_0^{\mathsf{T}} \mathbf{Z}_i) h(\mathbf{Z}_i) I(Z_{ij} \le z) (\mathbf{Z}_i - \bar{\mathbf{Z}}(\boldsymbol{\beta}_0,u))^{\mathsf{T}} \alpha_0(u) du$$

$$\times (-\boldsymbol{DU}(\boldsymbol{\beta}_0))^{-1} \sum_k \int_0^\infty (\mathbf{Z}_k - \bar{\mathbf{Z}}(\boldsymbol{\beta}_0,u)) dM_k(u),$$

where

$$g_j(\boldsymbol{\beta},u,z) = \frac{\sum_i Y_i(u) \exp(\boldsymbol{\beta}^{\mathsf{T}} \mathbf{Z}_i) h(\mathbf{Z}_i) I(Z_{ij} \le z)}{S_0(\boldsymbol{\beta},u)}.$$

This asymptotic distribution is approximated by replacing $\boldsymbol{\beta}_0$ by $\widehat{\boldsymbol{\beta}}$, $\alpha_0(u)du$ by the Breslow estimator, and $dM_i(t)$ by $dN_i(t)U_i$ with $U_1,\ldots,U_n$ i.i.d standard normal variables.

Lin et al. (1993) suggested to use $W_j(t,z)$ with $h(\cdot) = 1$ and $t = \infty$ to check the functional form for a quantitative Z_{ij}, i.e., to plot *cumulative martingale residuals*

$$\sum_i I(Z_{ij} \le z) \widehat{M}_i(\infty)$$

against z, together with a large number of paths generated from the approximate asymptotic distribution.

To examine proportional hazards, Lin et al. (1993) proposed to let $h(\mathbf{Z}) = Z_{ij}$ and $z = \infty$, i.e., cumulative *Schoenfeld* or *'score' residuals*

$$\sum_i \int_0^t (Z_{ij} - \bar{Z}_j(\widehat{\boldsymbol{\beta}}, u))dN_i(u)$$

are plotted against t, where the jth Schoenfeld residual for subject i failing at time X_i is its contribution $Z_{ij} - \bar{Z}_j(\widehat{\boldsymbol{\beta}}, X_i)$ to the Cox score. The observed path for the goodness-of-fit process is plotted together with a large number of paths generated from the approximate asymptotic distribution.

5.7.4 Direct regression models (*)

The general idea has also been applied to a number of other special regression models, such as those discussed in Sections 5.5.2-5.5.5. Thus, Li et al. (2015) used the technique for the Fine-Gray regression model (5.19), Lin et al. (2000) did it for the multiplicative mean model for recurrent events without competing risks (5.25), and Martinussen and Scheike (2006, ch. 5) for the Aalen additive hazards model (3.23). The technique was also used in connection with analyses based on *pseudo-values* by Pavlič et al. (2019), see Section 6.4.

5.8 Examples

In this section, we will exemplify some of the new methods that have been introduced in the current chapter.

5.8.1 Non-Markov transition probabilities

PROVA trial in liver cirrhosis

In Section 3.7.6, models for the transition intensities in the three-state illness-death model for the PROVA data (Example 1.1.4) were studied and one important conclusion from these analyses was that the process did not fulfill the Markov property. This was seen in Table 3.8 where the mortality rate $\alpha_{12}(\cdot)$ after bleeding depended on the duration $d = d(t) = t - T_1$ in the bleeding state. In this example, we will study to what extent this deviation from the Markov assumption affects estimation of the transition probability $P_{01}(s,t)$, i.e., the probability of being alive in the bleeding state at time t given alive without bleeding at the earlier time point s. Some of the analyses to be reported, as well as some further analyses of the PROVA data were presented by Andersen et al. (2022).

Under a Markov assumption, the probability $P_{01}(s,t)$ may be estimated using the Aalen-Johansen estimator which is the plug-in estimator of $\int_s^t P_{00}(s,u)\alpha_{01}(u)P_{11}(u,t)du$ (Section 5.1.3). This estimate (based on the entire data set of 286 observations, disregarding treatment and other covariates) is shown in Figure 5.8 for $s = 1$ year. This figure also shows three landmark based estimators, namely the landmark Aalen-Johansen estimator suggested by Putter and Spitoni (2018), the Titman (2015) estimator for a transient state, and the Pepe estimator also discussed by Titman (2015) (see Section 5.2.2). At the time point $s = 1$ year, there were 190 patients still at risk in state 0, and the three landmark estimators are based on those subjects. It is seen that the curves are quite different, with the Markov-based estimator throughout over-estimating the probability compared to the landmark Aalen-Johansen and Pepe estimators, whereas Titman's estimator is close to the other landmark based

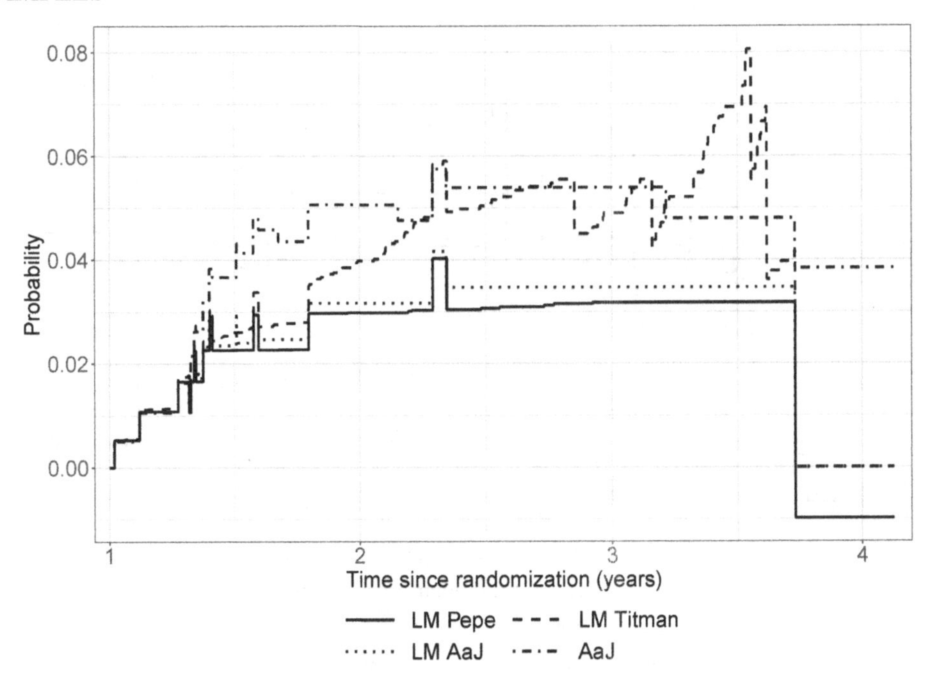

Figure 5.8 *PROVA trial in liver cirrhosis: Estimates for the transition probability* $P_{01}(s,t), t > s$ *for* $s = 1$ *year (LM: Landmark, AaJ: Aalen-Johansen).*

estimators for $t < 1.8$ years while, for larger values of t, it approaches the Aalen-Johansen estimator.

We next compare with plug-in estimators using the expression

$$P_{01}(s,t) = \int_s^t P_{00}(s,u)\alpha_{01}(u)P_{11}(u,t \mid u)du,$$

and basing the probability $P_{11}(u,t \mid u) = \exp\left(-\int_u^t \alpha_{12}(x,x-u)dx\right)$ of staying in state 1 until time t given entry into that state at the earlier time u on different models for the $1 \to 2$ transition intensity. We consider the following models

$$\alpha_{12}(t,d) = \alpha_{12,0}(d), \tag{5.39}$$

$$\alpha_{12}(t,d) = \alpha_{12,0}(t)\exp(\text{LP}(d)). \tag{5.40}$$

Equation (5.39) is the special semi-Markov model where the intensity only depends on d. In (5.40), the baseline $1 \to 2$ intensity depends on t and functions of d are used as time-dependent covariates (as in Table 3.8). In (5.40), the linear predictor is either chosen as

$$LP(d) = \beta_1 I(d < 5 \text{ days}) + \beta_2 I(5 \text{ days} \leq d < 10 \text{ days})$$

or $\text{LP}(d) = \beta \cdot d$. Figure 5.9 shows the resulting estimates $\widehat{P}_{01}(s,t)$ for $s = 1$ year together with the Aalen-Johansen and landmark Aalen-Johansen estimates. It is seen that the estimate based on a semi-Markov model with d as the baseline time-variable is close to the landmark Aalen-Johansen estimate. On the other hand, the models with t as baseline time-variable and duration-dependent covariates differ according to the way in which the effect of duration is modeled: With a piece-wise constant effect it is closer to the Markov-based Aalen-Johansen estimator, and with a linear effect it is closer to the landmark estimate.

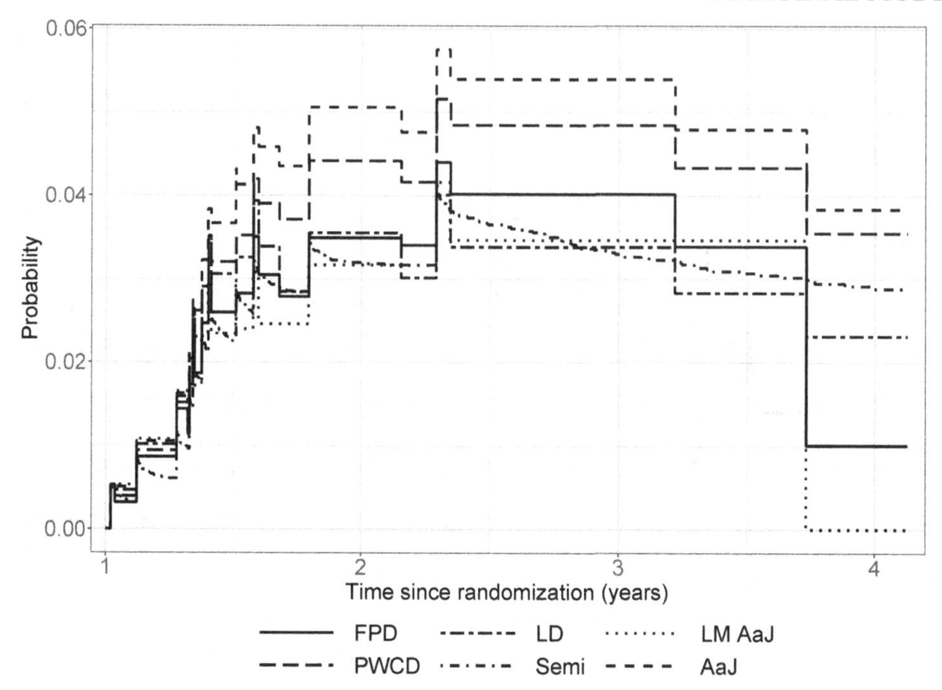

Figure 5.9 *PROVA trial in liver cirrhosis: Estimates for the transition probability* $P_{01}(s,t), t > s$ *for* $s = 1$ *year (FPD: Fractional polynomial duration effect in (5.40), LD: Linear duration effect in (5.40), LM: Landmark, AaJ: Aalen-Johansen, PWCD: Piece-wise constant duration effect in (5.40), Semi: Semi-Markov (5.39)).*

To study whether a more detailed model for the duration effect $\text{LP}(d)$ in (5.40) would provide a better fit to the data, a model with a duration effect modeled using a fractional polynomial

$$\text{LP}(d) = \beta_1 d + \beta_2 d^2 + \beta_3 d^3 + \beta_4 \log(d)$$

(e.g., Andersen and Skovgaard, 2010, ch. 4) was also studied, see Figure 5.9. It is seen that the latter estimate is close to that using a linear duration effect.

To assess the variability of the estimators, Andersen et al. (2022) also conducted a bootstrap experiment by sampling $B = 1,000$ times with replacement from the PROVA data set and repeating the analyses on each bootstrap sample. It was found that the Aalen-Johansen estimator has a relatively large SD; however, since this estimator tends to be upwards biased as seen in Figures 5.8-5.9, a more fair comparison between the estimated variabilities is obtained by studying the relative SD, i.e., the coefficient of variation $\text{SD}(\widehat{P})/\widehat{P}$. This showed that the estimators based on sub-sampling (landmark Aalen-Johansen, Pepe, Titman) have relatively large relative SD-values. On the other hand, the Aalen-Johansen estimator and the plug-in estimators (5.39) and (5.40) (with a linear duration effect as covariate) have smaller relative SD.

In conclusion, the estimators based on sub-sampling are truly non-parametric and hence reliable; however, being based on fewer subjects, they are likely to be more variable. On the other hand, the plug-in estimators are based on the full sample and hence less variable though it may be a challenge to correctly model the effect of 'the other time-variable' using time-dependent covariates.

Table 5.5 *Bone marrow transplantation in acute leukemia: Estimated coefficients (and SD) for duration effects in Cox models for the transition intensities $\alpha_{12}(\cdot), \alpha_{13}(\cdot)$ and $\alpha_{23}(\cdot)$ (GvHD: Graft versus host disease).*

Transition	Duration in		$\widehat{\beta}$	SD
$1 \to 2$	GvHD state	$(t - T_1)$	0.074	0.046
$1 \to 3$	GvHD state	$(t - T_1)$	0.050	0.021
$2 \to 3$	Relapse state	$(t - T_2)$	-0.066	0.016

Bone marrow transplantation in acute leukemia

We will here illustrate analyses on the bone marrow transplantation data (Example 1.1.7) similar to those conducted for the PROVA trial in the previous section. Following Andersen et al. (2022), we will be focusing on the probability $P_{02}(s,t)$ of being alive in the relapse state at time t given alive in the initial state at an earlier time s, see Figure 1.6. For this example, a Markov model does not fit the data well; see Table 5.5 where results from Cox models for the transition intensities $\alpha_{12}(\cdot), \alpha_{13}(\cdot)$ and $\alpha_{23}(\cdot)$ allowing for duration dependence in states 1 or 2 are summarized. It is seen that the two death intensities depend on (a linear effect of) duration, $t - T_1$ or $t - T_2$ in states 1 or 2, respectively. The former increases with duration ($\widehat{\beta} > 0$), while the latter decreases ($\widehat{\beta} < 0$).

Figures 5.10 and 5.11 show estimates of $P_{02}(s,t), t > s$ for $s = 3$ and 9 months using various estimators: The Markov-based Aalen-Johansen estimator, the landmark Aalen-Johansen and Pepe estimators, and three plug-in estimators. The first plug-in estimator uses duration in states 1 or 2 as baseline time-variables with no adjustment for time t since transplantation, and the two others model the intensities out of states 1 or 2 using t as baseline time-variable and adjusting for duration in states 1 or 2 using the models with estimates given in Table 5.5. It is seen that, for this example, the deviations from the Markov assumption are less severe for the estimation of transition probabilities, and no big differences between the various estimators are apparent. However, it does seem as if that based on the semi-Markov model for $s = 9$ months gives somewhat lower estimates for $t > 2.5$ years. A possible explanation is that the semi-Markov model does not take time t since transplantation into account for the transition intensities out of states 1 and 2 and, as seen in Section 3.7.8, this time-variable does have an effect. So, the example illustrates that when using plug-in estimators modeling duration effects explicitly, great care must be exercised when setting up these models.

5.8.2 *Direct binomial regression*

We will illustrate the use of direct binomial regression (Section 5.5.5) for estimating covariate effects on the cumulative incidence of death without transplantation in the PBC3 trial. In Section 4.2.2, these data were analyzed using the Fine-Gray model and estimates with a sub-distribution hazard ratio interpretation were obtained. Using the estimating equations (5.32), it is possible to apply other link functions for linking the cumulative incidence to covariates, e.g., a logistic link yielding estimates with an odds ratio interpretation. Table 5.6 shows estimates obtained by fitting such models to the cumulative incidence for

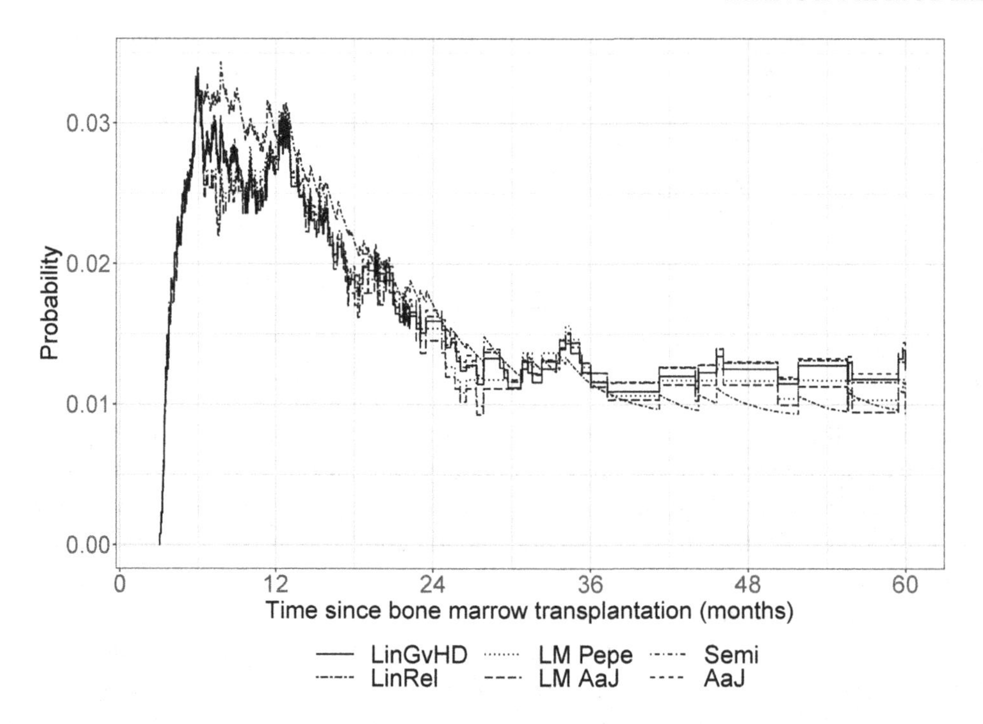

Figure 5.10 *Bone marrow transplantation in acute leukemia: Estimates for the transition probability* $P_{02}(s,t), t > s$ *for* $s = 3$ *months (LinGvHD: Linear duration effect in state 1, LM: Landmark, Semi: Semi-Markov, LinRel: Linear duration effect in state 2, AaJ: Aalen-Johansen).*

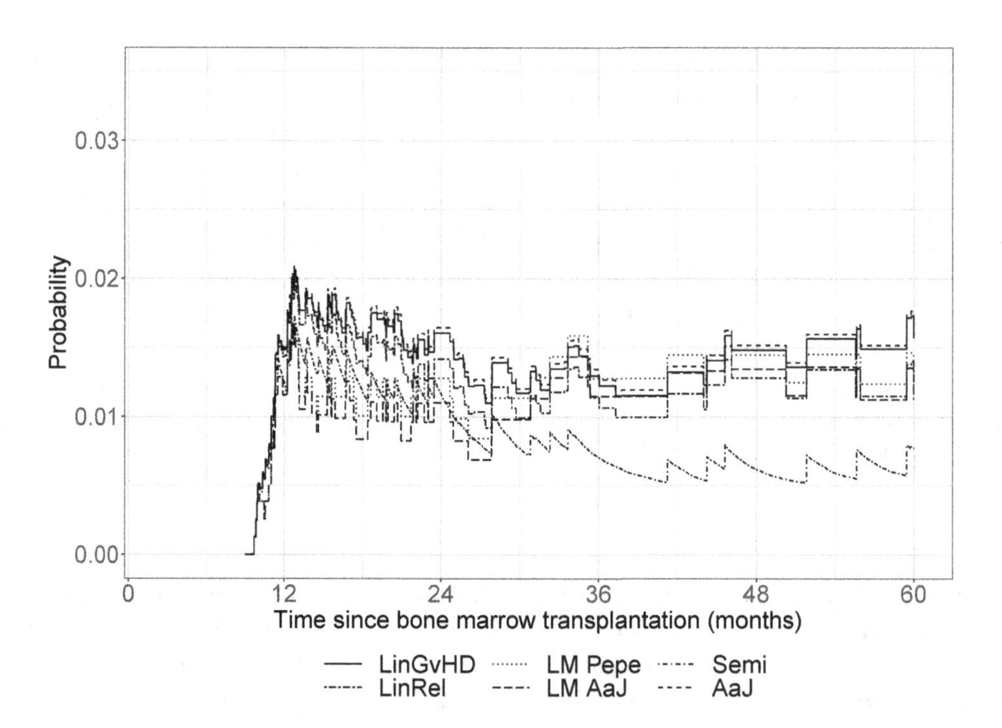

Figure 5.11 *Bone marrow transplantation in acute leukemia: Estimates for the transition probability* $P_{02}(s,t), t > s$ *for* $s = 9$ *months (LinGvHD: Linear duration effect in state 1, LM: Landmark, Semi: Semi-Markov, LinRel: Linear duration effect in state 2, AaJ: Aalen-Johansen).*

death without a liver transplantation ($F_2(\cdot)$), either at $t_0 = 2$ years or simultaneously at $(t_1, t_2, t_3) = (1, 2, 3)$ years. It is seen that, unadjusted, the odds of dying without transplantation before 2 years is $1.098 = \exp(0.093)$ times higher for a CyA-treated person compared to placebo with 95% confidence interval from 0.531 to 2.27, and after adjustment for albumin and $\log_2$(bilirubin) the corresponding odds ratio is $0.630 = \exp(-0.463)$ (95% confidence interval (0.266, 1.488)). Moving from analyzing $F_2(t_0 \mid Z)$ to jointly analyzing $F_2(t_j \mid Z), j = 1, 2, 3$ and assuming time-constant effects, the estimated SD become smaller.

Table 5.6 *PBC3 trial in liver cirrhosis: Estimated coefficients (and SD) from direct binomial (logistic) models for the cumulative incidence of death without transplantation at $t_0 = 2$ years or simultaneously at $(t_1, t_2, t_3) = (1, 2, 3)$ years.*

(a) $t_0 = 2$ *years*

Covariate		$\widehat{\beta}$	SD	$\widehat{\beta}$	SD
Treatment	CyA vs. placebo	0.093	0.371	-0.463	0.439
Albumin	per 1 g/L			-0.147	0.037
$\log_2$(bilirubin)	per doubling			0.639	0.151

(b) $(t_1, t_2, t_3) = (1, 2, 3)$ *years*

Covariate		$\widehat{\beta}$	SD	$\widehat{\beta}$	SD
Treatment	CyA vs. placebo	-0.030	0.323	0.520	0.373
Albumin	per 1 g/L			-0.125	0.035
$\log_2$(bilirubin)	per doubling			0.579	0.128

5.8.3 *Extended models for recurrent events*

Furberg et al. (2022) applied versions of the Mao-Lin (2016) model for recurrent events with competing risks to data from the LEADER trial (Example 1.1.6). Recall from Section 5.5.4 that this model (5.31) concerns the mean of a weighted recurrent end-point counting both recurrent events and death. For recurrent MI including all-cause death (with all severity weights $c_h = 1$), the estimated log(mean ratio) was $\widehat{\beta} = -0.159$ ($SD = 0.057$). Furberg et al. also studied 'recurrent 3-p MACE', thus counting both recurrent myocardial infarctions, recurrent strokes, and cardiovascular deaths as events (giving all events a severity weight of $c_h = 1$). Non-cardiovascular death was here (somewhat incorrectly) treated as censoring. This yielded a mean ratio of $\exp(-0.183) = 0.833$ between liraglutide and placebo with a 95% confidence interval from 0.742 to 0.935. Adjusting properly for the competing risk of non-cardiovascular death (in a model combining the Mao-Lin model with a Ghosh-Lin model – see appendix in Furberg et al., 2022) had almost no impact on the estimate which was a mean ratio of 0.832 (0.741, 0.934). A possible explanation for this similarity is the lack of difference between non-cardiovascular death rates in the two treatment groups.

5.8.4 *Goodness-of-fit based on cumulative residuals*

We will show how plots of cumulative residuals may be used for assessing goodness-of-fit for the Cox regression models fitted to the data from the PBC3 trial (Example 1.1.1).

Checking linearity

The *martingale residual* from a Cox model is the difference

$$\widehat{M}_i = D_i - \widehat{A}_0(X_i)\exp(\widehat{\mathrm{LP}}_i)$$

between the failure indicator D_i (= 1 for a failure and = 0 for a censored observation) for subject i and the estimated cumulative hazard evaluated at the time, X_i, of failure/censoring. The latter has an interpretation as an 'expected value' of D_i at time X_i. If Z_j is a quantitative covariate, then, according to Lin et al. (1993), a plot of the cumulative sum of martingale residuals for subjects with $Z_{ij} \leq z$ against z is sensitive to non-linearity of the effect of the covariate on the linear predictor. If linearity provides a good description of the effect, then the resulting curve should vary non-systematically around 0. A formal test for linearity may be obtained by comparing the curve with a large number of random realizations of how the curve should look like under linearity, e.g., focusing on the maximum value of the observed curve compared to the maxima of the random realizations.

In Section 2.2, a model for the rate of failure of medical treatment including the covariates treatment, albumin, and bilirubin was fitted to the PBC3 data, see Table 2.4. To assess linearity of the two quantitative covariates albumin and bilirubin, Figures 5.12 and 5.13 show cumulative martingale residuals plotted against the covariate. While, for albumin, linearity is not contra-indicated, the plot for bilirubin shows clear departures from 'random variation around 0'. This is supported by P-values from a formal significance test (0.459 for albumin and extremely small for bilirubin). The curve for bilirubin gets negative for small values of the covariate indicating that the 'expected' number of failures is too large

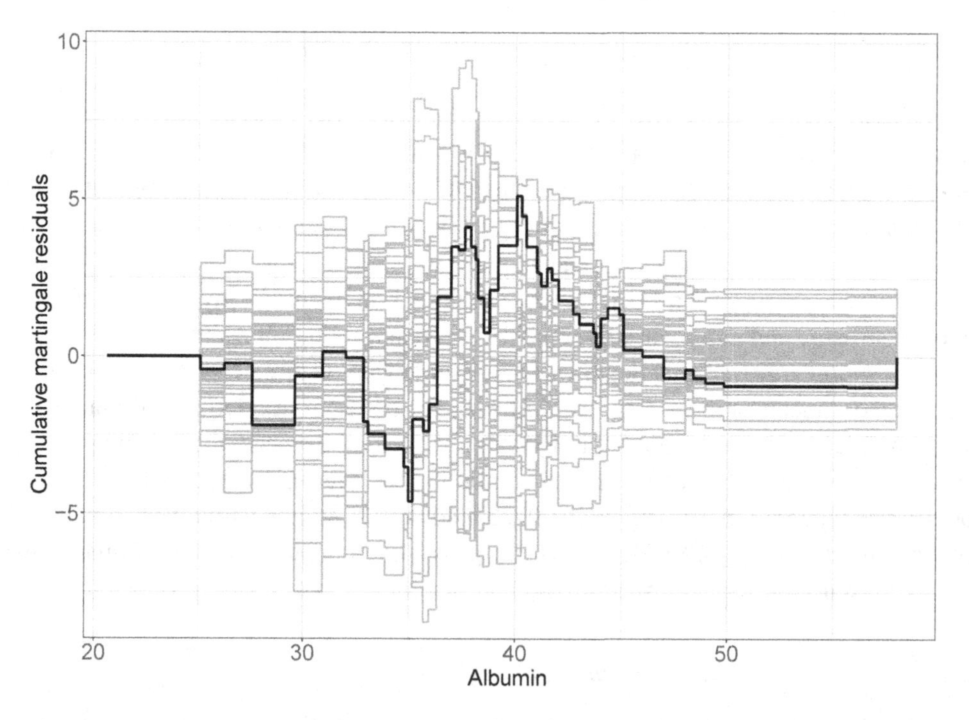

Figure 5.12 *PBC3 trial in liver cirrhosis: Checking linearity using cumulative martingale residuals plotted against albumin.*

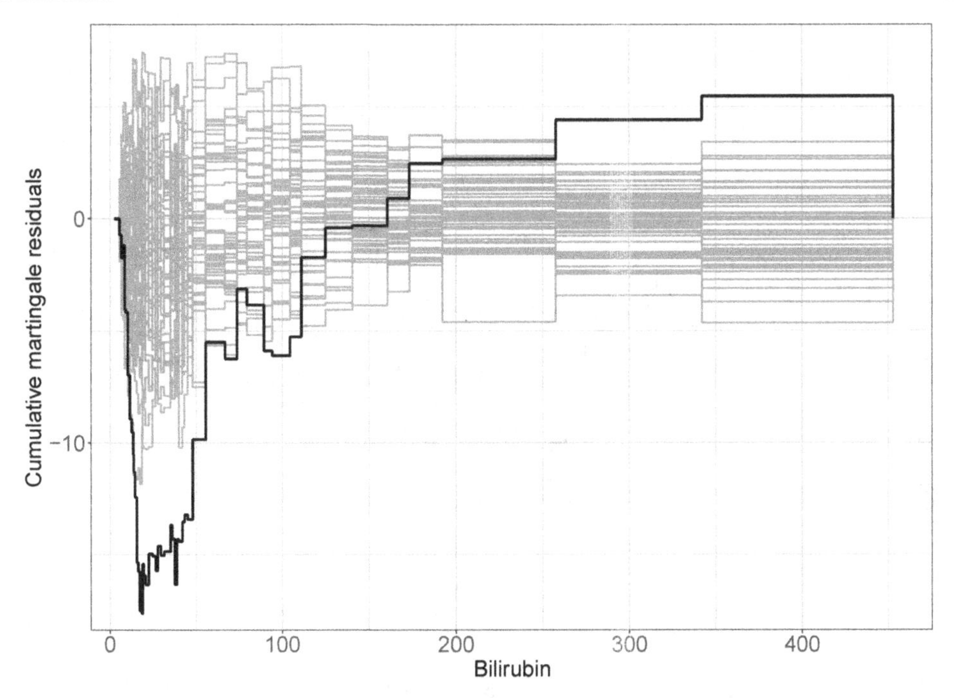

Figure 5.13 *PBC3 trial in liver cirrhosis: Checking linearity using cumulative martingale residuals plotted against bilirubin.*

for low values of bilirubin compared to the observed (the latter is often equal to 0). This suggests that relatively more weight should be given to low values of bilirubin in the linear predictor and relatively less weight to high values – something that may be achieved by a transformation of the covariate with a concave ('downward bending') function, such as the logarithm. Figure 5.14 shows the plot after transformation and now the curve is more in accordance with what would be expected under linearity. This is supported by the P-value of 0.481. The estimates in this model were given in Table 2.7. The plot for albumin in this model (not shown) is not much different from what was seen in Figure 5.12.

Checking proportional hazards

The *Schoenfeld* (or *score*) *residuals* from a Cox model are the differences

$$D_i(Z_{ij} - \bar{Z}_j(X_i)),$$

between the observed covariate Z_{ij} for a subject failing at time X_i and an expected average value for covariate j, $\bar{Z}_j(X_i)$, among subjects at risk at that time. According to Lin et al. (1993), a plot of the cumulative sum of Schoenfeld residuals for subjects with $X_i \leq t$ against time t is sensitive to departures from proportional hazards. If the proportional hazards assumption fits the data well, then the resulting curve should vary non-systematically around 0, and a formal goodness-of-fit test may be obtained along the same lines as for the plot of cumulative martingale residuals.

Figures 5.15-5.17 show plots of cumulative Schoenfeld residuals (standardized by division by $\mathrm{SD}(\widehat{\beta}_j)$) against time for the three covariates in the model: Treatment, albumin, and

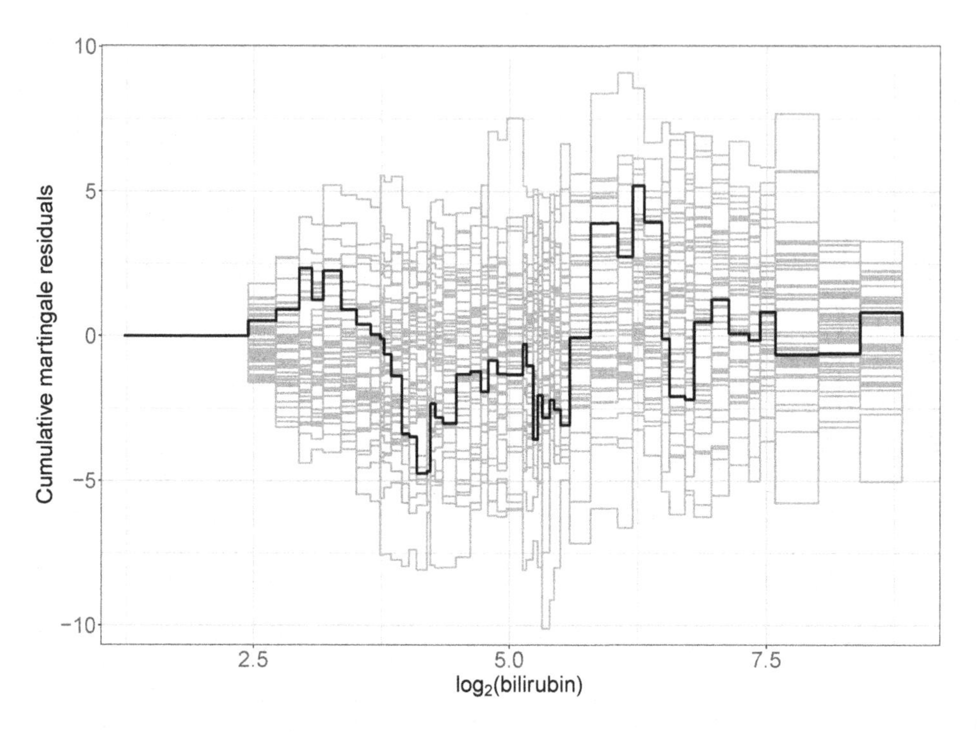

Figure 5.14 *PBC3 trial in liver cirrhosis: Checking linearity using cumulative martingale residuals plotted against* $\log_2$*(bilirubin).*

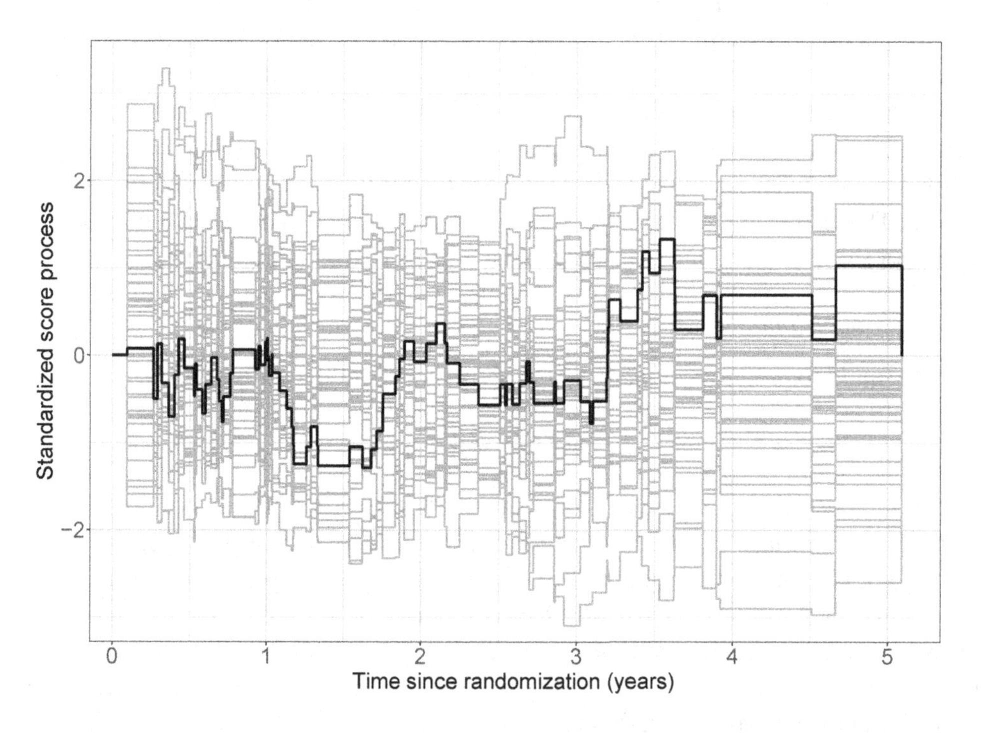

Figure 5.15 *PBC3 trial in liver cirrhosis: Checking proportional hazards using cumulative Schoenfeld residuals for treatment (standardized) plotted against the time-variable.*

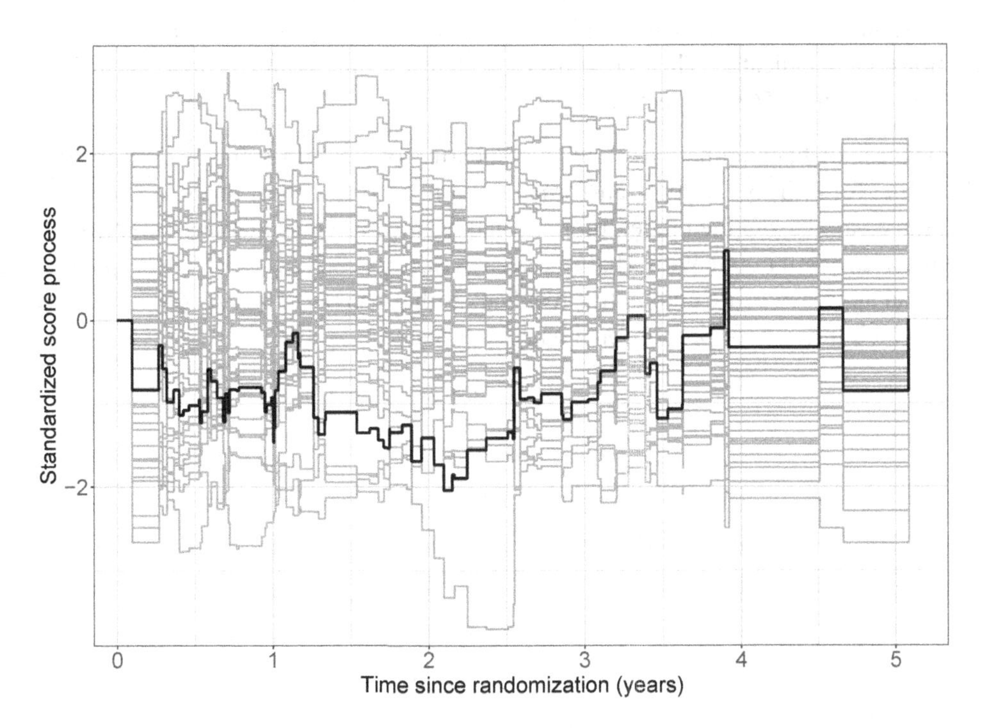

Figure 5.16 *PBC3 trial in liver cirrhosis: Checking proportional hazards using cumulative Schoenfeld residuals for albumin (standardized) plotted against the time-variable.*

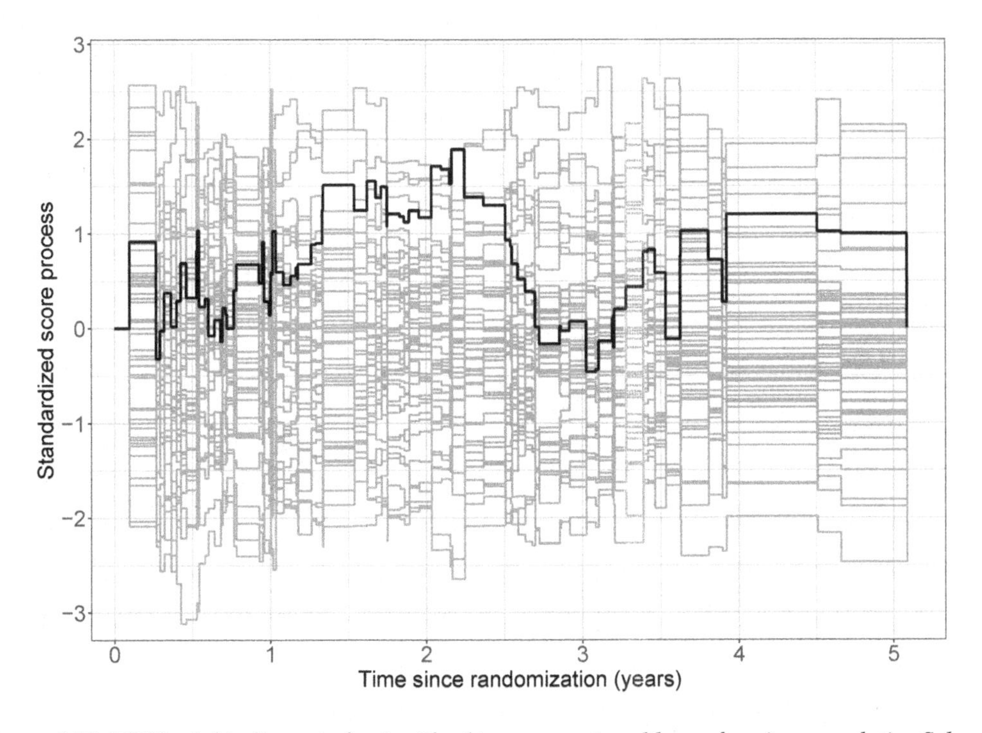

Figure 5.17 *PBC3 trial in liver cirrhosis: Checking proportional hazards using cumulative Schoenfeld residuals for* $\log_2$*(bilirubin) (standardized) plotted against the time-variable.*

$\log_2$(bilirubin). It is seen that for neither of the covariates is the proportional hazards assumption contra-indicated. This is confirmed both by the curves and the associated P-values (0.919, 0.418, and 0.568, respectively).

The goodness-of-fit examinations show that linearity for albumin seems to describe the data well, while, for bilirubin, a log-transformation is needed to obtain linearity. Furthermore, for all three variables in the model, proportional hazards is a reasonable assumption. These conclusions are well in line with what was seen in Sections 2.2.2 and 3.7.7. An advantage of the general approach using cumulative residuals is that one needs not specify an alternative against which linearity or proportional hazards is tested.

5.9 Exercises

Exercise 5.1 (*) Consider the two-state reversible Markov model

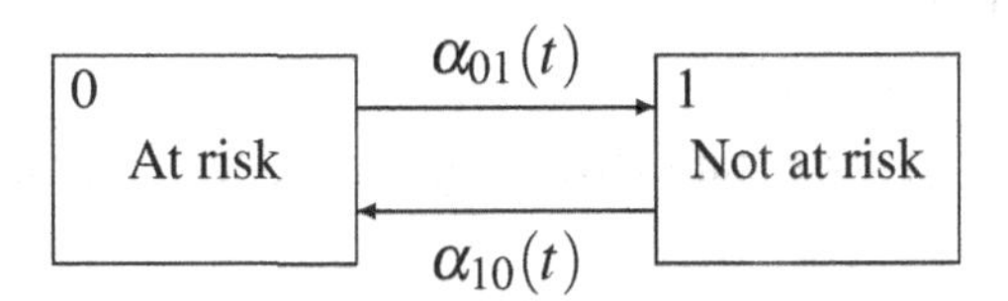

Set up the $\boldsymbol{A}(t)$ and $\boldsymbol{P}(s,t)$ matrices and express, using the Kolmogorov forward differential equations, the transition probabilities in terms of the transition intensities (Section 5.1).

Exercise 5.2 (*) Consider the four-state model for the bone marrow transplantation study, Figure 1.6. Set up the $\boldsymbol{A}(t)$ and $\boldsymbol{P}(s,t)$ matrices and express, using the Kolmogorov forward differential equations, the transition probabilities in terms of the transition intensities (Section 5.1).

Exercise 5.3 (*)

1. Consider the competing risks model and show that the ratio between the cause h sub-distribution hazard and the corresponding cause-specific hazard is

$$\frac{\widetilde{\alpha}_h(t)}{\alpha_h(t)} = \frac{S(t)}{1 - F_h(t)}.$$

2. Show that, thereby, proportional sub-distribution hazards and proportional cause-specific hazards are incompatible.

Exercise 5.4 (*)

Consider the competing risks model and direct binomial regression for $Q_h(t_0)$, the cause-h cumulative incidence at time t_0 (Section 5.5.5). The estimating equation (5.18) is

$$\sum_i D_i(t_0)\widehat{W}_i(t_0)\boldsymbol{A}(\boldsymbol{\beta}^\mathsf{T}\mathbf{Z}_i)\big(N_{hi}(t_0) - Q_h(t_0 \mid \mathbf{Z}_1)\big)$$

with $D_i(t_0)$ the indicator $I(T_i \wedge t_0 \leq C_i)$ of observing the state occupied at t_0 and $\widehat{W}_i(t_0) = 1/\widehat{G}((t_0 \wedge X_i)-)$ the estimated inverse probability of no censoring (strictly) before the minimum of t_0 and the observation time X_i for subject i. The alternative estimating equation (5.32) is

$$\sum_i \boldsymbol{A}(\boldsymbol{\beta}^\mathsf{T}\mathbf{Z}_i)\big(N_{hi}(t_0)D_i(t_0)\widehat{W}_i(t_0) - Q_h(t_0 \mid \mathbf{Z}_i)\big).$$

Show that, replacing $\widehat{G}$ by the true G, both estimating equations are unbiased.

Exercise 5.5 (*) Derive the estimating equations for the landmark model (5.11).

Exercise 5.6 Consider an illness-death model for the Copenhagen Holter study with states '0: Alive without AF or stroke', '1: Alive with AF and no stroke', '2: Dead or stroke', see Figures 1.3 and 1.7. Examine, using a time-dependent covariate, whether this process may be modeled as being Markovian.

Exercise 5.7 Consider the four-state model for the Copenhagen Holter study, see Figure 1.7.

1. Fit separate landmark models at times 3, 6, and 9 years for the mortality rate, including AF, stroke, ESVEA, sex, age, and systolic blood pressure.
2. Fit landmark 'super models' where the coefficients vary smoothly among landmarks but with separate baseline hazards at each landmark.
3. Fit a landmark 'super model' where both the coefficients and the baseline hazards vary smoothly among landmarks.

Exercise 5.8 Consider a competing risks model for the Copenhagen Holter study with states '0: Alive without AF or stroke', '1: Alive with AF and no stroke', '2: Dead or stroke', see Figures 1.2 and 1.7.

Fit, using direct binomial regression, a model for being in state 1 at time 3 years including the covariates ESVEA, sex, age, and systolic blood pressure.

Exercise 5.9 Consider the Cox model for stroke-free survival in the Copenhagen Holter study including the covariates ESVEA, sex, age, and systolic blood pressure (Exercises 2.4 and 3.7).

1. Investigate, using cumulative Schoenfeld residuals, whether the effects of the covariates may be described as time-constant hazard ratios.
2. Investigate, using cumulative martingale residuals, whether the effects of age and systolic blood pressure can be considered linear on the log(hazard) scale.

Exercise 5.10 Consider the data on recurrent episodes in affective disorder, Example 1.1.5. Fit a Mao-Lin regression model (5.31) for the mean of the composite end-point recurrent episode or death, including initial diagnosis as the only covariate and using severity weights equal to 1.

Chapter 6

Pseudo-values

In Sections 4.2 and 5.5, we discussed how direct regression models for marginal parameters in a multi-state model could be set up and fitted using generalized estimating equations (GEEs). It turned out that this could be done on a case-by-case basis and that, furthermore, it was typically necessary to explicitly address the censoring distribution. This is because the uncensored observations had to be re-weighted to also represent those who were censored and this required estimation of the probability of being uncensored at the times of observed failures. One might ask whether it would be possible to apply a more general technique when fitting marginal models for multi-state processes. The answer to this question is 'yes', under the proviso that one is content with a model for a single or a finite number of time points. A way to do this is to apply *pseudo-values* (or *pseudo-observations* – we will use these notions interchangeably in what follows).

The idea is as follows: With *complete* data, i.e., in the absence of censoring, a regression model could be set up and fitted using standard GEE using the relevant aspect of the complete data as response variable as explained in Section 5.5.1. To model the survival probability $S(t_0 \mid Z)$ in the point t_0, the survival indicator $I(T_i > t_0)$ would be observed for all subjects $i = 1, \ldots, n$ and could thus be used as outcome variable in the GEE. With *incomplete* data this is not possible, and in this case the pseudo-values are calculated based on the available data and they replace the incompletely observed response variables (e.g., Andersen et al., 2003; Andersen and Pohar Perme, 2010). This is doable because they, under suitable assumptions on the censoring distribution, have the correct expected value for given covariates (Graw et al., 2009; Jacobsen and Martinussen, 2016; Overgaard et al., 2017). The pseudo-values typically build on a non-parametric estimator for the marginal parameter, such as the Kaplan-Meier estimator for the survival function in the two-state model (Sections 4.1.1 or 5.1.1) or the Aalen-Johansen estimator for the competing risks cumulative incidence (Sections 4.1.2 or 5.1.2). Thereby, censoring is dealt with once and for all leaving us with a set of n observations which are approximately independent and identically distributed (*i.i.d.*). Note that, while *right-censoring* may be handled in this way, data with *left-truncation* are typically harder to deal with (Parner et al., 2023).

In Section 6.1, the basic idea is presented in an intuitive way with several examples and in Section 6.2, more mathematical details are provided. Section 6.3 presents a fast approximation to calculation of pseudo-values, and Section 6.4 gives a brief account of how to use cumulative residuals when assessing goodness-of-fit of models fitted to pseudo-values.

6.1 Intuition

6.1.1 Introduction

The set-up is as follows: $V(t)$ is a multi-state process, and interest focuses on a marginal parameter which is the expected value, $E(f(V)) = \theta$, say, of some function f of the process. Examples include the following:

- $V(t)$ is the two-state process for survival data, Figure 1.1, and θ is the state 0 occupation probability $Q_0(t_0)$ at a fixed time point t_0, i.e., the survival probability $S(t_0) = P(T > t_0)$ at that time.
- $V(t)$ is the competing risks process, Figure 1.2, and θ is the state $h, h > 0$ occupation probability $Q_h(t_0)$ at a fixed time point t_0, i.e., the cause-h cumulative incidence $F_h(t_0) = P(T \leq t_0, D = h)$ at that time.
- $V(t)$ is the two-state process for survival data, Figure 1.1, and θ is the expected time $\varepsilon_0(\tau)$ spent in state 0 before a fixed time point τ, i.e., the τ-restricted mean survival time.
- $V(t)$ is the competing risks process, Figure 1.2, and θ is the expected time $\varepsilon_h(\tau)$ spent in state $h > 0$ up to a fixed time point τ, i.e., the cause-h specific time lost due to that cause before time τ.
- $V(t)$ is a recurrent events process, Figures 1.4-1.5, and θ is the expected number, $\mu(t_0) = E(N(t_0))$ of events at a fixed time point t_0.

In this section, we present the idea for the first of these examples. The other examples follow the same lines and more discussion is provided in Sections 6.1.2-6.1.7. We are interested in a regression model for the survival function at time t_0, $S(t_0 \mid Z)$, that is, the expected value of the survival indicator $f(T) = I(T > t_0)$ given covariates Z

$$S(t_0 \mid Z) = E(I(T > t_0) \mid Z).$$

One typical model for this could be what corresponds to a Cox model, i.e.,

$$\log(-\log S(t_0 \mid Z)) = \beta_0 + \text{LP},$$

where the intercept is $\beta_0 = \log(A_0(t_0))$, the log(cumulative baseline hazard) at time t_0, and LP is the linear predictor $\text{LP} = \beta_1 Z_1 + \cdots + \beta_p Z_p$. Another would correspond to an additive hazard model

$$-\log(S(t_0 \mid Z))/t_0 = \beta_0/t_0 + \text{LP}$$

with $\beta_0 = A_0(t_0)$, the cumulative baseline hazard at t_0. In general, some function g, the *link function*, of the marginal parameter θ is the linear predictor. Note that such a model is required to hold only at time t_0 and not at all time points.

We first consider the unrealistic situation without censoring, i.e., survival times $T_1, \ldots, T_n$ are observed and so are the t_0-survival indicators $f(T_i) = I(T_i > t_0)$, $i = 1, \ldots, n$. This situation serves as motivation for the way in which pseudo-observations are defined and, in this situation, two facts can be noted:

1. The marginal mean $E(f(T)) = E(I(T > t_0)) = S(t_0)$ can be estimated as a simple average

$$\widehat{S}(t_0) = \frac{1}{n}\sum_i I(T_i > t_0).$$

2. A regression model for $\theta = S(t_0 \mid Z)$ with link function g can be analyzed using GEE with $f(T_1) = I(T_1 > t_0), \dots, f(T_n) = I(T_n > t_0)$ as responses. This is a standard generalized linear model for a binary outcome with link function g.

Let $\widehat{S}^{-i}$ be the estimator for S *without observation* i, i.e.,

$$\widehat{S}^{-i}(t_0) = \frac{1}{n-1}\sum_{j \neq i} I(T_j > t_0).$$

We now have that

$$\begin{aligned} n \cdot \widehat{S}(t_0) &= f(T_1) + \cdots + f(T_{i-1}) + f(T_i) + f(T_{i+1}) + \cdots + f(T_n), \\ (n-1) \cdot \widehat{S}^{-i}(t_0) &= f(T_1) + \cdots + f(T_{i-1}) + \qquad\quad f(T_{i+1}) + \cdots + f(T_n), \end{aligned}$$

i.e.,

$$n \cdot \widehat{S}(t_0) - (n-1) \cdot \widehat{S}^{-i}(t_0) = f(T_i).$$

Thus, the ith observation can be re-constructed by combining the marginal estimator based on all observations and that obtained without observation no. i.

We now turn to the realistic scenario where some survival times are incompletely observed because of right-censoring, i.e., the available data are (X_i, D_i), $i = 1, \dots, n$ where X_i is the ith time of observation, the smaller of the true survival time T_i and the censoring time C_i, and D_i is 1 if T_i is observed and 0 if the ith observation is censored. In this case it is still possible to estimate the marginal survival function, namely using the Kaplan-Meier estimator, $\widehat{S}$ given by Equation (4.3). Based on this, we can calculate the quantity

$$\theta_i = n \cdot \widehat{S}(t_0) - (n-1) \cdot \widehat{S}^{-i}(t_0), \tag{6.1}$$

where $\widehat{S}^{-i}(t_0)$ is the estimator (now Kaplan-Meier) applied to the sample of size $n-1$ obtained by eliminating observation no. i from the full sample. The θ_i, $i = 1, \dots, n$ given by (6.1) are the *pseudo-observations* for the incompletely observed survival indicators $f(T_i) = I(T_i > t_0)$, $i = 1, \dots, n$.

Note that *pseudo-values are computed for all subjects* – whether the survival time was observed or only a censored observation was available.

The idea is now, first, to transform the data (X_i, D_i), $i = 1, \dots, n$ into $\theta_i, i = 1, \dots, n$ using (6.1), i.e., to add one more variable to each of the n lines in the data set and, next, to analyze a regression model for θ by using θ_i, $i = 1, \dots, n$ as responses in a GEE with the desired link function, g. Here, typically, a *normal* error distribution is specified since this will enable the correct estimating equations to be set up for the mean value (in spite of the fact that the distribution of pseudo-values is typically far from normal). Such a procedure will provide estimators for the parameters in the regression model $S(t_0 \mid Z)$ that have been shown to be mathematically well-behaved if the distribution of the censoring times C_i

does not depend on the covariates, Z (Graw et al., 2009; Jacobsen and Martinussen, 2016; Overgaard et al., 2017, 2023). The situation with covariate-dependent censoring is more complex and will be discussed in Section 6.1.8. The standard sandwich estimator based on the GEE is most often used, though this may be slightly conservative, i.e., a bit too large, as will be explained in Section 6.2.

The use of pseudo-values for fitting marginal models for multi-state parameters has a number of attractive features:

1. It can be used quite generally for marginal multi-state parameters whenever a suitable estimator $\widehat{\theta}$ for the marginal mean $\theta = E(f(V))$ is available.
2. It provides us with a set of new variables $\theta_1, \dots, \theta_n$ for which standard models for complete data can be analyzed.
3. It provides us with a set of new variables $\theta_1, \dots, \theta_n$ to which various plotting techniques are applicable.
4. If interest focuses on a single time point t_0, then a specification of a model for other time points is not needed.

A number of difficulties should also be mentioned:

1. If censoring depends on covariates, then modifications of the method are necessary.
2. It only provides a model at a fixed point in time t_0, or as we shall see just below, at a number of fixed points in time $t_1, \dots, t_m$, and these time points need to be specified.
3. The base estimator needs to be re-calculated $n+1$ times, and if this computation is involved and/or n is large, then obtaining the n pseudo-values may be cumbersome.

A multivariate model for $S(t_1 \mid Z), \dots, S(t_m \mid Z)$ at a number, m of time points $t_1, \dots, t_m$ can be analyzed in a similar way. The response in the resulting GEE is m-dimensional and a joint model for all time points is considered. The model could be what corresponds to a Cox model, i.e.,

$$\log(-\log S(t_j \mid Z)) = \beta_{0j} + \text{LP},$$

with $\beta_{0j} = \log(A_0(t_j))$, $j = 1, \dots, m$, the log(cumulative baseline hazard) at t_j.

PBC3 trial in liver cirrhosis

As an example we will study the PBC3 trial (Example 1.1.1) and pseudo-observations for the indicator of no failure from medical treatment. For illustration, we first fix two subjects and plot the pseudo-observations for those subjects as a function of time. We choose one subject with a failure at 1 year ($X = 366$ days, $D = 1$) and one with a censored observation at 1 year ($X = 365$ days, $D = 0$), see Figure 6.1. It is seen that, for $t < 1$ year, the pseudo-observations for the two subjects coincide. This is because an observation time at X (here 1 year) has the same impact on the Kaplan-Meier estimator $\widehat{S}(t)$ for $t < X$ whether or not the observation time corresponds to a failure or a censoring – in both cases, the subject is a member of the risk set $R(t)$. The pseudo-values before 1 year are (slightly) above 1. For $t \geq X$, however, the pseudo-values for the two subjects differ. The resulting curve for the failure is seen to be a 'caricature' of the indicator $I(1 \text{ year} > t)$ (however, with

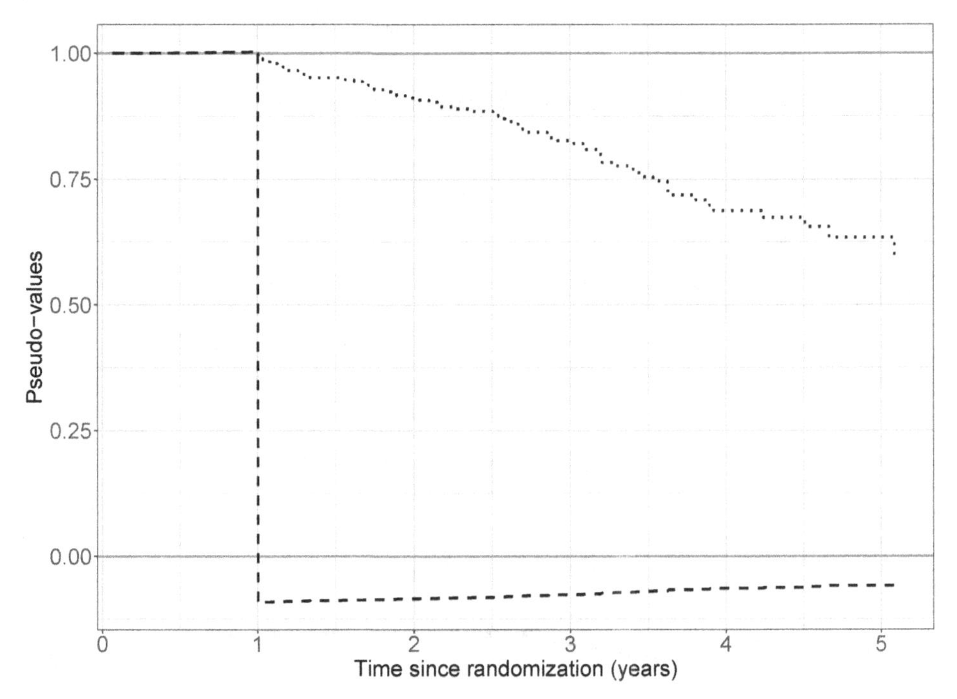

Figure 6.1 *PBC3 trial in liver cirrhosis: Pseudo-values for the survival indicator $I(T > t)$ as a function of follow-up time t for two subjects: A failure at $T = 1$ year (dashed) and a censoring at $C = 1$ year (dotted).*

negative values for $t > 1$ that increase towards 0), whereas, for the censored observation, the pseudo-values decrease (without reaching 0). Even though the pseudo-values for $I(T > t)$ go beyond the interval $[0, 1]$, they have approximately the correct conditional expectation given covariates, i.e.,

$$E(I(T_i > t) \mid Z_i) \approx E(\theta_i \mid Z_i)$$

if the censoring distribution is independent of covariates. This is why they can be used as responses in a GEE for $S(t \mid Z)$.

We next fix time and show how the pseudo-observations for all subjects in the data set look like at those time points. For illustration, we compute pseudo-values at times $(t_1, t_2, t_3) = (1, 2, 3)$ years. Figures 6.2a-6.2c show the results which are equivalent to what was seen in Figure 6.1. For an observed failure time at $T \leq t_j$ the pseudo-value is negative, for an observed censoring time at $C \leq t_j$ the pseudo-value is between 0 and 1, and for an observation time $X > t_j$ (both failures and censorings) the pseudo-value is slightly above 1.

We next show how pseudo-observations can be used when fitting models for $S(t_0 \mid Z)$ and look at $t_0 = 2$ years as an example. To assess models including only a single quantitative covariate like bilirubin or $\log_2$(bilirubin), *scatter-plots* may be used much like in simple linear regression models (Andersen and Skovgaard, 2010, ch. 4). Figure 6.3 (left panel) shows pseudo-values plotted against bilirubin. Note that adding a scatter-plot smoother to the plot is crucial – much like when plotting binary outcome data. The resulting curve

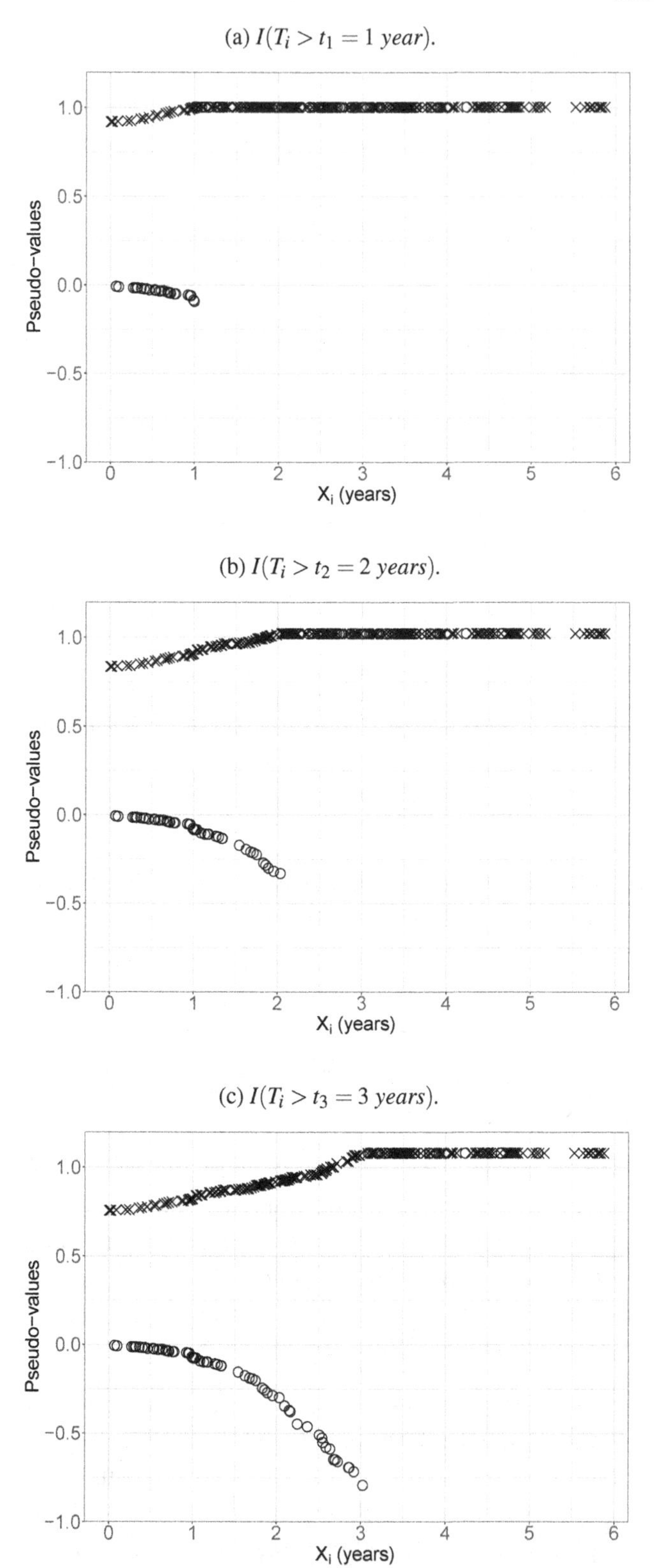

Figure 6.2 *PBC3 trial in liver cirrhosis: Pseudo-values for the survival indicator $I(T_i > t_l = l$ years), $l = 1, 2, 3$ for all subjects, i, plotted against the observation time X_i (failures: o, censorings: x).*

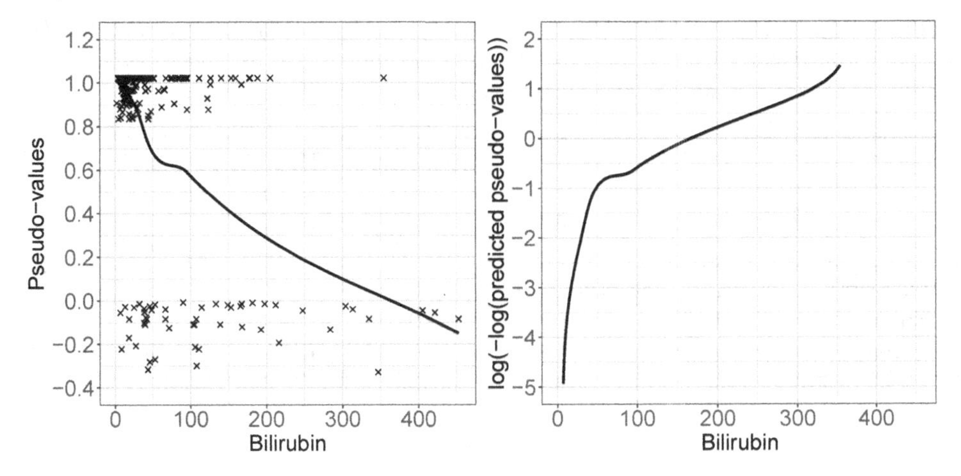

Figure 6.3 *PBC3 trial in liver cirrhosis: Pseudo-values for the survival indicator $I(T_i > 2$ years) for all subjects, i, plotted against the covariate Z_i = bilirubin with a scatter-plot smoother super-imposed (left); in the right panel, the smoother is transformed with the cloglog link function.*

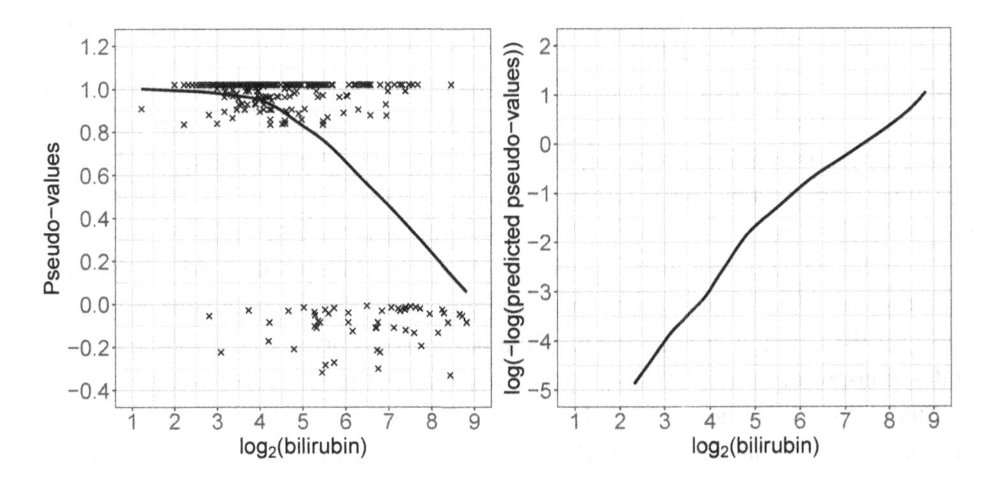

Figure 6.4 *PBC3 trial in liver cirrhosis: Pseudo-values for the survival indicator $I(T_i > 2$ years) for all subjects, i, plotted against the covariate $Z_i = \log_2$(bilirubin) with a scatter-plot smoother super-imposed (left); in the right panel, the smoother is transformed with the cloglog link function.*

should be linear on the scale of the link function

$$\text{cloglog}(1 - S(t_0)) = \log(-\log(S(t_0)))$$

and Figure 6.3 (right panel) shows the smoother after this transformation. It is seen that linearity does not describe the association well. Plotting, instead, against $\log_2$(bilirubin) (Figure 6.4) shows that using a linear model in this scale is not contra-indicated.

We can then fit a model for $S(t_0 \mid Z_1, Z_2, \log_2(Z_3))$ with Z_1, the indicator for CyA treatment, Z_2 = albumin, and Z_3 = bilirubin using the pseudo-values at $t_0 = 2$ years as the outcome variable and using the cloglog link. Table 6.1 (left panel) shows the results. Compared

Table 6.1 *PBC3 trial in liver cirrhosis: Estimated coefficients (and robust SD) from models for the survival function with linear effects of albumin and* $\log_2$*(bilirubin) based on pseudo-values. The cloglog link function was used, and the SD values are based on the sandwich formula.*

		A single time point $t_0 = 2$		Three time points $(t_1, t_2, t_3) = (1, 2, 3)$	
Covariate		$\widehat{\beta}$	SD	$\widehat{\beta}$	SD
Treatment	CyA vs. placebo	-0.705	0.369	-0.599	0.287
Albumin	per 1 g/L	-0.105	0.034	-0.094	0.026
$\log_2$(bilirubin)	per doubling	0.836	0.140	0.684	0.092

with the Cox model results in Table 2.7, it is seen that the estimated coefficients based on the pseudo-values are similar. The SD values are somewhat larger which should be no surprise since the Cox models use all data, whereas the pseudo-values concentrate on a single point in time. A potential advantage of using the pseudo-observations is that if interest does focus on a single time point, then they, in contrast to a Cox model, avoid making modeling assumptions about the behavior at other time points. In Table 6.1 (right panel), results are shown for a joint model for pseudo-values at times $(t_1, t_2, t_3) = (1, 2, 3)$ years: $\log(-\log S(t_j \mid Z)) = \beta_{0j} + \text{LP}$, $j = 1, 2, 3$. Now, the results are closer to those based on the Cox model in Table 2.7. In particular, values of SD are smaller than when based on pseudo-values at a single point in time and simulation studies (e.g., Andersen and Pohar Perme, 2010) have shown that the SD does tend to get smaller when based on more time points; however, more than $m \sim 5$ time points will typically not add much to the precision. The model for more time points is fitted by adding the time points at which pseudo-values are computed as a categorical covariate and the output then also includes estimates $(\widehat{\beta}_{01}, \widehat{\beta}_{02}, \widehat{\beta}_{03})$ of the Cox log(cumulative baseline hazard) at times (t_1, t_2, t_3). Note that, in such a model, non-proportional hazards (at the chosen time points) corresponds to interactions with this categorical time-variable. For the models based on pseudo-values, the SD values are obtained using *sandwich estimators* from the GEE. These have been shown to be slightly conservative, however, typically not seriously biased, see also Section 6.2.

Another type of plot which is applicable when assessing a regression model based on pseudo-observations is a *residual plot*. Figure 6.5 shows residuals from the model in Table 6.1 (right panel), i.e.,

$$r_{ij} = \theta_{ij} - \exp(-\exp(\widehat{\beta}_{0j} + \widehat{\text{LP}}_i))$$

plotted against $\log_2$(bilirubin) for subject i. Here, $j = 1, 2, 3$ refers to the three time points (t_1, t_2, t_3). A smoother has been superimposed for each j and it is seen that the residuals vary roughly randomly around 0 indicating a suitable fit of the model. Note that residual plots are applicable also for multiple regression models in contrast to the scatter-plots in Figure 6.3 (left panel). In Section 6.4 we will briefly discuss how formal significance tests for the goodness-of-fit of regression models based on pseudo-observations may be devised using *cumulative pseudo-residuals*.

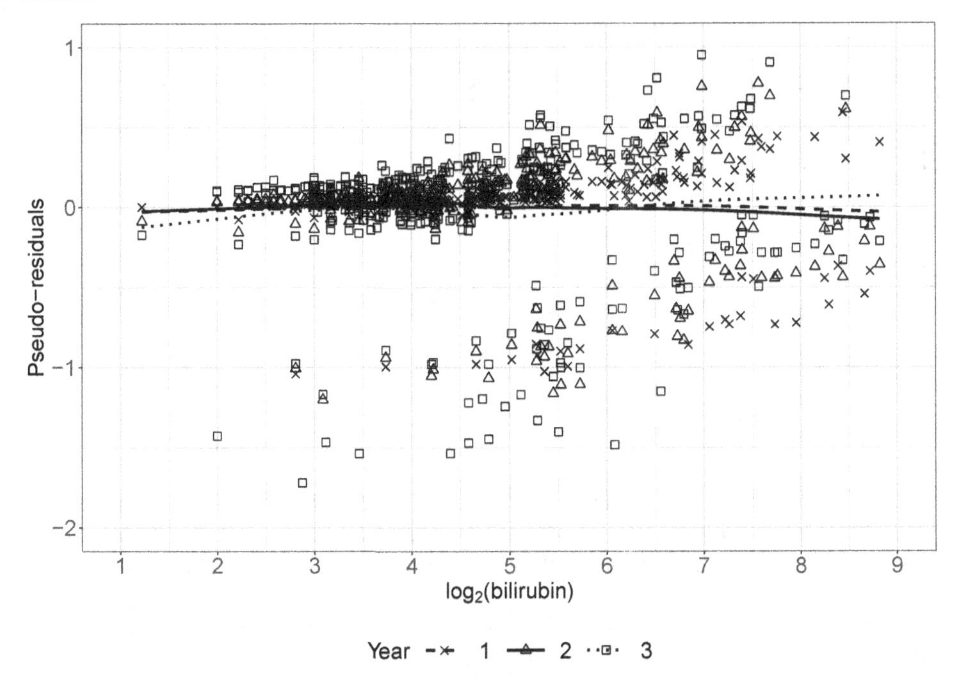

Figure 6.5 *PBC3 trial in liver cirrhosis: Pseudo-residuals for the survival indicator* $I(T_i > t_j)$ *for all subjects, i, and for* $(t_1, t_2, t_3) = (1, 2, 3)$ *years plotted against* $\log_2$*(bilirubin). The cloglog link function was used.*

> **Pseudo-values**
>
> Pseudo-observations are computed once and for all at the chosen time points. This takes care of censoring and provides us with a set of new observations that can be used as response variables in a GEE with the desired link function. This also enables application of various graphical techniques for data presentation and model assessment. In the example, this provided estimates comparable to those obtained with a Cox model. This, however, is only a 'poor man's Cox model' since fitting the full Cox model is both easier and more efficient. So, the main argument for using pseudo-observations is the generality of the approach: The same basic ideas apply for a number of marginal parameters in multi-state models and for several link functions. We will demonstrate these features in Sections 6.1.2-6.1.7.

6.1.2 *Hazard difference*

In Section 6.1.1, the idea of pseudo-values was introduced via the PBC3 trial in liver cirrhosis, and regression models for the survival function at one or a few time points were studied using the cloglog link function. The resulting regression coefficients were then comparable to log(hazard ratios) estimated using the Cox model. If, instead, one wishes to estimate parameters with a *hazard difference* interpretation, then this may be achieved by modeling the exact same pseudo-values but using instead a (minus) log link. Using bilirubin as an example, linearity may be investigated (in a univariate model) by plotting pseudo-values against

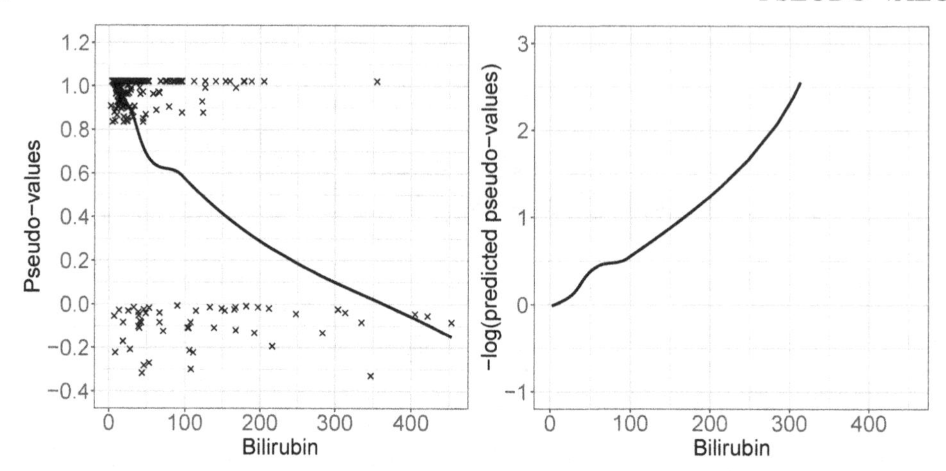

Figure 6.6 *PBC3 trial in liver cirrhosis: Pseudo-values for the survival indicator $I(T_i > 2$ years$)$ for all subjects, i, plotted against the covariate $Z_i =$ bilirubin with a scatter-plot smoother superimposed (left); in the right panel, the smoother is transformed with the (minus) log link function.*

Table 6.2 *PBC3 trial in liver cirrhosis: Estimated coefficients (and robust SD) from a model for the survival indicator $I(T_i > t_j)$ with linear effects of albumin and bilirubin based on pseudo-values at $(t_1,t_2,t_3) = (1,2,3)$ years. The (minus) log link function was used.*

Covariate		$\widehat{\beta}$	SD
Treatment	CyA vs. placebo	-0.048	0.031
Albumin	per 1 g/L	-0.0097	0.0032
Bilirubin	per 1 μmol/L	0.0042	0.0008

bilirubin. This is done in Figure 6.6 where a smoother has been superimposed (left), and in the right-hand panel this smoother is log transformed. There seems to be a problem with the fit for large values of bilirubin (where the smoother gets negative, thereby preventing a log transform). Table 6.2 shows the estimated coefficients in a model for $S(t_j \mid Z)$ for the three time points $(t_1,t_2,t_3) = (1,2,3)$ years, using the (minus) log link function and including the covariates treatment, albumin and bilirubin. The coefficients have hazard difference interpretations and may be compared to those seen in Table 2.10. The estimates based on pseudo-values are seen to be similar, however, with larger SD values.

A residual plot may be used to assess the model fit, and Figure 6.7 shows the pseudo-residuals from the model in Table 6.2 plotted against bilirubin. Judged from the smoothers, the fit is not quite as good as that using the cloglog link.

6.1.3 *Restricted mean*

Turning now to the restricted mean life time, $\varepsilon_0(\tau) = E(\min(T, \tau))$, pseudo-values are based on the integrated Kaplan-Meier estimator

$$\widehat{\varepsilon}_0(\tau) = \int_0^\tau \widehat{S}(t)dt,$$

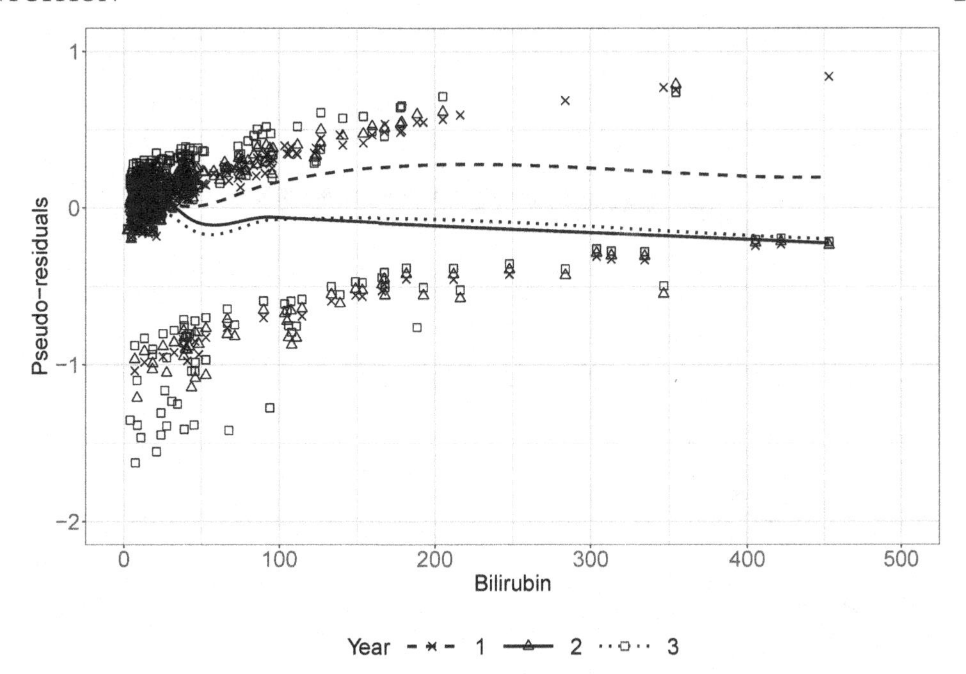

Figure 6.7 *PBC3 trial in liver cirrhosis: Pseudo-residuals for the survival indicator $I(T_i > t_j)$ for all subjects, i, and for $(t_1,t_2,t_3) = (1,2,3)$ years plotted against bilirubin. The model used the (minus) log link function (Table 6.2).*

Table 6.3 *PBC3 trial in liver cirrhosis: Estimated coefficients (and robust SD) from a linear model (identity link function) for the τ-restricted mean life time for $\tau = 3$ years based on pseudo-values.*

Covariate		$\widehat{\beta}$	SD
Intercept		2.83	0.35
Treatment	CyA vs placebo	0.148	0.073
Albumin	per 1 g/L	0.023	0.0068
$\log_2$(bilirubin)	per doubling	-0.243	0.032

that is,

$$\theta_i = n\int_0^\tau \widehat{S}(t)dt - (n-1)\int_0^\tau \widehat{S}^{-i}(t)dt.$$

We consider the PBC3 trial and the value τ =3 years and compare with results using the model by Tian et al. (2014). Figure 6.8 shows the scatter-plot where the pseudo-values θ_i are plotted against the observation times X_i. It is seen that all observations $X_i > \tau$ give rise to identical pseudo-values slightly above τ while observed failures before τ have pseudo-values close to the observed $X_i = T_i$ and censored observations before τ have values that increase with X_i in the direction of τ.

Table 6.3 shows the results from a linear model (i.e., identity as link function) for $\varepsilon_0(\tau \mid Z), \tau = 3$ years, based on pseudo-observations. The results are seen to coincide quite well with those obtained using the Tian et al. model (Table 4.4). Figure 6.9 shows scatter-plots of pseudo-values against $\log_2$(bilirubin) and seems to not contra-indicate a linear model.

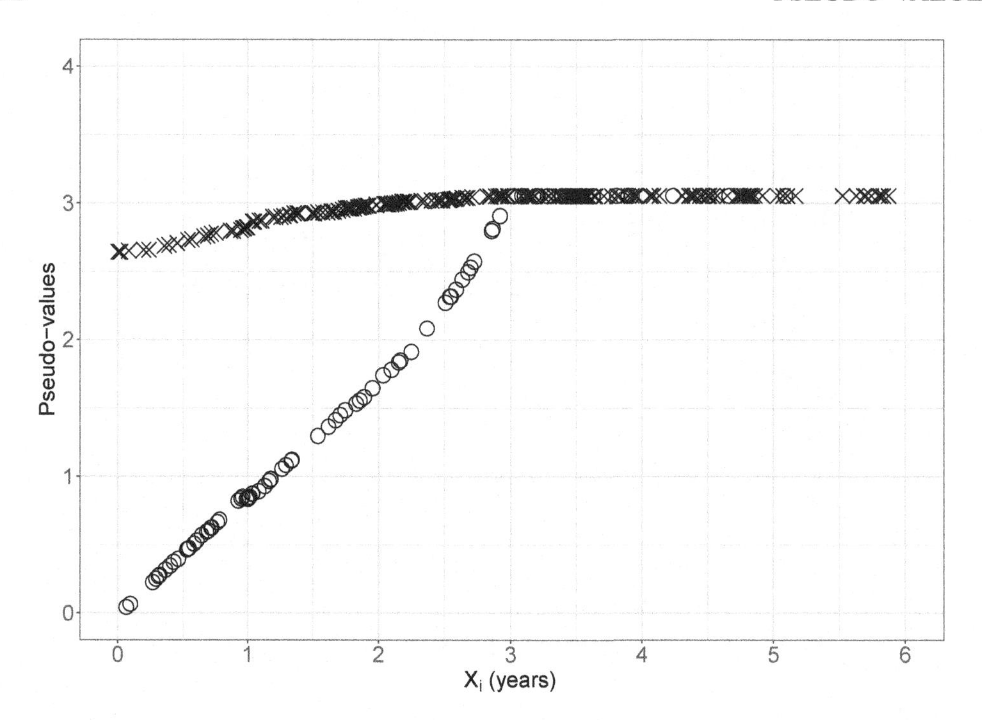

Figure 6.8 *PBC3 trial in liver cirrhosis: Pseudo-values for the restricted life time* $\min(T_i, \tau)$ *for all subjects, i, plotted against the observation time* X_i *for* $\tau = 3$ *years: Observed failures (o), censored observations (x).*

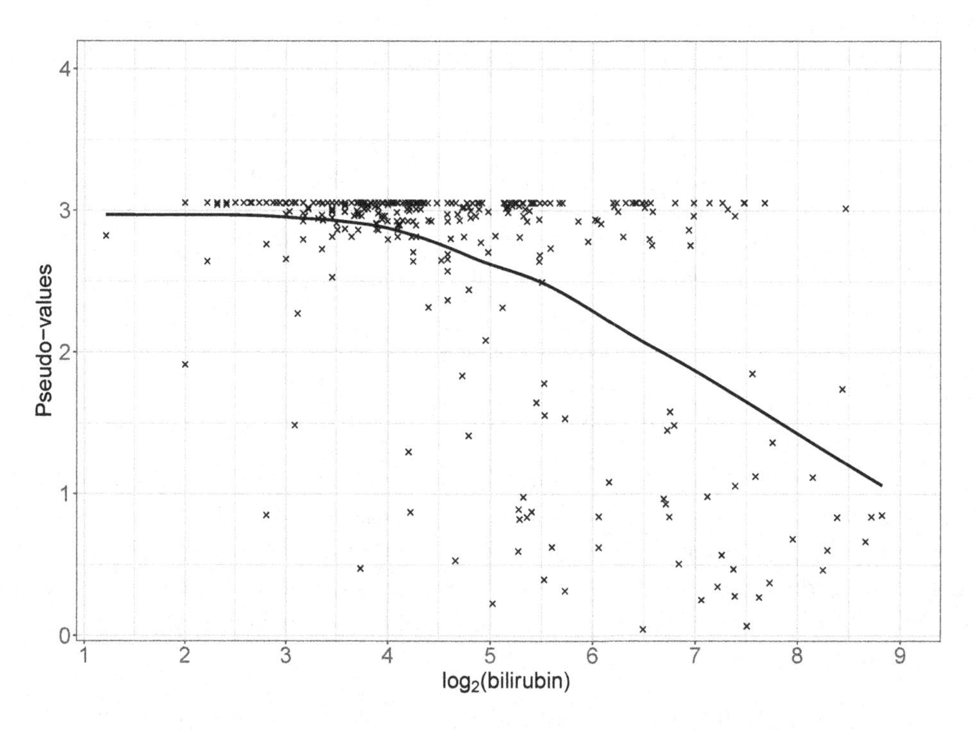

Figure 6.9 *PBC3 trial in liver cirrhosis: Pseudo-values for the restricted life time* $\min(T_i, \tau)$ *for all subjects, i, plotted against* $\log_2$*(bilirubin) for* $\tau = 3$ *years. A scatter-plot smoother has been superimposed.*

Table 6.4 *PBC3 trial in liver cirrhosis: Estimated coefficients (and robust SD) from models with logistic and cloglog link functions for the cumulative incidence of death without transplantation before $t_0 = 2$ years based on pseudo-values.*

(a) *logit link function*

Covariate		$\widehat{\beta}$	SD	$\widehat{\beta}$	SD
Treatment	CyA vs. placebo	0.112	0.370	-0.574	0.506
Albumin	per 1 g/L			-0.144	0.049
$\log_2$(bilirubin)	per doubling			0.713	0.188

(b) *cloglog link function*

Covariate		$\widehat{\beta}$	SD	$\widehat{\beta}$	SD
Treatment	CyA vs placebo	0.106	0.351	-0.519	0.425
Albumin	per 1 g/L			-0.114	0.037
$\log_2$(bilirubin)	per doubling			0.570	0.145

6.1.4 Cumulative incidence

We will now, for comparison with the direct binomial regression analyses in Section 5.8.2, present models for the 2-year cumulative incidence of death without transplantation in the PBC3 trial based on pseudo-observations. Table 6.4 shows the results. Comparing with Table 5.6 it is seen that the two approaches yield very similar results – both in terms of estimated log(odds ratios) and of their SD. The table also shows similar results using, instead of the logistic link function, a cloglog link function as in the Fine-Gray model. Though the coefficients are similar, they have a different interpretation, namely log(sub-distribution hazard ratios) rather than log(odds ratios). It is an advantage of the pseudo-value approach (and of direct binomial regression) that several link functions may be used. However, it should be kept in mind that the time point(s) at which pseudo-values are calculated must be selected. Choosing, as in the previous section, time points 1, 2, and 3 years the estimated coefficients (SD) using the cloglog link are, respectively, -0.511 (0.349) for treatment, -0.107 (0.032) for albumin, and 0.519 (0.117) for $\log_2$(bilirubin), i.e., similar coefficients but somewhat smaller SD. Other aspects to consider when choosing the link function are, as already discussed in Section 5.5.5, that predicted probabilities may exceed 1 (e.g., for identity and log links) or may be negative (e.g., for the identity link) and that, for the same mentioned link functions, models with a time-constant effect may be implausible.

For comparison with the Fine-Gray analyses presented in Table 4.5, we fitted pseudo-value based models including also the covariates sex and age. Both a model using three time points (1, 2, 3 years) and one using ten time points (0.5, 1.0,..., 4.5, 5.0 years) are shown in Table 6.5. Some efficiency gain is seen when using more time points and comparing with the Fine-Gray results, similar coefficients with somewhat increased SD are seen when using pseudo-values.

6.1.5 *Cause-specific time lost*

In Section 4.2.2, we showed models for the time lost before $\tau = 3$ years in the PBC3 trial due to transplantation or death without transplantation using estimating equations suggested by Conner and Trinquart (2021). We will repeat these analyses using pseudo-observations, see Table 6.6. Comparing with Table 4.6, it is seen that coefficients are quite similar with smaller SD when based on pseudo-values. As we did in Section 4.2.2, we may compare the coefficients from Tables 6.3 and 6.6 and notice that, for each explanatory variable, the coefficient from the former table equals minus the sum of the coefficients from the latter (e.g., for treatment we have $-(-0.063-0.085) = 0.148$).

6.1.6 *Non-Markov transition probabilities*

This example continues the example in Section 5.8.1 and reports some further results on data from the PROVA trial in liver cirrhosis (Example 1.1.4) presented by Andersen et al. (2022). We look at models for the probability $P_{01}(s,t)$ of being alive in the bleeding state 1 at time $t > 1$ year given alive without bleeding at time $s = 1$ year and base the analyses on pseudo-observations for the indicator $I(V_i(t) = 1)$. These are based on estimators $\widehat{P}_{01}(1,t)$ using landmarking or plug-in. As an example, we will study how this probability depends on whether sclerotherapy was given or not. Figure 6.10 shows the landmark Aalen-Johansen estimators for $P_{01}(1,t)$ in the two treatment groups and the two curves seem to be rather close. This tendency is also seen when computing pseudo-values at time $t = 2$ years (close to the median of observed transition times) based on different base estimators for the transition probability and fitting a model with a log link, including only an indicator for sclerotherapy. Table 6.7 shows the resulting estimates of the treatment effect at that time using as base estimators, respectively, the landmark Aalen-Johansen or Pepe estimators or different plug-in estimators. For the plug-in estimators, there is a choice between different data sets on which estimation can be based, and this could affect the efficiency. The model could be based on fitting the intensity models to the entire data set, the landmark data set, or to the data set consisting of all patients who were still at risk at time $s = 1$ year. All estimated coefficients provide a log(relative risk) close to 0. Using the 'at-risk' data set appears to be associated with the smallest SD.

Table 6.5 *PBC3 trial in liver cirrhosis: Estimated coefficients (and robust SD) from models for the cumulative incidence of death without transplantation. Models use the cloglog link function based on pseudo-values at either 3 or 10 time points.*

		3 time points		10 time points	
Covariate		$\widehat{\beta}$	SD	$\widehat{\beta}$	SD
Treatment	CyA vs. placebo	-0.272	0.337	-0.413	0.318
Albumin	per 1 g/L	-0.076	0.033	-0.038	0.032
$\log_2$(Bilirubin)	per doubling	0.666	0.121	0.669	0.118
Sex	male vs. female	-0.502	0.400	-0.855	0.388
Age	per year	0.073	0.022	0.096	0.022

Table 6.6 *PBC3 trial in liver cirrhosis: Estimated coefficients (and robust SD) from linear models for time lost (in years) due to transplantation or to death without transplantation before $\tau = 3$ years. Analyses are based on pseudo-values.*

(a) *Transplantation*

Covariate		$\widehat{\beta}$	SD	$\widehat{\beta}$	SD
Treatment	CyA vs. placebo	-0.056	0.051	-0.063	0.046
Albumin	per 1 g/L			-0.001	0.004
$\log_2$(Bilirubin)	per doubling			0.100	0.026

(b) *Death without transplantation*

Covariate		$\widehat{\beta}$	SD	$\widehat{\beta}$	SD
Treatment	CyA vs. placebo	-0.015	0.073	-0.085	0.069
Albumin	per 1 g/L			-0.022	0.007
$\log_2$(Bilirubin)	per doubling			0.143	0.032

6.1.7 *Recurrent events*

Furberg et al. (2023) studied bivariate pseudo-values in the context of recurrent events with competing risks, with examples from the LEADER trial (Example 1.1.6). That is, a joint model based on pseudo-values for both the mean of the number, $N(t)$, of recurrent myocardial infarctions and that of the survival indicator $I(T_D > t)$ was proposed. This enables joint inference for the treatment effects on both components. For the time points $(t_1, t_2, t_3) = (20, 30, 40)$ months, the following models were studied

$$\log(E(N(t_j) \mid Z)) = \log(\mu_0(t_j)) + \beta_R Z$$

and

$$\log(-\log(E(I(T_D > t_j) \mid Z))) = \log(A_0(t_j)) + \beta_S Z,$$

where Z is the indicator for treatment with liraglutide. Joint estimation of the treatment effects (β_R, β_S) was based on pseudo-values for $(N(t_j), I(T_D > t_j))$, $j = 1, 2, 3$, using the

Table 6.7 *PROVA trial in liver cirrhosis: Estimated coefficients (log(relative risk)), (with robust SD) of sclerotherapy (yes vs. no) on the probability $P_{01}(1,t)$ of being alive in the bleeding state at time $t = 2$ years given alive in the initial state at time $s = 1$ year based on pseudo-values using different base estimators.*

Base estimator	$\widehat{\beta}$	SD
Landmark Pepe	-0.151	0.925
Landmark Aalen-Johansen	-0.261	0.882
Plug-in linear, complete	-0.849	1.061
Plug-in duration scale, complete data	-0.636	0.832
Plug-in linear, at-risk data	-0.650	0.674
Plug-in linear, landmark data	-0.079	0.920

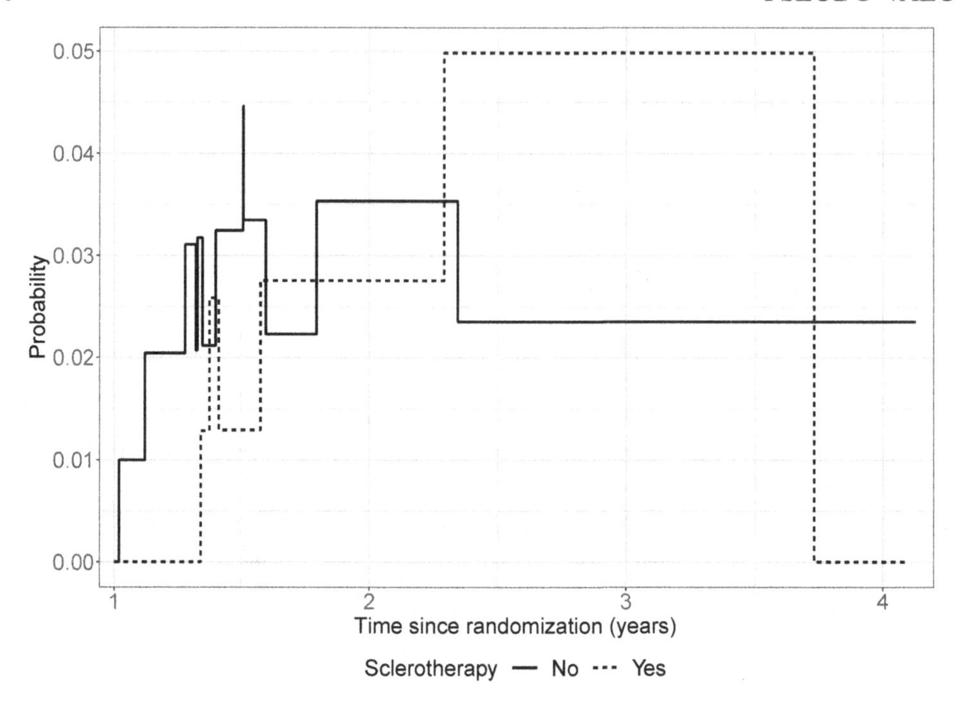

Figure 6.10 *PROVA trial in liver cirrhosis: Landmark Aalen-Johansen estimators for the probability $P_{01}(1,t)$ of being alive in the bleeding state at time $t > 1$ year among patients observed to be in the initial state at time $s = 1$ year. Separate curves are estimated for patients treated or not with sclerotherapy.*

sandwich estimator to estimate the SD and correlations of $(\widehat{\beta}_R, \widehat{\beta}_S)$. Figure 6.11 shows the non-parametric Cook-Lawless estimates for $E(N(t) \mid Z)$ and the Kaplan-Meier estimates for $S(t \mid Z)$, and Table 6.8 shows the results from the bivariate pseudo-value regression. The estimated regression coefficients are close to those based on separate Ghosh-Lin and Cox models quoted in Section 4.2.3, i.e., -0.159 (SD $= 0.088$) for the log(mean ratio) and -0.166 (SD $= 0.070$) for the log(hazard ratio). From the estimated joint distribution of $(\widehat{\beta}_R, \widehat{\beta}_S)$, i.e., a bivariate normal distribution with SD and correlation equal to the values from Table 6.8, it is possible to conduct a bivariate Wald test for the hypothesis $(\beta_R, \beta_S) = (0,0)$. The 2 DF Wald statistic takes the value 8.138 corresponding to $P = 0.017$.

Table 6.8 *LEADER cardiovascular trial in type 2 diabetes: Parameter estimates (with robust SD) for treatment (liraglutide vs. placebo) from a bivariate pseudo-value model with recurrent myocardial infarctions (R) and overall survival (S) at three time points at (20,30,40) months.*

$\widehat{\beta}_R$	$\widehat{\beta}_S$	SD($\widehat{\beta}_R$)	SD($\widehat{\beta}_S$)	Corr($\widehat{\beta}_R$, $\widehat{\beta}_S$)
-0.218	-0.163	0.097	0.082	0.100

6.1.8 *Covariate-dependent censoring*

In this section, we will briefly explain the needed modifications to the approach (for the survival indicator $I(T > t_0)$) when there is covariate-dependent censoring. It can be shown

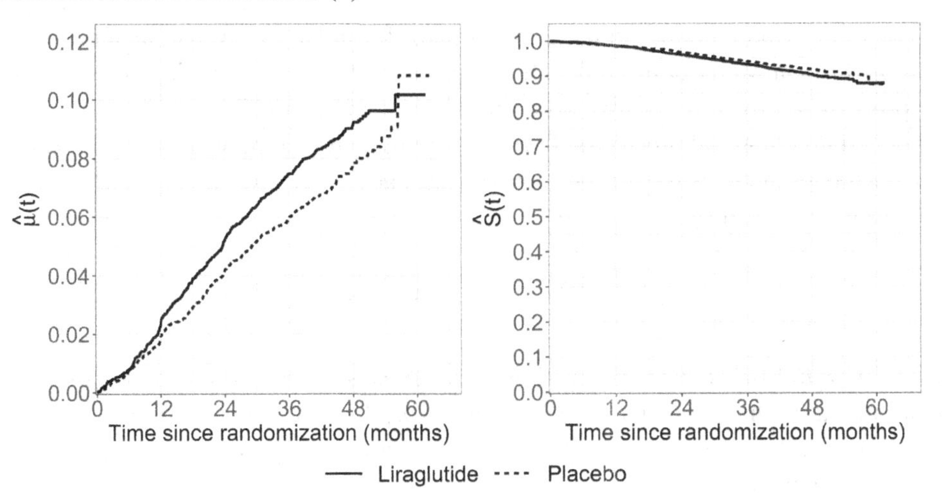

Figure 6.11 *LEADER cardiovascular trial in type 2 diabetes: Cook-Lawless estimates of the mean number of recurrent myocardial infarctions (left) and Kaplan-Meier estimates of the survival function (right), by treatment group.*

(see, e.g., Binder et al., 2014) that when survival data (X_i, D_i), $i = 1, \ldots, n$ are available then the Kaplan-Meier estimator may, alternatively, be written using IPCW

$$\widehat{S}(t) = 1 - \frac{1}{n} \sum_{i=1}^{n} \frac{N_i(t)}{\widehat{G}(X_i-)}, \tag{6.2}$$

where, as previously, the ith counting process is $N_i(t) = I(X_i \leq t, D_i = 1)$ and $\widehat{G}$ is the Kaplan-Meier estimator for the distribution of censoring times. When censoring times depend on covariates Z, this motivates another estimator for $S(t)$, namely

$$\widehat{S}_c(t) = 1 - \frac{1}{n} \sum_{i=1}^{n} \frac{N_i(t)}{\widehat{G}(X_i- \mid Z_i)}, \tag{6.3}$$

with $\widehat{G}(t \mid Z)$ now based on a regression model for the censoring distribution, e.g., a Cox model. For this situation, pseudo-values θ_i for the survival indicator can be based on $\widehat{S}_c(t)$

$$\theta_i = n \cdot \widehat{S}_c(t) - (n-1) \cdot \widehat{S}_c^{-i}(t)$$

and used as explained in Section 6.1.1. If the model for the censoring distribution is correctly specified, then the resulting estimators have the desired properties (Overgaard et al., 2019). Similar modifications may be applied to other estimators on which pseudo-values are based.

6.2 Theoretical properties (*)

Theoretical properties of methods based on pseudo-observations have been derived by, among others, Graw et al. (2009), Jacobsen and Martinussen (2016), and by Overgaard et al. (2017). In this section we will give a sketch of how these properties may be derived using the Aalen-Johansen estimator as an example.

Recall that the pseudo-values build on an estimator for the marginal parameter of interest, such as the Kaplan-Meier or the Aalen-Johansen estimator. The basic idea in the derivation of the properties is to consider this basic estimator as a functional (i.e., a function of functions) of certain empirical processes. In the following, we will indicate how this works for the Aalen-Johansen estimator for the cause-h cumulative incidence $F_h(t) = \int_0^t S(u)dA_h(u)$. The estimator is (Equation (5.4))

$$\widehat{F}_h(t) = \int_0^t \widehat{S}(u-)d\widehat{A}_h(u)$$

with $d\widehat{A}_h(u) = dN_h(u)/Y(u)$ being the jumps in the Nelson-Aalen estimator of the cause-h specific cumulative hazard (Equation (3.10)) and $\widehat{S}(u)$ the all-cause Kaplan-Meier estimator (Equation (4.3)). Now, by Equation (4.15), the fraction of subjects still at risk at time $t-$, i.e., $Y(t)/n$, can be re-written as

$$\frac{1}{n}Y(t) = \widehat{S}(t-)\widehat{G}(t-),$$

leading to the *IPCW version* of the Aalen-Johansen estimator

$$\widehat{F}_h(t) = \int_0^t \frac{1}{\widehat{G}(u-)}dN_h(u)/n,$$

compare Section 6.1.8. Here, as previously, $\widehat{G}$ is the Kaplan-Meier estimator for the censoring distribution. The empirical processes on which the estimator is based are

$$\widehat{H}_Y(t) = \frac{1}{n}\sum_i Y_i(t)$$

and

$$\widehat{H}_h(t) = \frac{1}{n}\sum_i N_{hi}(t), \, h = 0, 1, \ldots, k,$$

where N_{0i} is the counting process for censoring, $N_{0i}(t) = I(X_i \leq t, D_i = 0)$. With this notation, the cumulative censoring hazard is estimated by the Nelson-Aalen estimator

$$\widehat{A}_0(t) = \int_0^t \frac{1}{\widehat{H}_Y(u)}d\widehat{H}_0(u)$$

and the corresponding survival function G by the product-integral

$$\widehat{G}(t) = \prod_{[0,t]}\left(1 - d\widehat{A}_0(u)\right).$$

Since observations for $i = 1, \ldots, n$ are independent, it follows by the law of large numbers that $\widehat{\boldsymbol{H}} = (\widehat{H}_Y, \widehat{H}_0, \widehat{H}_1, \ldots, \widehat{H}_k)$ converges to a certain limit $\boldsymbol{\eta} = (\eta_Y, \eta_0, \eta_1, \ldots, \eta_k)$. The Aalen-Johansen estimator $\widehat{F}_h$ is a certain functional, say ϕ, of $\widehat{\boldsymbol{H}}$ and the true value $F_h(t) = \theta$ is the same functional applied to $\boldsymbol{\eta}$.

A smooth functional such as ϕ may be Taylor (*von Mises*) expanded

$$\begin{aligned}\phi(\widehat{\boldsymbol{H}}) &\approx \phi(\boldsymbol{\eta})+\frac{1}{n}\sum_i\dot{\phi}(X_i^*)\\ &= \theta+\frac{1}{n}\sum_i\dot{\phi}(X_i^*)\end{aligned}$$

where $X_i^*=(X_i,D_i)$ is the data point for subject i and $\dot{\phi}$ is the *first order influence function* for $\phi(\cdot)$. This is defined by

$$\begin{aligned}\dot{\phi}(x) &= \frac{d}{du}\phi\left((1-u)\boldsymbol{\eta}+u\delta_x\right)_{|u=0}\\ &= \phi'_{\boldsymbol{\eta}}(\delta_x-\boldsymbol{\eta}),\end{aligned}$$

i.e., the derivative of ϕ at $\boldsymbol{\eta}$ in the direction $\delta_x-\boldsymbol{\eta}$ where δ_x is Dirac's delta, $\delta_x(y)=I(y=x)$ (e.g., Overgaard et al., 2017).

We can now approximate the pseudo-observation for $I(T_i\leq t,D_i=h)$

$$\begin{aligned}\theta_i &= n\widehat{F}_h(t)-(n-1)\widehat{F}_h^{-i}(t)\\ &= n\phi(\widehat{\boldsymbol{H}})-(n-1)\phi(\widehat{\boldsymbol{H}}^{-i})\\ &\approx n\Big(\phi(\boldsymbol{\eta})+\frac{1}{n}\sum_i\dot{\phi}(X_i^*)\Big)-(n-1)\Big(\phi(\boldsymbol{\eta})+\frac{1}{n-1}\sum_{\ell\neq i}\dot{\phi}(X_\ell^*)\Big),\end{aligned}$$

i.e.,

$$\theta_i\approx\theta+\dot{\phi}(X_i^*). \tag{6.4}$$

We assume a model for the cumulative incidence of the form

$$g(E(I(T\leq t,D=h)\mid \mathbf{Z}))=\boldsymbol{\beta}^{\mathsf{T}}\mathbf{Z},$$

i.e., with link function g and where $\mathbf{Z}$ contains the constant '1' and $\boldsymbol{\beta}$ the corresponding intercept, and estimates of $\boldsymbol{\beta}$ are obtained by solving the GEE

$$\boldsymbol{U}(\boldsymbol{\beta})=\sum_i\boldsymbol{A}(\boldsymbol{\beta},\boldsymbol{Z}_i)(\theta_i-g^{-1}(\boldsymbol{\beta}^{\mathsf{T}}\boldsymbol{Z}_i))=\mathbf{0}.$$

Now by (6.4), these GEEs are seen to be (approximately) unbiased if

$$E(\dot{\phi}(X_i^*)\mid \mathbf{Z}_i)=g^{-1}(\boldsymbol{\beta}^{\mathsf{T}}\mathbf{Z}_i)-\theta \tag{6.5}$$

and this must be verified on a case-by-case basis by explicit calculation of the influence function. This has been done by Graw et al. (2009) for the cumulative incidence and more generally by Overgaard et al. (2017) under the assumption that censoring is independent of covariates. For the cumulative incidence the influence function is

$$\dot{\phi}(X_i^*)=\int_0^t\frac{dN_{hi}(u)}{G(u-)}-F_h(t)+\int_0^t\frac{F_h(t)-F_h(u)}{S(u)G(u)}dM_{0i}(u), \tag{6.6}$$

where N_{hi} counts h-events for subject i and $M_{0i}(t) = N_{0i}(t) - \int_0^t Y_i(u)dA_0(u)$ is the martingale for the process N_{0i} counting censorings for subject i. From this expression, (6.5) may be shown using the martingale property of $M_{0i}(t)$.

By the first order von Mises expansion, unbiasedness of the GEE was established and, had the pseudo-values $\theta_1, \ldots, \theta_n$ been independent, the standard sandwich variance estimator would apply for $\widehat{\boldsymbol{\beta}}$. A second order von Mises expansion gives the approximation

$$\theta_i \approx \theta + \dot{\phi}(X_i^*) + \frac{1}{n-1}\sum_{j \neq i} \ddot{\phi}(X_i^*, X_j^*) \tag{6.7}$$

where $\ddot{\phi}$ is the *second-order influence function*. This may be shown to have expectation zero (Overgaard et al., 2017); however, the presence of the second order terms shows that $\theta_1, \ldots, \theta_n$ are *not* independent meaning that the GEEs are not a sum of independent terms even when inserting the true value $\boldsymbol{\beta}_0$, and, therefore, the sandwich estimator needs to be modified to properly describe the variability of $\widehat{\boldsymbol{\beta}}$. The details were presented by Jacobsen and Martinussen (2016) for the Kaplan-Meier estimator and more generally by Overgaard et al. (2017). However, use of the standard sandwich variance estimator based on the GEE for pseudo-values from the Aalen-Johansen estimator turns out to be only slightly *conservative* because the extra term in the correct variance estimator arising from the second order terms in the expansion is negative and tends to be numerically small.

6.3 Approximation of pseudo-values (*)

In this section, we will first discuss how a less computer-intensive approximation to pseudo-values may be computed and, next, show how well this approximation works in the PBC3 trial. Computation of the pseudo-values may be time-consuming because the estimator $\widehat{\theta}$ needs to be re-calculated n times. Equation (6.4) suggests a plug-in approximation to the pseudo-value for subject i, namely,

$$\widehat{\theta}_i = \widehat{\theta} + \widehat{\dot{\phi}}(X_i^*), \tag{6.8}$$

where, in the latter term, estimates for the quantities appearing in the expression for the influence function are inserted (Parner et al., 2023; Bouaziz, 2023). To illustrate this idea, we study the Aalen-Johansen estimator $\widehat{F}_h(t)$ for which the influence function is given by (6.6). The corresponding approximation in (6.8) is then

$$\widehat{\theta}_i = \int_0^t \frac{dN_{hi}(u)}{\widehat{G}(u-)} + \int_0^t \frac{\widehat{F}_h(t) - \widehat{F}_h(u)}{Y(u)/n} d\widehat{M}_{0i}(u),$$

inserting estimates for F_h, G and the cumulative censoring hazard A_0. An advantage is that these estimates can be calculated once based on the full sample and, next, the estimated influence function is evaluated at each observation X_i^*. Thus,

$$\begin{aligned} \widehat{\theta}_i &= \frac{N_{hi}(t)}{\widehat{G}((X_i \wedge t)-)} + \frac{N_{0i}(t)(\widehat{F}_h(t) - \widehat{F}_h(X_i \wedge t))}{Y(X_i \wedge t)/n} \\ &- \int_0^{X_i \wedge t} \frac{(\widehat{F}_h(t) - \widehat{F}_h(u))dN_0(u)}{Y(u)^2/n} \end{aligned} \tag{6.9}$$

where the first term contributes if i has a cause-h event before time t, the second if i is censored before time t, and the last term contributes in all cases. Note that the first term corresponds to the outcome variable in direct binomial regression, Section 5.5.5.

The empirical influence function may be obtained in the following, alternative way, known as the *infinitesimal jackknife* (IJ) (Jaeckel, 1972; Efron, 1982, ch. 6). Using results from Lu and Tsiatis (2008), the influence function (6.6) may be re-written as

$$\phi(X_i^*) = \int_0^t \frac{1}{G(u)} dM_{hi}(u) - \int_0^t S(u) \int_0^u \frac{1}{S(s)G(s)} dM_i(s) dA_h(u), \tag{6.10}$$

where, for subject i, M_{hi} is the cause-h event martingale and $M_i = \sum_h M_{hi}$ the martingale for all events. The estimator $\widehat{F}_h(t) = \int_0^t \widehat{S}(u-) d\widehat{A}_h(u)$ is written as a function of *weights* where the actual estimator is obtained when all weights equal $1/n$

$$\widehat{F}_h^w(t) = \int_0^t \exp\left(-\sum_\ell \widehat{A}_\ell^w(u)\right) d\widehat{A}_h^w(u), \tag{6.11}$$

with $\widehat{A}_\ell^w(t) = \frac{\sum_i w_i dN_{\ell i}(u)}{\sum_i w_i Y_i(s)}$. The IJ influence function is obtained as

$$\dot{\hat{\phi}}(X_i^*) = \frac{\partial \widehat{F}_h^w(t)}{\partial w_i}|_{\mathbf{w}=1/n}$$

and the approximate pseudo-value is

$$\widehat{\theta}_i = \widehat{\theta} + \dot{\hat{\phi}}(X_i^*),$$

see Exercise 6.2.

PBC3 trial in liver cirrhosis

As an illustration, we will study the closeness of the IJ-approximation (for the survival function, see Exercise 6.1) using the PBC3 trial (Example 1.1.1) as example. Pseudo-values for the indicator $I(T_i > t_0)$ of no failure of medical treatment before time $t_0 = 2$ years were computed using the direct formula $n \cdot \widehat{S}(t_0) - (n-1) \cdot \widehat{S}^{-i}(t_0)$ and compared with the IJ approximation, see Figure 6.12. The approximation seems to work remarkably well (notice the vertical scale on the right-hand plot).

6.4 Goodness-of-fit (*)

If the response variable T_i in Section 5.7.2 is a pseudo-value θ_i, then the Taylor expansion of the GEE becomes the (second order) von Mises expansion discussed in Section 6.2 in order to take into account that the pseudo-values θ_i, $i = 1, \ldots, n$ are not independent. Therefore, some extra arguments are required when using this idea for pseudo-residuals (Pavlič et al., 2019). However, since the second order terms tend to be numerically small, the first order expansion may provide a satisfactory approximation.

Pavlič et al. (2019) used data from the PBC3 trial to show how plots of cumulative pseudo-residuals were applicable when assessing goodness-of-fit of regression models based on

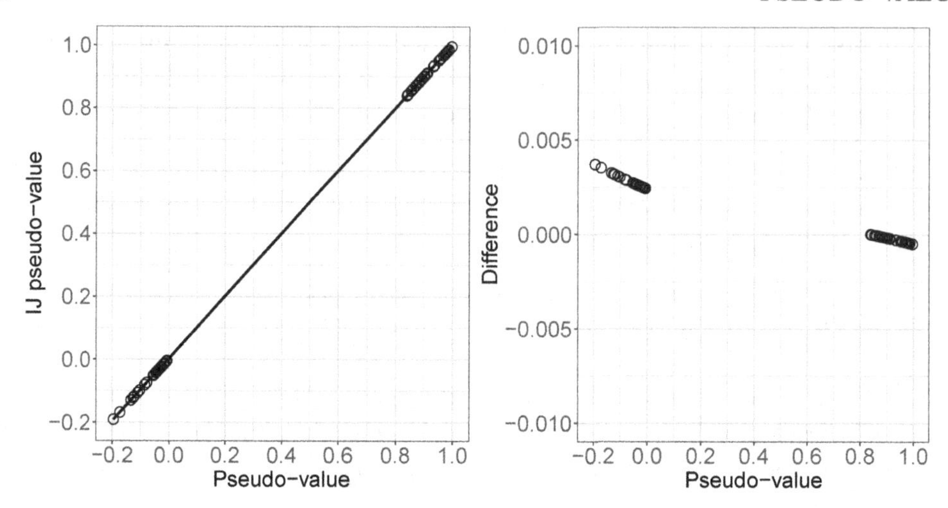

Figure 6.12 *PBC3 trial in liver cirrhosis: Infinitesimal jackknife (IJ) pseudo-values for the survival indicator $I(T_i > 2$ years) for all subjects, i, (left) and difference between IJ pseudo-values and ordinary pseudo-values (6.1) (right) plotted against the ordinary pseudo-values. An identity line has been added to the left-hand plot.*

pseudo-observations. Models for the probability of no medical failure before chosen time points were fitted to pseudo-values using the cloglog link and including only the covariate bilirubin or only log(bilirubin). The conclusion was that linearity is rejected without transforming the covariate, whereas no deviations from linearity were detected after a log-transformation.

6.5 Exercises

Exercise 6.1 (*) The influence function for the survival function $S(t)$ is

$$\dot{\phi}(X_i^*) = -S(t)\int_0^t \frac{dM_i(u)}{S(u)G(u)}$$

with $M_i(u) = N_i(u) - \int_0^u Y_i(s)dA(s)$ being the martingale for the failure counting process for subject i (Overgaard et al. 2017). The corresponding 'plug-in' approximation is then

$$\hat{\dot{\phi}}(X_i^*) = -\widehat{S}(t)\left(\frac{N_i(t)}{Y(t\wedge X_i)/n} - \int_0^{t\wedge X_i}\frac{dN(u)}{Y(u)^2/n}\right).$$

1. Show that, writing the estimator in the 'exp(−Nelson-Aalen)' form $\widehat{S}^w(t) = \exp(-\int_0^t(\sum_i w_i dN_i(u)/\sum_i w_i Y_i(u))$, this expression is obtained as

$$\hat{\dot{\phi}}(X_i^*) = \frac{\partial \widehat{S}^w(t)}{\partial w_i}|_{w=1/n}.$$

2. Show that for the standard Kaplan-Meier estimator

$$\widehat{S}^w(t) = \prod_{[0,t]}(1-\sum_i w_i dN_i(u)/\sum_i w_i Y(u))$$

it holds that

$$\frac{\partial \widehat{S}^w(t)}{\partial w_i}|_{w=1/n} = -\widehat{S}(t)\left(\frac{N_i(t)}{Y(t\wedge X_i) - dN(t\wedge X_i)} - \int_0^{t\wedge X_i}\frac{dN(u)}{Y(u)(Y(u)-dN(u))}\right).$$

3. Show that, in the case of no censoring, the influence function reduces to

$$\dot{\phi}(X_i^*) = I(T_i > t) - S(t).$$

Exercise 6.2 (*)

1. Show that, for the Aalen-Johansen estimator, the calculation

$$\hat{\dot{\phi}}(X_i^*) = \frac{\partial \widehat{F}_h^w(t)}{\partial w_i}|_{w=1/n},$$

with $\widehat{F}_h^w(t)$ given by (6.11), leads to the pseudo-value approximation obtained by plugging-in estimators into (6.10).

2. Show that, in the case of no censoring, the influence function reduces to

$$\dot{\phi}(X_i^*) = I(T_i \le t, D = h) - F_h(t).$$

Exercise 6.3 Consider the Copenhagen Holter study and the composite end-point stroke-free survival.

1. Fit, using pseudo-values, a cloglog model for experiencing that end-point before time 3 years including the covariates ESVEA, sex, age, and systolic blood pressure.
2. Compare the results with those of Exercise 2.4.

Exercise 6.4 Consider the Copenhagen Holter study and the composite end-point stroke-free survival.

1. Fit, using pseudo-values a linear model for the 3-year restricted mean time to the composite event including the covariates ESVEA, sex, age, and systolic blood pressure.
2. Compare with the results of Exercise 4.3.

Exercise 6.5 Consider the competing outcomes stroke and death without stroke in the Copenhagen Holter study.

1. Fit, using pseudo-values, a cloglog-model for the cumulative incidences at 3 years including ESVEA, sex, age, and systolic blood pressure.
2. Compare with the results of Exercises 4.4 and 5.8.

Chapter 7

Further topics

In previous chapters, we have discussed a number of methods for analyzing statistical models for multi-state survival data based on *rates* or on marginal parameters, such as *risks* of being in certain states at certain time-points – the latter type of models sometimes based on *pseudo-values*. In this final chapter, we will introduce a number of possible extensions to these methods. For these further topics, entire books and review papers have been written, e.g., Sun (2006) and van den Hout (2020) on interval-censored data (see also Cook and Lawless, 2018, ch. 5), Hougaard (2000) and Prentice and Zhao (2020) on non-independent data, Hernán and Robins (2020) on causal inference, Rizopoulos (2012) on joint models, and Borgan and Samuelsen (2014) on cohort sampling. This means that our exposition will be brief and we provide references for further reading.

7.1 Interval-censoring

So far, we have assumed that the multi-state process $V_i(t)$ was observed *continuously*, i.e., exact transition times were observed up till the time $X_i = T_i \wedge C_i$ – the minimum of the time of reaching an absorbing state and the time of right-censoring. Such an observation scheme is not always possible. Sometimes, $V_i(t)$ is only observed *intermittently*, that is, only the values $V_i(J_0^i), V_i(J_1^i), \ldots, V_i(J_{N_i}^i)$ at a number $(N_i + 1)$ of *inspection times* $\boldsymbol{J}^i = (J_0^i, J_1^i, \ldots, J_{N_i}^i)$ are ascertained. Typically the first time, J_0^i equals 0 for all subjects, i but, in general, the inspection times may vary among subjects. The resulting observations of $V_i(t)$ are said to be *interval-censored*. The data arising when $\boldsymbol{J}^i$ is the same for all i are known as *panel data* (e.g., Kalbfleisch and Lawless, 1985). There may also be situations where the very concept of an exact transition time is not meaningful, e.g., the time of onset of a slowly developing disease such as dementia. In such a case, typically only a last time seen without the disease and a first time seen with the disease are available for any subject who develops the disease, once more giving rise to interval-censoring.

An assumption that will be made throughout, similar to that of independent censoring (Section 1.3), is that the inspection process $\boldsymbol{J}^i$ is independent of $V_i(t)$ (e.g., Sun, 2006, ch. 1; see also Cook and Lawless, 2018, ch. 7).

In this section, we will give a brief account of some techniques that have been developed for analysis of interval-censored multi-state survival data.

7.1.1 *Markov processes (*)*

Intermittent observation of the process $V(t)$ gives rise to a likelihood contribution from subject i that is a product of factors

$$P(V_i(J^i_\ell) = s_\ell \mid V_i(J^i_{\ell-1}) = s_{\ell-1}), \ell = 1, \dots, N_i,$$

each corresponding to the probability of moving from the state $s_{\ell-1}$ occupied at time $J^i_{\ell-1}$ to the state s_ℓ occupied at the next inspection time, J^i_ℓ. The resulting likelihood is tractable if $V(t)$ is a *Markov process* and transition hazards are assumed to be piece-wise constant because then the transition probabilities are explicit functions of the transition hazards, see Equation (5.5). Piece-wise constant hazard models for general Markov multi-state processes were discussed by Jackson (2011) (see also van den Hout, 2017, ch. 4) and may also be used in the special models to be discussed in the next sections.

7.1.2 *Two-state model (*)*

For the two-state model (Figure 1.1), at most one transition can be observed for any given subject and interval-censored observations then reduce to an interval $(J_L, J_R]$ where that transition took place, i.e., $T \in (J_L, J_R]$. General interval-censoring is $0 < J_L < J_R < \infty$ with $V(J_L) = 0, V(J_R) = 1$, while the special case $J_L = 0, J_R < \infty, V(J_R) = 1$ is *left-censoring*. The situation with $J_L = 0$ or $J_R = \infty$ with either $V(J_R)$ or $V(J_L) \in \{0, 1\}$ is known as *current status data* in which case there is a single inspection time where it is ascertained whether or not the event has already happened. The special cases $J_L > 0, J_R = \infty$ with $V(J_L) = 0$ and $J_L = J_R = T$ are, respectively, *right-censoring* (at $C = J_L$) and exact observation of T.

A simple, but not recommendable approach (e.g., Sun, 2006, ch. 2) is *mid-point imputation* where, for $J_R < \infty$, the 'exact' survival time is set to $T = (J_R + J_L)/2$ and, for $J_R = \infty$, observation of T is considered right-censored at $C = J_L$. With this approach, analysis proceeds as if data were exactly observed, except from right-censoring.

Observation of the interval $(J_L, J_R]$, with $T \in (J_L, J_R]$, gives rise to the following special case of the likelihood discussed in Section 7.1.1

$$\prod_i (S(J^i_L) - S(J^i_R)),$$

where S is the survival function $S(t) = P(T > t)$. Analysis of parametric models, including the piece-wise exponential model, based on this likelihood is simple and asymptotic properties follow from standard likelihood theory. Non-parametric maximization leads to the Turnbull (1976) estimator for which the large-sample distribution is more complex (e.g., Sun, 2006, ch. 2 and 3). Pseudo-values based on a parametric model were discussed by Bouaziz (2023).

Regression analysis of interval-censored survival data via transformation models based on this likelihood were studied by Zeng et al. (2016). This class of models includes the Cox model, previously studied by Finkelstein (1986). Regression analysis of *current status data* using an additive hazards model was discussed by Lin et al. (1998). For the two-state model, *panel data* give rise to *grouped survival data.* In this situation, non-parametric estimation of the survival function reduces to the classical *life-table* (e.g., Preston et al., 2000). The Cox model for grouped survival data was studied by Prentice and Gloeckler (1978).

7.1.3 *Competing risks (*)*

Also for the competing risks model (Figure 1.2), interval-censored data reduce to an interval $(J_L, J_R]$ with $T \in (J_L, J_R]$. Following Section 7.1.1, we will assume that, when $J_R < \infty$, we also observe the state $V(J_R)$, i.e., the cause of death. When $J_R = \infty$, observation of T is right-censored at J_L and the cause of death is unknown.

Similar to the two-state model, Section 7.1.2, mid-point imputation is a possibility but it is generally not recommended. The likelihood contribution from subject i based on the interval-censored competing risks data is

$$\left(\prod_h (F_h(J_R^i) - F_h(J_L^i))^{I(D_i=h)}\right) S(J_L^i)^{I(D_i=0)}$$

where D_i denotes the cause of death $h = 1, \dots, k$ with $D_i = 0$ if observation of T_i is right-censored, F_h is the cause-h cumulative incidence, and $S = 1 - \sum_h F_h$ the overall survival function. Following Section 7.1.1, a model with piece-wise constant cause-specific hazards is tractable via the resulting likelihood. Non-parametric estimation of the cumulative incidences $F_h(t)$ was discussed by Hudgens et al. (2004), generalizing the Turnbull (1976) estimator, while Frydman and Liu (2013) focused on non-parametric estimation of the cumulative cause-specific hazards. Competing risks *panel data* give rise to *multiple decrement life-tables* (e.g., Preston et al., 2000).

Regression modeling of the cumulative incidences using transformation models (including the Fine-Gray model) were studied by Mao et al. (2017).

7.1.4 *Progressive illness-death model (*)*

For the illness-death model without recovery (Figure 1.3), data are more complex. It is usually assumed that time of death, T is either exactly observed or right-censored at C. In addition, there may be earlier inspection times, $J_L < J_R \leq X = T \wedge C$ where the subject is last observed to be healthy, $V(J_L) = 0$, respectively, first observed to be diseased, $V(J_R) = 1$. When the subject is censored at C or observed to die at T and no prior time J_R with $V(J_R) = 1$ is observed (but there may be a time $J_L \geq 0$ last seen healthy), the disease status at C or T is typically not known. If a subject is observed with the disease at a time point J_R then, as mentioned in Sections 7.1.2-7.1.3, mid-point imputation is an option. The general likelihood contributions for these four situations are shown in Table 7.1 and we refer to Section 5.1.3 for equations giving the transition probabilities as functions of the transition intensities.

Table 7.1 *Likelihood contributions for four types of interval-censored observation of an irreversible illness-death model.*

Observation ($V(J_L) = 0$)	Likelihood contribution
$V(J_R) = 1, V(C) = 1$	$P_{00}(0, J_L) P_{01}(J_L, J_R) P_{11}(J_R, C)$
$V(J_R) = 1, V(T) = 2$	$P_{00}(0, J_L) P_{01}(J_L, J_R) P_{11}(J_R, T-) \alpha_{12}(T)$
$V(C) \in \{0, 1\}$	$P_{00}(0, J_L)(P_{00}(J_L, C) + P_{01}(J_L, C))$
$V(T) = 2$	$P_{00}(0, J_L)(P_{00}(J_L, T-) \alpha_{02}(T) + P_{01}(J_L, T-) \alpha_{12}(t))$

If the disease status ($h = 0, 1$) at the time of failure or censoring, $X = T \wedge C$, is known, then the likelihood contribution is instead $P_{00}(0, J_L) \sum_{h=0,1} I(V(X-) = h) P_{0h}(J_L, X-) \alpha_{h2}(X)^D$ where $D = I(T < C)$ is the usual failure indicator.

A model with piece-wise constant transition hazards is tractable (e.g., Lindsey and Ryan, 1993). Frydman (1995) studied non-parametric maximum likelihood estimation of the transition hazards for the case where disease status at time of death is always known and Frydman and Szarek (2009) extended the discussion to the situation where this is not the case – both papers generalizing the Turnbull estimator. Joly et al. (2002) studied the situation with possible unknown disease status at death and estimated the transition hazards using penalized likelihood.

Pseudo-values for state occupation indicators, in particular for being diseased, $I(V(t) = 1)$, were discussed by Sabathé et al. (2020) based on the penalized likelihood estimator and by Johansen et al. (2020) based on parametric models, including the piece-wise constant hazards model.

7.2 Models for dependent data

In Sections 3.9, 4.3, and 5.6, the situation with dependent failure times was studied, and two different scenarios were discussed. These are analysis of *clustered data* and inference for *times of entry into different states* in the multi-state model, a special case of the latter being times to recurrence no. $h = 1, \dots, K$ in a model for recurrent events. Either, a *frailty* was used to explicitly model the dependence, or the dependence was treated as a nuisance parameter, and focus was on inference for the *marginal distributions*, taking a potential dependence into account by using robust SD estimation. Reviews of these situations have been given by Lin (1994) with most emphasis on marginal hazard models, by Wei and Glidden (1997), and by Glidden and Vittinghoff (2004), the latter with much emphasis on the shared frailty model. We concluded that, for clustered data, both modeling approaches were useful, the main difference being that frailty models provide regression coefficients with a *within-cluster* interpretation, whereas marginal hazards make comparisons *among clusters*. At the same time, the frailty SD also gave a measure of the within-cluster association. We will return to a discussion of clustered data in Section 7.2.2 and discuss how a hybrid approach, *two-stage estimation* may sometimes provide the best of two worlds by, at the same time, providing regression parameters with a marginal interpretation and a measure of within-cluster association. We first return to the second situation concerning times of entry into different states.

7.2.1 *Times of entry into states*

For the situation concerning times of entry into different states, we concluded previously that this is not well treated by either of the two methods, a major problem being that these times are often *improper random variables* for which a marginal hazard (see Equation (5.36)) is not a straightforward quantity. As discussed in Section 5.6, it either refers to a hypothetical world where other events do not occur, or it should be taken to be a subdistribution hazard with its associated difficulties in interpretation (see Section 4.2.2). We also refer to Sections 4.4.3 and 4.4.4 for a discussion of difficulties in connection with the related concept of latent failure times for competing risks (and semi-competing risks). *Recurrent events* provide a special case where, as discussed in Section 3.9.3, a frailty model is useful if within-subject parameters are relevant and, as shown in Sections 4.2.3 and 5.5.4, also marginal models for recurrent events are, indeed, useful, however not models for marginal hazards but rather models for marginal means (or, equivalently, marginal rate functions), at least in the case where a terminating event (such as death) stops the recurrent events process. Without terminating events, the WLW model may be applicable. For general multi-state models, we find that transition (e.g., cause-specific) hazards or state occupation probabilities and associated expected lengths of stay are typically more relevant target parameters than marginal distributions of times of entry into states.

7.2.2 *Shared frailty model – two-stage estimation (*)*

We now return to the shared frailty model for clustered data, Equation (3.37). Recall from Section 3.9 that we assume that subjects have conditionally independent survival times given frailty, that the frailty distribution is independent of covariates, and the censoring is independent of frailty. We will concentrate on *bivariate* survival data, i.e., clusters of size $n_i = 2$, $i = 1, \dots, n$; however, all that follows goes through for general cluster sizes. We also concentrate on the shared gamma frailty model and only provide brief remarks on other frailty distributions. For any i (which we omit from the notation), we have that the conditional survival functions given frailty are

$$S_h^c(t \mid A) = P(T_h > t \mid A) = \exp\bigl(-A \int_0^t \alpha_h^c(u)du\bigr), \quad h = 1,2$$

and the marginal survival functions $E_A(S_h^c(t \mid A))$ are

$$S_h(t) = \int_0^\infty \exp\bigl(-a \int_0^t \alpha_h^c(u)du\bigr) f_\theta(a)da, \quad h = 1,2,$$

where $f_\theta(\cdot)$ is the density function for the gamma distribution with mean 1 and $\mathrm{SD}^2 = \theta$. The survival function can be evaluated to be

$$S_h(t) = \bigl(1 + \frac{1}{\theta} A_h^c(t)\bigr)^{-\theta}$$

with $A_h^c(t) = \int_0^t \alpha_h^c(u)du$ (e.g., Hougaard, 2000, ch. 7).

From the assumption of conditional independence of (T_1, T_2) given frailty, A, it follows that the *bivariate survival function* is

$$S(t_1,t_2) = P(T_1 > t_1, T_2 > t_2) = \bigl(S_1(t_1)^{-1/\theta} + S_2(t_2)^{-1/\theta} - 1\bigr)^{-\theta}, \tag{7.1}$$

(e.g., Hougaard, 2000, ch. 7; see also Cook and Lawless, 2018, ch. 6). Based on this result, the intensity process for the counting process $N_h(t)$ for subject $h = 1,2$ can be seen to equal

$$\lim_{\Delta \to 0} \frac{1}{\Delta} P(N_h(t+\Delta) - N_h(t) = 1 \mid \mathscr{H}_{t-}) = Y_h(t)\alpha_h^c(t) \frac{\theta + N_1(t-) + N_2(t-)}{\theta + A_1(t \wedge X_1) + A_1(t \wedge X_2)},$$

where $\mathscr{H}_t$, as introduced in Section 1.4, denotes the observed *past* in $[0,t]$ (see, e.g., Nielsen et al., 1992). This shows that the shared gamma frailty model induces an intensity for members of a given cluster that, at time t, depends on the number of previously observed events in the cluster.

These derivations show how the marginal and joint distributions of T_1 and T_2 follow from the shared (gamma) frailty specification where the joint distribution is expressed in terms of the marginals by (7.1). Based on this expression, Glidden (2000), following earlier work by Hougaard (1987) and Shih and Louis (1995), showed that it is possible to go the other way around, i.e., first to specify the margins $S_1(t)$ and $S_2(t)$ and, subsequently, to combine them into a joint survival function using (7.1). This equation is an example of a *copula*, i.e., a joint distribution on the unit square $[0,1] \times [0,1]$ with uniform margins. It is seen that a shared frailty model induces a copula – other examples were given by, among others, Andersen (2005). Glidden (2000), for the gamma distribution, and Andersen (2005) for other frailty distributions studied *two-stage estimation*, as follows. First, the marginal survival functions $S_1(t)$ and $S_2(t)$ are specified and analyzed, e.g., using marginal Cox models as in Sections 4.3 and 5.6. Next, estimates from these marginal models are inserted into (7.1) (or another specified copula) to obtain a profile-like likelihood for the parameter(s) in the copula, i.e., for θ in the case of the gamma distribution. This is maximized to obtain an estimate, $\widehat{\theta}$, for the association parameter. These authors derived asymptotic results for the estimators (building, to a large extent, on Spiekerman and Lin, 1999). With a two-stage approach, it is possible to get regression coefficients with a marginal interpretation (based on a model for which goodness-of-fit examinations are simple) and at the same time get a quantification of the within-cluster association, e.g., using Kendall's coefficient of concordance as exemplified in Section 3.9.2. Methods for evaluating the fit to the data of the chosen copula are, however, not well developed (see, e.g., Andersen et al., 2005).

7.3 Causal inference

The g-formula was introduced in Section 1.2.5 as a means by which a single marginal treatment difference can be estimated based on a regression model including both treatment and other explanatory variables (*confounders*), and the method was later illustrated for a number of marginal parameters in Chapter 4. In this section, we will give a brief discussion of the circumstances under which a marginal treatment difference can be given a causal interpretation. This requires a definition of causality, and we will follow the approach of Hernán and Robins (2020) based on *potential outcomes* (or *counterfactuals*) that is applicable whenever the treatment corresponds to a well-defined intervention, i.e., it should be possible to describe the target randomized trial in which the treatment effect would be estimable. We will define causality following these lines in Section 7.3.1 and demonstrate that, in this setting, data from a randomized trial do, indeed, allow estimation of average causal effects. In the more technical Sections 7.3.2-7.3.3, we will discuss the assumptions that are needed for causal inference in observational studies (including *consistency*, *exchangeability*, and *positivity*) and show that, under such assumptions, average causal effects may be estimated via the g-formula or via an alternative approach using *Inverse Probability of Treatment Weighting* (IPTW) based on *propensity scores*. The final Section 7.3.4 gives a less technical summary with further discussion.

7.3.1 *Definition of causality*

Let Z be a binary treatment variable, $Z = 1$ for treated, $Z = 0$ for controls, let $V(t)$ be a multi-state process, and let $\theta = E(f(V))$ be a marginal parameter of interest. Imagine a population in which all subjects receive the treatment; in this population we could observe the process $V^{z=1}(t)$. Imagine, similarly, the same population but now every subject receives the control, whereby we could observe $V^{z=0}(t)$. The processes $V^{z=z_0}(t)$, $z_0 = 0, 1$ are the *potential outcomes* or *counterfactuals*, so called because in reality, each subject, i will receive at most one of the treatments which means that at least one of $V_i^{z=0}(t)$ or $V_i^{z=1}(t)$ will never be observable. The *average causal effect* of treatment on θ is now defined as

$$\theta_Z = E(f(V^{z=1})) - E(f(V^{z=0})), \tag{7.2}$$

i.e., the difference between the means of what would be observed if every subject either receives the treatment or every subject receives the control. In (7.2), the average causal effect is defined as a difference; however, the causal treatment effect could equally well be a ratio or another contrast between $E(f(V^{z=1}))$ and $E(f(V^{z=0}))$.

Examples of functions $f(\cdot)$ are the state h indicator $f(V) = I(V(t_0) = h)$ at time t_0 in which case θ_Z would be the causal risk difference of occupying state h at time t_0, e.g., the t_0-year risk difference of failure in the two-state model (Figure 1.1). Another example based on the two-state model is $f(V) = \min(T, \tau)$, in which case θ_Z is the causal difference between the τ-restricted mean life times under treatment and control. Note that θ_Z could not be a hazard difference or a hazard ratio because the hazard functions

$$\alpha^{z=z_0}(t) \approx P(T^{z=z_0} \leq t + dt \mid T^{z=z_0} > t)/dt, \, z_0 = 0, 1,$$

do not contrast the same population under treatment and control but, rather, they are contrasting the two, possibly different, sub-populations who would survive until time t under either treatment or under control (e.g., Martinussen et al., 2020).

If treatment were randomized, then θ_Z would be estimable based on the observed data. This is because for $z_0 = 0, 1$ we have that

$$E(f(V^{z=z_0})) = E(f(V^{z=z_0}) \mid Z = z_0)$$

due to *exchangeability* – because of randomization, treatment allocation Z is independent of everything else, including the potential outcomes $V^{z=z_0}$, that is, computing the mean over all subjects, $E(f(V^{z=z_0}))$, results in the same as computing the mean, $E(f(V^{z=z_0}) \mid Z = z_0)$, over the subset of subjects who were randomized to treatment z_0. By assuming *consistency* $V_i^{z=z_0}(t) = V_i(t)$ if $Z_i = z_0$, i.e., what is observed for subject i if receiving treatment z_0 equals that subject's counterfactual outcome under treatment z_0, we have that

$$E(f(V^{z=z_0}) \mid Z = z_0) = E(f(V) \mid Z = z_0)$$

and the latter mean is estimable from the subset of subjects randomized to treatment z_0 – at least under an assumption of independent censoring. Note that, by the consistency assumption, counterfactual outcomes are linked to the observed outcomes.

7.3.2 *The g-formula (*)*

In the previous section, we presented a formal definition of causality based on counterfactuals and argued why average causal effects were estimable based on data from a randomized study under the assumption of consistency. We will now turn to observational data and discuss under which extra assumptions an average causal effect can be estimated using the g-formula. Recall from Section 1.2.5 that the g-formula computes the average prediction

$$\widehat{\theta}_{z_0} = \frac{1}{n}\sum_i \widehat{f}(V_i(t) \mid Z = z_0, \widetilde{\mathbf{Z}}_i) \tag{7.3}$$

based on some regression model for the parameter, θ of interest including treatment, Z and other covariates (confounders) $\widetilde{\mathbf{Z}}$. The prediction is performed by setting treatment to $z_0 \, (= 0, 1)$ for all subjects and keeping the observed confounders $\widetilde{\mathbf{Z}}_i$ for subject $i = 1, \ldots, n$. This estimates (under assumptions to be stated in the following) the mean $E(f(V^{z=z_0}))$. This is because we always have the identity

$$E(f(V^{z=z_0})) = E_{\widetilde{\mathbf{Z}}}\left(E(f(V^{z=z_0} \mid \widetilde{\mathbf{Z}})\right)$$

and, under an assumption of *conditional exchangeability*, this equals

$$E_{\widetilde{\mathbf{Z}}}\left(E(f(V^{z=z_0}) \mid \widetilde{\mathbf{Z}})\right) = E_{\widetilde{\mathbf{Z}}}\left(E(f(V^{z=z_0}) \mid \widetilde{\mathbf{Z}}, Z = z_0)\right).$$

That is, we assume that sufficiently many confounders are collected in $\widetilde{\mathbf{Z}}$ to obtain exchangeability for given value of $\widetilde{\mathbf{Z}}$ or, in other words, for given confounders those who get treatment 1 and those who get treatment 0 are exchangeable. This assumption is also known as *no unmeasured confounders*. Finally, *consistency*, i.e.,

$$E_{\widetilde{\mathbf{Z}}}\left(E(f(V^{z=z_0}) \mid \widetilde{\mathbf{Z}}, Z = z_0)\right) = E_{\widetilde{\mathbf{Z}}}\left(E(f(V) \mid \widetilde{\mathbf{Z}}, Z = z_0)\right),$$

is assumed, where the right-hand side is the quantity that is estimated by the g-formula. In addition, an assumption of *positivity* should be imposed, meaning that for all values of $\widetilde{\mathbf{Z}}$ the probability of receiving either treatment should be positive. By this assumption, prediction of the outcome based on the regression model for θ is feasible for all confounder values under both treatments and, therefore, 'every corner of the population is reached by the predictions'. The g-formula estimate of (7.2) then becomes

$$\widehat{\theta}_Z = \widehat{\theta}_1 - \widehat{\theta}_0 \tag{7.4}$$

with $\widehat{\theta}_{z_0}$, $z_0 = 0, 1$ given by (7.3).

7.3.3 *Inverse probability of treatment weighting (*)*

In the previous section, we argued that an average causal effect, under suitable assumptions, is estimable using the g-formula via modeling of certain features of the data – namely the expected outcome for given treatment and confounders. The same average causal effect is estimable under the same assumptions by modeling a completely different feature of the data, namely the probability of treatment assignment for given values of the confounders, $\widetilde{\mathbf{Z}}$. The conditional probability

$$\text{PS}(\widetilde{\mathbf{Z}}_i) = P(Z_i = 1 \mid \widetilde{\mathbf{Z}}_i) \tag{7.5}$$

of subject i receiving treatment 1 is known as the *propensity score* and the idea in *inverse probability of treatment weighting*, IPTW, is to construct a re-weighted data set, replacing the outcome for subject i by a weighted outcome using the weights

$$\widehat{W}_i = \frac{Z_i}{\widehat{\text{PS}}(\widetilde{\mathbf{Z}}_i)} + \frac{1 - Z_i}{1 - \widehat{\text{PS}}(\widetilde{\mathbf{Z}}_i)}, \tag{7.6}$$

where the propensity score has been estimated. That is, the outcome for subject i is weighted by the inverse probability of receiving the treatment that was actually received and, by this, the re-weighted data set becomes *free of confounding* because $\widetilde{\mathbf{Z}}$ has the same distribution among treated ($Z = 1$) and controls ($Z = 0$) (e.g., Rosenbaum and Rubin, 1983).

Therefore, a simple model including only treatment can be fitted to the re-weighted data set to estimate θ_Z. This could be any of the models discussed in previous chapters from which $\theta = E(f(V))$ can be estimated, e.g., Cox or Fine-Gray models for risk parameters at some time point, or direct models for the expected length of stay in a state in $[0, \tau]$.

In the situation where the outcome is represented by a pseudo-value θ_i (Andersen et al., 2017) or with complete data, i.e., with no censoring, whereby $f(V_i(t))$ is completely observable and equals θ_i, see Section 6.1, the estimate is a simple difference between weighted averages

$$\widehat{\theta}_Z = \frac{1}{n} \sum_{i:Z_i=1} \widehat{W}_i \theta_i - \frac{1}{n} \sum_{i:Z_i=0} \widehat{W}_i \theta_i.$$

In this case, it can be seen that this actually estimates the average causal effect, as follows. The mean of the estimate in treatment group 1 (inserting the true propensity score) is

$$E\Big(\frac{1}{n} \sum_{i:Z_i=1} W_i \theta_i\Big) = \frac{1}{n} E_{\widetilde{\mathbf{Z}}} \sum_i E\Big(\frac{Z_i \theta_i}{\text{PS}(\widetilde{\mathbf{Z}}_i)} \mid \widetilde{\mathbf{Z}}_i\Big)$$

and, assuming consistency, this is

$$\frac{1}{n} E_{\widetilde{\mathbf{Z}}} \sum_i \frac{1}{\text{PS}(\widetilde{\mathbf{Z}}_i)} E\big(Z_i \theta_i \mid \widetilde{\mathbf{Z}}_i\big) = \frac{1}{n} E_{\widetilde{\mathbf{Z}}} \sum_i \frac{1}{\text{PS}(\widetilde{\mathbf{Z}}_i)} E\big(Z_i f(V_i^{z=1}) \mid \widetilde{\mathbf{Z}}_i\big)$$

and, finally, by conditional exchangeability, this is

$$\frac{1}{n} E_{\widetilde{\mathbf{Z}}} \sum_i \frac{1}{\text{PS}(\widetilde{\mathbf{Z}}_i)} E\big(Z_i \mid \widetilde{\mathbf{Z}}_i\big) E\big(f(V_i^{z=1}) \mid \widetilde{\mathbf{Z}}_i\big) = \frac{1}{n} \sum_i E(f(V_i^{z=1})).$$

An identical calculation for the control group gives the desired result. It is seen that, because we divide by $\text{PS}(\widetilde{\mathbf{Z}})$ or $1 - \text{PS}(\widetilde{\mathbf{Z}})$ in (7.6), the assumption of positivity is needed.

7.3.4 Summary and discussion

Sections 7.3.2 and 7.3.3 demonstrated that the average causal effect (7.2) may be estimated in two different ways under a certain set of assumptions. The g-formula (Equations (7.3) and (7.4)) builds on an *outcome model*, i.e., a model by which the marginal parameter, θ of interest may be predicted for given values of treatment Z and confounders $\widetilde{\mathbf{Z}}$. On the other hand, IPTW builds on a model for treatment assignment (the *propensity score*, Equation (7.5)) from which weights (7.6) are calculated and a re-weighted data is constructed. The re-weighted data set is free of confounding from $\widetilde{\mathbf{Z}}$ and, therefore, the average causal effect (7.2) may be estimated by fitting a simple model including only treatment Z to this data set. The assumptions needed for a causal interpretation of the resulting $\widehat{\theta}_Z$ include: Consistency that links the observed outcomes to the counterfactuals, see Section 7.3.1, positivity, i.e., a probability different from both 0 and 1 for any subject in the population of receiving either treatment, and no unmeasured confounders – sufficiently many confounders are collected in $\widetilde{\mathbf{Z}}$ to ensure, for given confounder values, that those who get treatment 1 and those who get treatment 0 are exchangeable. It is an important part of any causal inference endeavor to discuss to what extent these conditions are likely to be fulfilled. In addition to these assumptions, the g-formula rests on the outcome model being correctly specified and IPTW on the propensity score model being correctly specified. *Doubly robust* methods have been devised that only require one of these models to be correct, as well as even less model-dependent techniques based on *targeted maximum likelihood estimation* (TMLE), see, e.g., van der Laan and Rose (2011).

Causal inference is frequently used – also for the analysis of multi-state survival data, see, e.g., Gran et al. (2015), while Janvin et al. (2023) discussed causal inference for recurrent events with competing risks. In this connection, analysis with time-dependent covariates poses particular challenges because these may, at the same time, be affected by previous treatment allocation and be predictive for both future treatment allocation and for the outcome (known as *time-dependent confounding*, see, e.g., Daniel et al., 2013).

7.4 Joint models with time-dependent covariates

We have previously discussed difficulties in connection with estimating marginal parameters by plug-in based on intensity models with *time-dependent covariates*. The problem was announced in Section 3.7.3 when introducing inference in models with time-dependent covariates and further discussed in Section 3.7.8 in connection with considering the role of GvHD in the multi-state model for the bone marrow transplantation data (Example 1.1.7). Finally, landmarking was introduced in Section 5.3 as a simple way to circumvent the difficulty. In the bone marrow transplantation study, estimation of state occupation probabilities in a model that accounts for GvHD became possible when considering it as a *state* in the multi-state model (Figure 1.6) rather than as a non-adapted time-dependent covariate in a competing risks model with end-points relapse and non-relapse mortality. Thus, the solution used was to study a *joint model* for the time-dependent covariate and the events of interest. This approach is, in principle, available whenever the time-dependent covariate is categorical, though the resulting multi-state models quickly get complicated if several categorical time-dependent covariates need consideration and/or have several categories.

For a quantitative time-dependent covariate, the approach is not attractive, as it would require a categorization of the covariate; however, a solution is still to consider a *joint model* for the multi-state process and the time-dependent covariate. The model for the covariate should now be of an entirely different type, namely a model for repeated measurements of a quantitative random variable. A large literature about this topic has evolved, some of it summarized in the book by Rizopoulos (2012), and earlier review articles in the area include Henderson et al. (2000) and Tsiatis and Davidian (2004). In this section, we will give a very brief introduction to the topic, concentrating on a random effects model for the evolvement of the time-dependent covariate and how this may be used as a basis for estimating (conditional) survival probabilities in the framework of the two-state model (Figure 1.1). Competing risks and recurrent events in this setting were discussed by Rizopoulos (2012, ch. 5) and will not be further considered in our brief account here.

7.4.1 Random effects model

In the joint model to be discussed in this section, the time-dependent covariate, $Z_i(t)$ for subject i is assumed to be a sum of a true value at time t, $m_i(t)$ and a *measurement error*, $\varepsilon_i(t)$,

$$Z_i(t) = m_i(t) + \varepsilon_i(t),$$

where the error terms are assumed independent of everything else and normally distributed with mean zero and a certain SD. The true value follows a linear mixed model

$$m_i(t) = \gamma_0 + \mathrm{LP}_i^m(t) + \sum_{\ell=1}^{k} f_\ell(t)\log(A_{\ell i})$$

with a fixed-effects linear predictor $\mathrm{LP}_i^m(t) = \sum_\ell \gamma_\ell \widetilde{Z}_{\ell i}(t)$ depending on covariates $\widetilde{Z}$ that are either time-fixed or deterministic functions of time. The random effects, $\log(A_{1i}),\dots,\log(A_{ki})$ enter via k fixed functions of time, $f_\ell(t), \ell = 1,\dots,k$, where k is often taken to be 2 with $(f_1(t), f_2(t)) = (1,t)$, corresponding to random intercept and random slope. The random effects are assumed to follow a $k-$variate normal distribution with mean zero and some covariance. The hazard function

$$\alpha_i(t \mid \boldsymbol{A}_i) = \alpha_0(t)\exp(\beta_0 m_i(t) + \mathrm{LP}_i^\alpha)$$

is assumed to depend on the true value of the time-dependent covariate and, thereby, on the random effects, and possibly on other time-fixed covariates via the linear predictor $\mathrm{LP}_i^\alpha = \sum_\ell \beta_\ell Z_{\ell i}$. Some components, $Z, \widetilde{Z}$ may appear in both linear predictors $\mathrm{LP}^m, \mathrm{LP}^\alpha$. The baseline hazard is typically modeled parametrically, e.g., by assuming it to be piecewise constant. In this model, the random effects $(A_1,\dots,A_k)$ serve as *frailties* (Section 3.9) which, at the same time, affect the longitudinal development of the time-dependent covariate. The survival time, T and the time-dependent covariate are assumed to be conditionally independent given the frailties, and measurements $Z_i(t_{i\ell_1})$ and $Z_i(t_{i\ell_2})$ taken at different time points are also conditionally independent for given frailties. Thus, the correlation among repeated measurements of $Z_i(\cdot)$ is given entirely by the random effects. These assumptions are utilized when setting up the likelihood in the next section. A careful discussion of the assumptions was given by Tsiatis and Davidian (2004).

7.4.2 *Likelihood (*)*

The data for each of n independent subjects include the censored event time information (X_i, D_i), covariates $(\mathbf{Z}_i, \widetilde{\mathbf{Z}}_i(t))$, and measurements of the time-dependent covariate $\mathbf{Z}_i(t) = (Z_i(t_{i1}), \dots, Z_i(t_{in_i}))$ taken at n_i time points (typically with $t_{i1} = 0$). The likelihood contribution from the event time information (X_i, D_i) *for given frailties* and given $(\mathbf{Z}_i, \widetilde{\mathbf{Z}}_i(t))$ is

$$L_i^{\alpha}(\boldsymbol{\theta} \mid \boldsymbol{A}_i) = (\alpha_i(X_i \mid \boldsymbol{A}_i))^{D_i} \exp\left(-\int_0^{X_i} \alpha_i(t \mid \boldsymbol{A}_i)dt\right).$$

From the conditional independence assumptions summarized in Section 7.4.1, it follows that, for given frailties, the likelihood contribution from observation of the time-dependent covariate is

$$L_i^{\mathbf{Z}}(\boldsymbol{\theta} \mid \boldsymbol{A}_i) = \prod_{\ell=1}^{n_i} \varphi(Z_i(t_{\ell i})),$$

where φ is the relevant normal density function. The observed-data likelihood is now obtained by integrating over the unobserved frailties

$$L_i(\boldsymbol{\theta}) = \int L_i^{\alpha}(\boldsymbol{\theta} \mid \boldsymbol{A}_i) L_i^{\mathbf{Z}}(\boldsymbol{\theta} \mid \boldsymbol{A}_i) \varphi(\boldsymbol{A}_i) d\boldsymbol{A}_i$$

with normal density $\varphi(\boldsymbol{A}_i)$. Maximization of $L(\boldsymbol{\theta}) = \prod_i L_i(\boldsymbol{\theta})$ over the set of all parameters (denoted $\boldsymbol{\theta}$) involves numerical challenges which may be approached, e.g., using the EM-algorithm (Rizopoulos, 2012; ch. 4). Also, variance estimation for the parameter estimates, $\widehat{\boldsymbol{\theta}}$ may be challenging though, in principle, these estimates may be obtained from the second derivative of $\log(L(\boldsymbol{\theta}))$.

7.4.3 *Prediction of survival probabilities*

Our goal with the joint model was estimation of survival probabilities based on a model with time-dependent covariates. The model was described in Section 7.4.1 and inference for the model parameters in Section 7.4.2. Estimation of (conditional) survival probabilities $P(T_i > t \mid T_i > s), t > s$ for given values of time-fixed covariates and given observed time-dependent covariate up till time s, additionally, requires prediction of the random effects for the subject in question. A large literature exists on prediction in random effects models and details go beyond what can be covered here. It is, indeed, possible to make a prediction, $\widehat{\boldsymbol{A}}_{is}$ based on observation of time-fixed covariates $(\mathbf{Z}_i, \widetilde{\mathbf{Z}}_i)$ and of the time-dependent covariate at times $t_{i1}, \dots, t_{in_s}$ $(< s)$ (Rizopoulos, 2012; ch. 7). The estimated conditional survival function is given by

$$\widehat{S}_i(t \mid s) = \exp\left(-\int_s^t \alpha_i(u \mid \widehat{\boldsymbol{A}}_{is}; \widehat{\boldsymbol{\theta}})du\right),$$

where $\widehat{\boldsymbol{\theta}}$ is the maximum likelihood estimator for all model parameters.

The same information also enables prediction of future values of the time-dependent covariate – even beyond the survival time for the subject in question. The joint modeling approach has been criticized for being able to do this (briefly discussed, e.g., by Tsiatis and Davidian, 2004); however, for the purpose of estimating survival probabilities, which was our goal with the joint model, this point of criticism is less relevant.

7.4.4 *Landmarking and joint models*

We have discussed two ways of obtaining estimates of marginal parameters based on models with time-dependent covariates: *Landmarking* (Section 5.3) and *joint models* (current section). Following Putter and van Houwelingen (2022), it can be concluded that the former is a 'pragmatic approach that avoids specifying a model for the time-dependent covariate' while the latter is 'quite efficient when the model is well specified' but 'quite sensitive to misspecification of the longitudinal trajectory'. As a compromise, Putter and van Houwelingen (2022) suggested a hybrid method – still based on landmarking, but also involving a working model for $Z(\cdot)$ by which the conditional expectation $E(Z(t) \mid \{Z(u), u \leq s\}, T > s)$ may be approximated and used for prediction at the landmark time s. The details go beyond what we can cover here; however, this 'landmarking 2.0' idea seems to be a viable compromise that addresses the bias-variance trade-off between the two approaches.

7.5 Cohort sampling

In Sections 2.2.1 and 3.3 (see also Section 5.6), it was shown how the Cox regression model could be fitted to a sample of n subjects (the full cohort) by solving estimating equations based on the Cox partial likelihood (e.g., Equations (2.1) and (3.17)). Furthermore, the cumulative baseline hazard could be estimated using the Breslow estimator (Equations (2.2) or (3.18)). To compute these estimates, information on covariates, at any time t, was needed for all subjects at risk at that time.

In practical analyses of large cohorts, covariate ascertainment may be unnecessarily costly, in particular when relatively few subjects actually experience the event for which the hazard is being modeled. In such cases, ways of *sampling from the cohort* may provide considerable savings of resources without seriously compromising statistical efficiency, and this section discusses two such cohort sampling methods – *nested case-control sampling* and *case-cohort sampling*. As an example, Josefson et al. (2000) studied the association between cervical carcinoma in situ (CIN) and HPV-16 viral load. They applied a nested case-control study including all 468 cases of CIN and 1-2 controls per case sampled from the cohort consisting of 146,889 Swedish women. This cohort was screened between 1969 and 1995 and generated the cases. The purpose of applying this design was to reduce the costs in connection with doing the cytological analyses needed to ascertain the viral load from the smears, taken from the screened woman and, subsequently, stored. As a second example, Petersen et al. (2005) used a case-cohort design in a study of the association between cause-specific mortality rates among Danish adoptees and cause of death information for their biological and adoptive parents. Data on all 1,403 adoptees who were observed to die before 1993 (the cases) were ascertained together with data on a random sub-cohort sampled from the entire Danish Adoption Register. The sub-cohort consisted of 1,683 adoptees among whom 203 were also cases. In that study, ascertainment of data on cause-specific mortality for the biological and adoptive parents was time-consuming, as it involved scrutiny of non-computerized mortality records.

The discussion in this section will be relatively brief and concentrates on the main ideas. Sections 7.5.1-7.5.2 give some technical results with a broader summary and examples in Section 7.5.3.

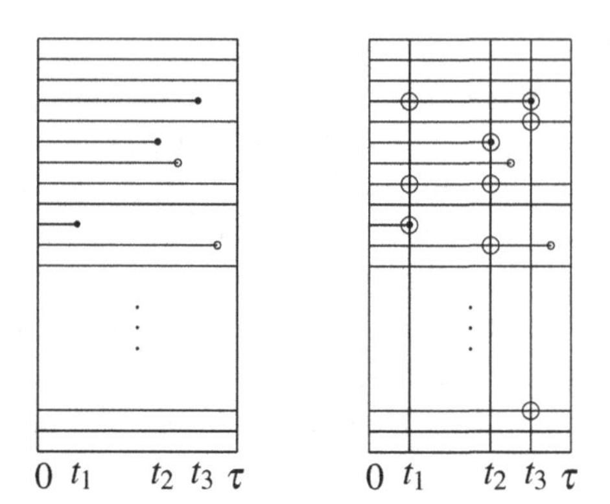

Figure 7.1 *A cohort observed from time $t = 0$ to τ with $D = 3$ cases observed at times t_1, t_2, t_3, at which $m - 1 = 2$ controls are sampled from the respective risk sets.*

7.5.1 *Nested case-control studies (*)*

The nested case-control study is a case-control study *matched on time* in the sense that the data set for analysis consists of all the cases observed in the cohort and a set of controls randomly sampled from each risk set at the times at which cases are observed. Covariates, which for sake of simplicity are assumed to be time-constant, are ascertained, first of all, for all subjects from the full cohort $i = 1, \ldots, n$ for whom $N_i(X_i) = 1$. For simplicity, we concentrate on models for the hazard function in the two-state model for survival data, Figure 1.1, though the same ideas apply for general transition intensities. These are the *cases*, say $\ell = 1, \ldots, D = \sum_i N_i(X_i)$ occurring at times $t_1, \ldots, t_D$. In addition, at each failure time t_ℓ, a number, $m - 1$ of *controls* are randomly sampled from the risk set at that time and their covariate values are also ascertained. The *sampled risk set* at time t_ℓ, $\widetilde{R}(t_\ell)$ consists of the case and the sampled controls. Figure 7.1 depicts the situation with $m = 3$.

The nested case-control study was discussed by Thomas (1977) with full mathematical details by Borgan et al. (1995). A survey of both this design and the case-cohort study was given by Borgan and Samuelsen (2014). Estimation of regression coefficients $\boldsymbol{\beta}$ in a Cox model for the hazard function $\alpha_i(t) = \alpha_0(t)\exp(\boldsymbol{\beta}^\mathsf{T}\mathbf{Z}_i)$ in the cohort proceeds by solving the score equations based on a partial likelihood

$$\widetilde{\mathrm{PL}}_{\mathrm{NCC}}(\boldsymbol{\beta}) = \prod_{\ell=1}^{D} \frac{\exp(\boldsymbol{\beta}^\mathsf{T}\mathbf{Z}_\ell)}{\sum_{j \in \widetilde{R}(t_\ell)} \exp(\boldsymbol{\beta}^\mathsf{T}\mathbf{Z}_j)}, \tag{7.7}$$

which is Equation (3.16) with the sum $\sum_j Y_j(t)\exp(\boldsymbol{\beta}^\mathsf{T}\mathbf{Z}_j)$ over the full risk set replaced by the corresponding sum over the sampled risk set $\widetilde{R}(t_\ell)$. Having estimated the regression coefficients, the cumulative baseline hazard function $A_0(t) = \int_0^t \alpha_0(u)du$ may be estimated by

$$\widehat{A}_{0,\mathrm{NCC}}(t) = \sum_{t_\ell \le t} \frac{1}{(Y(t_\ell)/m)\sum_{j \in \widetilde{R}(t_\ell)} \exp(\widehat{\boldsymbol{\beta}}^\mathsf{T}\mathbf{Z}_j)}, \tag{7.8}$$

which is the Breslow estimator (3.18) with the sum over the sampled risk set up-weighted by the ratio between the full risk set size, $Y(t_\ell)$ and that of the sampled risk set, m. Large

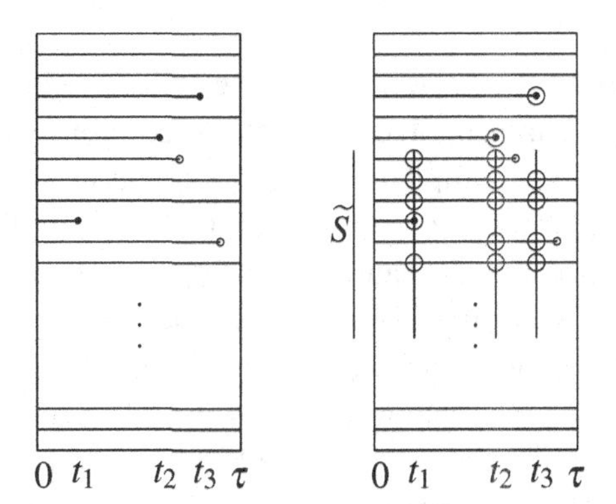

Figure 7.2 *A cohort observed from time $t = 0$ to τ with $D = 3$ cases observed at times t_1, t_2, t_3. A random sub-cohort, $\widetilde{S}$ is sampled at time $t = 0$.*

sample properties of (7.7) and (7.8), including estimation of SD, were discussed by Borgan et al. (1995) who also introduced other ways of sampling from the risk set than simple random sampling.

7.5.2 Case-cohort studies (*)

In the nested case-control study (Section 7.5.1), new controls are sampled at each failure time. In the case-cohort design a random *sub-cohort*, say $\widetilde{S}$ of size $\widetilde{m}$ is sampled at time $t = 0$ and used as a comparison group for all subsequent cases, see Prentice (1986) and Borgan and Samuelsen (2014). Figure 7.2 depicts the situation in the same cohort as in Figure 7.1.

Estimation of regression coefficients $\boldsymbol{\beta}$ in a Cox model for the hazard function $\alpha_i(t) = \alpha_0(t)\exp(\boldsymbol{\beta}^\mathsf{T}\mathbf{Z}_i)$ in the cohort may be carried out by solving the score equations based on the pseudo-likelihood

$$\widetilde{\mathrm{PL}}_{\mathrm{CC}}(\boldsymbol{\beta}) = \prod_{\ell=1}^{D} \frac{\exp(\boldsymbol{\beta}^\mathsf{T}\mathbf{Z}_\ell)}{\sum_{j\in\widetilde{S}\cup\{\ell\}} Y_j(t_\ell)\exp(\boldsymbol{\beta}^\mathsf{T}\mathbf{Z}_j)}. \tag{7.9}$$

Here, the comparison group at case time t_ℓ is the part of the sub-cohort $\widetilde{S}$ that is still at risk (i.e., with $Y_j(t_\ell) = 1$) – with the case $\{\ell\}$ added if this occurred outside the sub-cohort. Let $\widetilde{Y}(t_\ell)$ be the size of this comparison group. From $\widehat{\boldsymbol{\beta}}$, the cumulative baseline hazard may be estimated by

$$\widehat{A}_{0,\mathrm{CC}}(t) = \sum_{t_\ell\le t} \frac{1}{(Y(t_\ell)/\widetilde{Y}(t_\ell))\sum_{j\in\widetilde{S}\cup\{\ell\}} Y_j(t_\ell)\exp(\widehat{\boldsymbol{\beta}}^\mathsf{T}\mathbf{Z}_j)}, \tag{7.10}$$

which is the Breslow estimator (3.18) with the sum over the remaining sub-cohort at time t_ℓ up-weighted to represent the sum over the full risk set at that time. Large sample properties of (7.9) and (7.10) were discussed by Self and Prentice (1988), and modifications of the estimating equations by Borgan and Samuelsen (2014). Thus, all cases still at risk at t_ℓ may be included in the comparison group when equipped with suitable weights.

7.5.3 Discussion

Figures 7.1 and 7.2 illustrate the basic ideas in the two sampling designs. In the nested case-control design, controls are sampled at the observed failure times, while, in the case-cohort design, the sub-cohort is sampled at time $t = 0$ and used throughout the period of observation. It follows that the latter design is useful when more than one case series is of interest in a given cohort, because the same sub-cohort may be used as comparison group for all cases. This was the situation in the study by Petersen et al. (2005) where mortality rates from a number of different causes were analyzed. In the nested case-control design, controls are matched on time – a feature that was useful in the study by Josefsen et al. (2000) because the smears from cases and matched controls had the same storage time and, thereby, 'similar quality'. If, in a nested case-control study, more case series are studied then new controls must be sampled for each new series since the failure times will typically differ among case series. However, Støer et al. (2012) discussed how to re-use controls among case series in such a situation.

In situations where both designs are an option, one may wonder about their relative efficiencies. It appears that the efficiency of the two designs are quite similar when based on similar numbers of subjects for whom covariates are ascertained. The efficiency of a nested case-control study compared to a full cohort study has been shown to be of the order of magnitude of $(m-1)/m$, see, e.g., Borgan and Samuelsen (2014).

Guinea-Bissau childhood vaccination study

In this study (Example 1.1.2), relatively few children died during follow-up (222 or 4.2%, see Table 1.1), and cohort sampling could be an option, even though vaccination status and other covariates were, indeed, ascertained for all children in the study. To illustrate the techniques, a nested case-control study was conducted within the cohort of 5,274 children by sampling $m-1 = 3$ controls at each of the $D = 222$ observed failures. For comparison, a similar-sized case-cohort study was also conducted by sampling a 12.5% random sub-cohort (664 children) from the full cohort, resulting in 641 'new' children and 23 cases within the sub-cohort. Table 7.2 shows the estimated coefficients for BCG vaccination from Cox models with follow-up time as the time-variable, adjusted for age at recruitment as a categorical variable. It is seen that similar estimates are obtained in the three analyses with a somewhat smaller SD from the full cohort design and with similar values of SD for the two cohort sampling designs. The ratio between SD^2 for the full cohort and the nested case-control design, $(0.146/0.174)^2 = 0.71$, is well in line with the ratio $(m-1)/m = 0.75$.

Table 7.2 *Guinea-Bissau childhood vaccination study: Estimated coefficients (and SD) for BCG vaccination (yes vs. no) from Cox models using follow-up time as the time-variable. Adjustment for age at entry was made, and different sampling designs were used: Full cohort, nested case-control with $m-1 = 3$ controls per case, and case-cohort with $\widetilde{m} = 664$.*

Design	$\widehat{\beta}$	SD
Full cohort	-0.347	0.146
Nested case-control	-0.390	0.174
Case cohort	-0.389	0.166

Bibliography

Aalen, O. O. (1978). Nonparametric estimation of partial transition probabilities in multiple decrement models. *Ann. Statist.*, 6:534–545.

– (1989). A linear regression model for the analysis of life times. *Statist. in Med.*, 8:907–925.

Aalen, O. O., Borgan, Ø., Fekjær, H. (2001). Covariate adjustment of event histories estimated from Markov chains: The additive approach. *Biometrics*, 57:993–1001.

Aalen, O. O., Borgan, Ø., Gjessing, H. (2008). *Survival and Event History Analysis: A Process Point of View*. New York: Springer.

Aalen, O. O., Johansen, S. (1978). An empirical transition matrix for nonhomogeneous Markov chains based on censored observations. *Scand. J. Statist.*, 5:141–150.

Allignol, A., Beyersmann, J., Gerds, T. A., Latouche, A. (2014). A competing risks approach for nonparametric estimation of transition probabilities in a non-Markov illness-death model. *Lifetime Data Analysis*, 20:495–513.

Amorim, L. D. A. F., Cai, J. (2015). Modelling recurrent events: a tutorial for analysis in epidemiology. *Int. J. Epidemiol.*, 44:324–333.

Andersen, E. W. (2005). Two-stage estimation in copula models used in family studies. *Lifetime Data Analysis*, 11:333–350.

Andersen, P. K. (2013). Decomposition of number of years lost according to causes of death. *Statist. in Med.*, 32:5278–5285.

Andersen, P. K., Angst, J., Ravn, H. (2019). Modeling marginal features in studies of recurrent events in the presence of a terminal event. *Lifetime Data Analysis*, 25:681–695.

Andersen, P. K., Borgan, Ø., Gill, R. D., Keiding, N. (1993). *Statistical Models Based on Counting Processes*. New York: Springer.

Andersen, P. K., Ekstrøm, C. T., Klein, J. P., Shu, Y., Zhang, M.-J. (2005). A class of goodness of fit tests for a copula based on bivariate right-censored data. *Biom. J.*, 47:815–824.

Andersen, P. K., Geskus, R. B., de Witte, T., Putter, H. (2012). Competing risks in epidemiology: possibilities and pitfalls. *Int. J. Epidemiol.*, 41:861–870.

Andersen, P. K., Gill, R. D. (1982). Cox's regression model for counting processes: a large sample study. *Ann. Statist.*, 10:1100–1120.

Andersen, P. K., Hansen, L. S., Keiding, N. (1991). Non- and semi-parametric estimation of transition probabilities from censored observations of a non-homogeneous Markov process. *Scand. J. Statist.*, 18:153–167.

Andersen, P. K., Keiding, N. (2002). Multi-state models for event history analysis. *Statist. Meth. Med. Res.*, 11:91–115.

Andersen, P. K., Keiding, N. (2012). Interpretability and importance of functionals in competing risks and multistate models. *Statist. in Med.*, 31:1074–1088.

Andersen, P. K., Klein, J. P., Rosthøj, S. (2003). Generalized linear models for correlated pseudo-observations, with applications to multi-state models. *Biometrika*, 90:15–27.

Andersen, P. K., Liestøl, K. (2003). Attenuation caused by infrequently updated covariates in survival analysis. *Biostatistics*, 4:633–649.

Andersen, P. K., Pohar Perme, M. (2008). Inference for outcome probabilities in multi-state models. *Lifetime Data Analysis*, 14:405–431.

– (2010). Pseudo-observations in survival analysis. *Statist. Meth. Med. Res.*, 19:71–99.

Andersen, P. K., Pohar Perme, M., van Houwelingen, H. C., Cook, R. J., Joly, P., Martinussen, T., Taylor, J. M. G., Abrahamowicz, M., Therneau, T. M. (2021). Analysis of time-to-event for observational studies: Guidance to the use of intensity models. *Statist. in Med.*, 40:185–211.

Andersen, P. K., Skovgaard, L. T. (2006). *Regression with Linear Predictors*. New York: Springer.

Andersen, P. K., Syriopoulou, E., Parner, E. T. (2017). Causal inference in survival analysis using pseudo-observations. *Statist. in Med.*, 36:2669–2681.

Andersen, P. K., Wandall, E. N. S., Pohar Perme, M. (2022). Inference for transition probabilities in non-Markov multi-state models. *Lifetime Data Analysis*, 28:585–604.

Anderson, J. R., Cain, K. C., Gelber, R. D. (1983). Analysis of survival by tumor response. *J. Clin. Oncol.*, 1:710–719.

Austin, P. C., Steyerberg, E. W., Putter, H. (2021). Fine-Gray subdistribution hazard models to simultaneously estimate the absolute risk of different event types: Cumulative total failure probability may exceed 1. *Statist. in Med.*, 40:4200–4212.

Azarang, L., Scheike, T., Uña-Alvarez, J. (2017). Direct modeling of regression effects for transition probabilities in the progressive illness-death model. *Statist. in Med.*, 36:1964–1976.

Balan, T. A., Putter, H. (2020). A tutorial on frailty models. *Statist. Meth. Med. Res.*, 29:3424–3454.

Bellach, A., Kosorok, M. R., Rüschendorf, L., Fine, J. P. (2019). Weighted NPMLE for the subdistribution of a competing risk. *J. Amer. Statist. Assoc.*, 114:259–270.

Beyersmann, J., Allignol, A., Schumacher, M. (2012). *Competing Risks and Multistate Models with R*. New York: Springer.

Beyersmann, J., Latouche, A., Bucholz, A., Schumacher, M. (2009). Simulating competing risks data in survival analysis. *Statist. in Med.*, 28:956–971.

Binder, N., Gerds, T. A., Andersen, P. K. (2014). Pseudo-observations for competing risks with covariate dependent censoring. *Lifetime Data Analysis*, 20:303–315.

Blanche, P. F., Holt, H., Scheike, T. H. (2023). On logistic regression with right censored data, with or without competing risks, and its use for estimating treatment effects. *Lifetime Data Analysis*, 29:441–482.

Bluhmki, T., Schmoor, C., Dobler, D., Pauly, M., Finke, J., Schumacher, M., Beyersmann, J. (2018). A wild bootstrap approach for the Aalen–Johansen estimator. *Biometrics*, 74:977–985.

Borgan, Ø., Goldstein, L., Langholz, B. (1995). Methods for the analysis of sampled cohort data in the Cox proportional hazards model. *Ann. Statist.*, 23:1749–1778.

Borgan, Ø., Samuelsen, S. O. (2014). "Nested case-control and case-cohort studies". *Handbook of Survival Analysis*. Ed. by J. P. Klein, H. C. van Houwelingen, J. G. Ibrahim, T. H. Scheike. Boca Raton: CRC Press. Chap. 17:343–367.

Bouaziz, O. (2023). Fast approximations of pseudo-observations in the context of right-censoring and interval-censoring. *Biom. J.*, 65:22000714.

Breslow, N. E. (1974). Covariance analysis of censored survival data. *Biometrics*, 30:89–99.

Broström, G. (2012). *Event history analysis with R*. London: Chapman and Hall/CRC.

Bühler, A., Cook, R. J., Lawless, J. L. (2023). Multistate models as a framework for estimand specification in clinical trials of complex diseases. *Statist. in Med.*, 42:1368–1397.

Bycott, P., Taylor, J. M. G. (1998). A comparison of smoothing techniques for CD4 data measured with error in a time-dependent Cox proportional hazards model. *Statist. in Med.*, 17:2061–2077.

Clayton, D. G., Hills, M. (1993). *Statistical Models in Epidemiology*. Oxford: Oxford University Press.

Collett, D. (2015). *Modelling Survival Data in Medical Research (3rd ed.)* Boca Raton: Chapman and Hall/CRC.

Conner, S. C., Trinquart, L. (2021). Estimation and modeling of the restricted mean time lost in the presence of competing risks. *Statist. in Med.*, 40:2177–2196.

Cook, R. J., Lawless, J. F. (1997). Marginal analysis of recurrent events and a terminating event. *Statist. in Med.*, 16:911–924.

– (2007). *The Statistical Analysis of Recurrent Events*. New York: Springer.

– (2018). *Multistate Models for the Analysis of Life History Data*. Boca Raton: Chapman and Hall/CRC.

Cook, R. J., Lawless, J. F., Lakhal-Chaieb, L., Lee, K.-A. (2009). Robust estimation of mean functions and treatment effects for recurrent events under event-dependent censoring and termination: Application to skeletal complications in cancer metastatic to bone. *J. Amer. Statist. Assoc.*, 104:60–75.

Cox, D. R. (1972). Regression models and life-tables. *J. Roy. Statist. Soc., ser. B*, 34:187–220.

– (1975). Partial likelihood. *Biometrika*, 62:269–276.

Crowder, M. (2001). *Classical Competing Risks*. London: Chapman and Hall/CRC.

Daniel, R. M., Cousens, S. N., de Stavola, B. L., Kenward, M. G., Sterne, J. A. C. (2013). Methods for dealing with time-dependent confounding. *Statist. in Med.*, 32:1584–1618.

Daniel, R. M., Zhang, J., Farewell, D. (2021). Making apples from oranges: Comparing non collapsible effect estimators and their standard errors after adjustment for different covariate sets. *Biom. J.*, 63:528–557.

Datta, S., Satten, G. A. (2001). Validity of the Aalen-Johansen estimators of stage occupation probabilities and Nelson-Aalen estimators of integrated transition hazards for non-Markov models. *Stat. & Prob. Letters*, 55:403–411.

– (2002). Estimation of integrated transition hazards and stage occupation probabilities for non-Markov systems under dependent censoring. *Biometrics*, 58:792–802.

Efron, B. (1982). *The Jackknife, the Bootstrap and Other Resampling Plans*. SIAM, Philadelphia: CBMS-NSF Regional Conference Series in Applied Mathematics.

Efron, B., Tibshirani, R. (1993). *An Introduction to the Bootstrap*. Boca Raton: Chapman and Hall/CRC.

Fine, J. P., Gray, R. J. (1999). A proportional hazards model for the subdistribution of a competing risk. *J. Amer. Statist. Assoc.*, 94:496–509.

Fine, J. P., Jiang, H., Chappell, R. (2001). On semi-competing risks data. *Biometrika*, 88:907–919.

Finkelstein, D. M. (1986). A proportional hazards model for interval-censored failure time data. *Biometrics*, 42:845–854.

Fisher, L. D., Lin, D. Y. (1999). Time-dependent covariates in the Cox proportional-hazards regression model. *Ann. Rev. Public Health*, 20:145–157.

Fix, E., Neyman, J. (1951). A simple stochastic model of recovery, relapse, death and loss of patients. *Hum. Biol.*, 23:205–241.

Frydman, H. (1995). Nonparametric estimation of a Markov illness-death process from interval-censored observations, with applications to diabetes survival data. *Biometrika*, 82:773–789.

Frydman, H., Liu, J. (2013). Nonparametric estimation of the cumulative intensities in an interval censored competing risks model. *Lifetime Data Analysis*, 19:79–99.

Frydman, H., Szarek, M. (2009). Nonparametric estimation in a Markov illness-death process from interval censored observations with missing intermediate transition status. *Biometrics*, 65:143–151.

Furberg, J. K., Korn, S., Overgaard, M., Andersen, P. K., Ravn, H. (2023). Bivariate pseudo-observations for recurrent event analysis with terminal events. *Lifetime Data Analysis*, 29:256–287.

Furberg, J. K., Rasmussen, S., Andersen, P. K., Ravn, H. (2022). Methodological challenges in the analysis of recurrent events for randomised controlled trials with application to cardiovascular events in LEADER. *Pharmaceut. Statist.*, 21:241–267.

Gerds, T. A., Scheike, T. H., Andersen, P. K. (2012). Absolute risk regression for competing risks: interpretation, link functions, and prediction. *Statist. in Med.*, 31:1074–1088.

Geskus, R. (2016). *Data Analysis with Competing Risks and Intermediate States*. Boca Raton: Chapman and Hall/CRC.

Ghosh, D., Lin, D. Y. (2000). Nonparametric analysis of recurrent events and death. *Biometrics*, 56:554–562.

– (2002). Marginal regression models for recurrent and terminal events. *Statistica Sinica*, 12:663–688.

Gill, R. D., Johansen, S. (1990). A survey of product-integration with a view towards application in survival analysis. *Ann. Statist.*, 18:1501–1555.

Glidden, D. V. (2000). A two-stage estimator of the dependence parameter for the Clayton-Oakes model. *Lifetime Data Analysis*, 6:141–156.

– (2002). Robust inference for event probabilities with non-Markov event data. *Biometrics*, 58:361–368.

Glidden, D. V., Vittinghoff, E. (2004). Modelling clustered survival data from multicentre clinical trials. *Statist. in Med.*, 23:369–388.

Gran, J. M., Lie, S. A., Øyeflaten, I., Borgan, Ø., Aalen, O. O. (2015). Causal inference in multi-state models – Sickness absence and work for 1145 participants after work rehabilitation. *BMC Publ. Health*, 15:1082.

Graw, F., Gerds, T. A., Schumacher, M. (2009). On pseudo-values for regression analysis in competing risks models. *Lifetime Data Analysis*, 15:241–255.

Grøn, R., Gerds, T. A. (2014). "Binomial regression models". *Handbook of Survival Analysis*. Ed. by J. P. Klein, H. C. van Houwelingen, J. G. Ibrahim, T. H. Scheike. Boca Raton: CRCPress. Chap. 11:221–242.

Gunnes, N., Borgan, Ø., Aalen, O. O. (2007). Estimating stage occupation probabilities in non-Markov models. *Lifetime Data Analysis*, 13:211–240.

Henderson, R., Diggle, P., Dobson, A. (2000). Joint modelling of longitudinal measurements and event time data. *Biostatistics*, 1:465–480.

Hernán, M. A., Robins, J. M. (2020). *Causal Inference: What If*. Boca Raton: Chapman and Hall/CRC.

Hougaard, P. (1986). A class of multivariate failure time distributions. *Biometrika*, 73:671–678.

– (1999). Multi-state models: a review. *Lifetime Data Analysis*, 5:239–264.

– (2000). *Analysis of Multivariate Survival Data*. New York: Springer.

– (2022). Choice of time scale for analysis of recurrent events data. *Lifetime Data Analysis*, 28:700–722.

Huang, C., Wang, M. (2004). Joint modeling and estimation for recurrent event processes and failure time data. *J. Amer. Statist. Assoc.*, 99:1153–1165.

Hudgens, M. G., Satten, G. A., Longini, I. M. (2004). Nonparametric maximum likelihood estimation for competing risks survival data subject to interval censoring and truncation. *Biometrics*, 57:74–80.

Iacobelli, S., Carstensen, B. (2013). Multiple time scales in multi-state models. *Statist. in Med.*, 30:5315–5327.

Jackson, C. (2011). Multi-state models for panel data: the msm package for R. *J. Statist. Software*, 38:1–27.

Jacobsen, M., Martinussen, T. (2016). A note on the large sample properties of estimators based on generalized linear models for correlated pseudo-observations. *Scand. J. Statist.*, 43:845–862.

Jaeckel, L. A. (1972). *The Infinitesimal Jackknife*. Tech. rep. Bell Laboratories, MM 72-1215-11.

Janvin, M., Young, J. G., Ryalen, P. C., Stensrud, M. J. (2023). Causal inference with recurrent and competing events. *Lifetime Data Analysis*. (in press).

Jensen, H., Benn, C. S., Nielsen, J., Lisse, I. M., Rodrigues, A., Andersen, P. K., Aaby, P. (2007). Survival bias in observational studies of the effect of routine immunisations on childhood survival. *Trop. Med. Int. Health*, 12:5–14.

Johansen, M. N., Lundbye-Christensen, S., Parner, E. T. (2020). Regression models using parametric pseudo-observations. *Statist. Med.*, 39:2949–2961.

Joly, P., Commenges, D., Helmer, C., Letenneur, L. (2002). A penalized likelihood approach for an illness-death model with interval-censored data: application to age-specific incidence of dementia. *Biostatistics*, 3:433–443.

Josefson, A. M., Magnusson, P. K. E., Ylitalo, N., Sørensen, P., Qwarforth-Tubbin, P., Andersen, P. K., Melbye, M., Adami, H.-O., Gyllensten, U. B. (2000). Viral load of human papilloma virus 16 as a determinant for development of cervical carcinoma in situ: a nested case-control study. *The Lancet*, 355:2189–2193.

Kalbfleisch, J. D., Lawless, J. F. (1985). The analysis of panel data under a Markov assumption. *J. Amer. Statist. Assoc.*, 80:863–871.

Kalbfleisch, J. D., Prentice, R. L. (1980). *The Statistical Analysis of Failure Time Data.* (2nd ed. 2002). New York: Wiley.

Kaplan, E. L., Meier, P. (1958). Non-parametric estimation from incomplete observations. *J. Amer. Statist. Assoc.*, 53:457–481, 562–563.

Keiding, N. (1998). "Lexis diagram". *Encyclopedia of Biostatistics vol. 3*. New York: Wiley:2232–2234.

Kessing, L. V., Hansen, M. G., Andersen, P. K., Angst, J. (2004). The predictive effect of episodes on the risk of recurrence in depressive and bipolar disorder - a life-long perspective. *Acta Psych. Scand.*, 109:339–344.

Kessing, L. V., Olsen, E. W., Andersen, P. K. (1999). Recurrence in affective disorder: Analyses with frailty models. *Amer. J. Epidemiol.*, 149:404–411.

Kristensen, I., Aaby, P., Jensen, H. (2000). Routine vaccinations and child survival: follow up study in Guinea-Bissau, West Africa. *Br. Med. J.*, 321:1435–1438.

Larsen, B. S., Kumarathurai, P., Falkenberg, J., Nielsen, O. W., Sajadieh, A. (2015). Excessive atrial ectopy and short atrial runs increase the risk of stroke beyond atrial fibrillation. *J. Amer. College Cardiol.*, 66:232–241.

Latouche, A., Allignol, A., Beyersmann, J., Labopin, M., Fine, J. P. (2013). A competing risks analysis should report results on all cause-specific hazards and cumulative incidence functions. *J. Clin. Epidemiol.*, 66:648–653.

Lawless, J. F., Nadeau, J. C. (1995). Some simple robust methods for the analysis of recurrent events. *Technometrics*, 37:158–168.

Li, J., Scheike, T. H., Zhang, M.-J. (2015). Checking Fine and Gray subdistribution hazards model with cumulative sums of residuals. *Lifetime Data Analysis*, 21:197–217.

Li, Q. H., Lagakos, S. W. (1997). Use of the Wei-Lin-Weissfeld method for the analysis of a recurrent and a terminating event. *Statist. in Med.*, 16:925–940.

Lin, D. Y. (1994). Cox regression analysis of multivariate failure time data: the marginal approach. *Statist. in Med.*, 13:2233–2247.

Lin, D. Y., Oakes, D., Ying, Z. (1998). Additive hazards regression models with current status data. *Biometrika*, 85:289–298.

Lin, D. Y., Wei, L. J. (1989). The robust inference for the Cox proportional hazards model. *J. Amer. Statist. Assoc.*, 84:1074–1078.

Lin, D. Y., Wei, L. J., Yang, I., Ying, Z. (2000). Semiparametric regression for the mean and rate functions of recurrent events. *J. Roy. Statist. Soc., ser. B*, 62:711–730.

Lin, D. Y., Wei, L. J., Ying, Z. (1993). Checking the Cox model with cumulative sums of martingale-based residuals. *Biometrika*, 80:557–572.

– (2002). Model-checking techniques based on cumulative residuals. *Biometrics*, 58:1–12.

Lin, D. Y., Ying, Z. (1994). Semiparametric analysis of the additive risk model. *Biometrika*, 81:61–71.

Lindsey, J. C., Ryan, L. M. (1993). A three-state multiplicative model for rodent tumorigenicity experiments. *J. Roy. Statist. Soc., ser. C*, 42:283–300.

Liu, L., Wolfe, R. A., Huang, X. (2004). Shared frailty models for recurrent events and a terminal event. *Biometrics*, 60:747–756.

Lombard, M., Portmann, B., Neuberger, J., Williams, R., Tygstrup, N., Ranek, L., Ring-Larsen, H., Rodes, J., Navasa, M., Trepo, C., Pape, G., Schou, G., Badsberg, J. H., Andersen, P. K. (1993). Cyclosporin A treatment in primary biliary cirrhosis: results of a long-term placebo controlled trial. *Gastroenterology*, 104:519–526.

Lu, C., Goeman, J., Putter, H. (2023). Maximum likelihood estimation in the additive hazards model. *Biometrics*, 28:700–722.

Lu, X., Tsiatis, A. A. (2008). Improving the efficiency of the log-rank test using auxiliary covariates. *Biometrika*, 95:679–694.

Malzahn, N., Hoff, R., Aalen, O. O., Mehlum, I. S., Putter, H., Gran, J. M. (2021). A hybrid landmark Aalen-Johansen estimator for transition probabilities in partially non-Markov multi-state models. *Lifetime Data Analysis*, 27:737–760.

Mao, L., Lin, D. Y. (2016). Semiparametric regression for the weighted composite endpoint of recurrent and terminal events. *Biostatistics*, 17:390–403.

– (2017). Efficient estimation of semiparametric transformation models for the cumulative incidence of competing risk. *J. Roy. Statist. Soc., ser. B*, 79:573–587.

Mao, L., Lin, D. Y., Zeng, D. (2017). Semiparametric regression analysis of interval-censored competing risks data. *Biometrics*, 73:857–865.

Marso, S. P., Daniels, G. H., Brown-Frandsen, K., Kristensen, P., Mann, J. F. E., Nauck, M. A., Nissen, S. E., Pocock, S., Poulter, N. R., Ravn, L. S., Steinberg, W. M., Stockner, M., Zinman, B., Bergenstal, R. M., Buse, J. B., for the LEADER steering committee (2016). Liraglutide and Cardiovascular Outcomes in Type 2 Diabetes. *New Engl. J. Med.*, 375:311–322.

Martinussen, T., Scheike, T. H. (2006). *Dynamic Regression Models for Survival Data.* New York: Springer.

Martinussen, T., Vansteelandt, S., Andersen, P. K. (2020). Subtleties in the interpretation of hazard contrasts. *Lifetime Data Analysis*, 26:833–855.

Meira-Machado, L., J. Uña-Alvarez, Cadarso-Saurez, C. (2006). Nonparametric estimation of transition probabilities in a non-Markov illness-death model. *Lifetime Data Analysis*, 13:325–344.

Mitton, L., Sutherland, H., Week, M., (eds.) (2000). *Microsimulation Modelling for Policy Analysis. Challenges and Innovations*. Cambridge: Cambridge University Press.

Nielsen, G. G., Gill, R. D., Andersen, P. K., Sørensen, T. I. A. (1992). A counting process approach to maximum likelihood estimation in frailty models. *Scand. J. Statist.*, 19:25–43.

O'Hagan, A., Stevenson, M., Madan, J. (2007). Monte Carlo probabilistic sensitivity analysis for patient level simulation models: efficient estimation of mean and variance using ANOVA. *Health Economics*, 16:1009–1023.

O'Keefe, A. G., Su, L., Farewell, V. T. (2018). Correlated multistate models for multiple processes: An application to renal disease progression in systemic lupus erythematosus. *Appl. Statist.*, 67:841–860.

Overgaard, M. (2019). State occupation probabilities in non-Markov models. *Math. Meth. Statist.*, 28:279–290.

Overgaard, M., Andersen, P. K., Parner, E. T. (2023). Pseudo-observations in a multi-state setting. *The Stata Journal*, 23:491–517.

Overgaard, M., Parner, E. T., Pedersen, J. (2017). Asymptotic theory of generalized estimating equations based on jack-knife pseudo-observations. *Ann. Statist.*, 45:1988–2015.

– (2019). Pseudo-observations under covariate-dependent censoring. *J. Statist. Plan. and Inf.*, 202:112–122.

Parner, E. T., Andersen, P. K., Overgaard, M. (2023). Regression models for censored time-to-event data using infinitesimal jack-knife pseudo-observations, with applications to left-truncation. *Lifetime Data Analysis*, 29:654–671.

Pavlič, K., Martinussen, T., Andersen, P. K. (2019). Goodness of fit tests for estimating equations based on pseudo-observations. *Lifetime Data Analysis*, 25:189–205.

Pepe, M. S. (1991). Inference for events with dependent risks in multiple endpoint studies. *J. Amer. Statist. Assoc.*, 86:770–778.

Pepe, M. S., Longton, G., Thornquist, M. (1991). A qualifier Q for the survival function to describe the prevalence of a transient condition. *Statist. in Med.*, 10:413–421.

Petersen, L., Andersen, P. K., Sørensen, T. I. A. (2005). Premature death of adult adoptees: Analyses of a case-cohort sample. *Gen. Epidemiol.*, 28:376–382.

Prentice, R. L. (1986). A case-cohort design for epidemiologic cohort studies and disease prevention trials. *Biometrika*, 73:1–11.

Prentice, R. L., Gloeckler, L. A. (1978). Regression analysis of grouped survival data with application to breast cancer data. *Biometrics*, 34:57–67.

Prentice, R. L., Kalbfleisch, J. D., Peterson, A. V., Flournoy, N., Farewell, V. T., Breslow, N. E. (1978). The analysis of failure times in the presence of competing risks. *Biometrics*, 34:541–554.

Prentice, R. L., Williams, B. J., Peterson, A. V. (1981). On the regression analysis of multivariate failure time data. *Biometrika*, 68:373–379.

Prentice, R. L., Zhao, S. (2020). *The Statistical Analysis of Multivariate Failure Time Data*. Boca Raton: Chapman and Hall/CRC.

Preston, S., Heuveline, P., Guillot, M. (2000). *Demography: Measuring and Modeling Population Processes*. New York: Wiley.

PROVA study group (1991). Prophylaxis of first time hemorrage from esophageal varices by sclerotherapy, propranolol or both in cirrhotic patients: A randomized multicenter trial. *Hepatology*, 14:1016–1024.

Putter, H., Fiocco, M., Geskus, R. B. (2007). Tutorial in biostatistics: competing risks and multi-state models. *Statist. in Med.*, 26:2389–2430.

Putter, H., Schumacher, M., van Houwelingen, H. C. (2020). On the relation between the cause-specific hazard and the subdistribution rate for competing risks data: The Fine-Gray model revisited. *Biom. J*, 62:790–807.

Putter, H., Spitoni, C. (2018). Non-parametric estimation of transition probabilities in non-Markov multi-state models: The landmark Aalen-Johansen estimator. *Statist. Meth. Med. Res.*, 27:2081–2092.

Putter, H., van Houwelingen, H. C. (2015). Frailties in multi-state models: Are they identifiable? Do we need them? *Statist. Meth. Med. Res.*, 24:675–692.

– (2022). Landmarking 2.0: Bridging the gap between joint models and landmarking. *Statist. in Med.*, 41:1901–1917.

Rizopoulos, D. (2012). *Joint Models for Longitudinal and Time-to-Event Data*. Boca Raton: Chapman and Hall/CRC.

Rodriguez-Girondo, M., Uña-Alvarez, J. (2012). A nonparametric test for Markovianity in the illness-death model. *Statist. in Med.*, 31:4416–4427.

Rondeau, V., Mathoulin-Pelissier, S., Jacqmin-Gadda, H., Brouste, V., Soubeyran, P. (2007). Joint frailty models for recurring events and death using maximum penalized likelihood estimation: Application on cancer events. *Biostatistics*, 8:708–721.

Rosenbaum, P. R., Rubin, D. B. (1993). The central role of the propensity score in observational studies for causal effects. *Biometrika*, 70:41–55.

Royston, P., Parmar, M. K. B. (2002). Flexible parametric proportional-hazards and proportional-odds models for censored survival data, with application to prognostic modelling and estimation of treatment effects. *Statist. in Med.*, 21:2175–2197.

Rutter, C. M., Zaslavsky, A. M., Feuer, E. J. (2011). Dynamic microsimulation models for health outcomes: a review. *Med. Decision Making*, 31:10–18.

Sabathé, C., Andersen, P. K., Helmer, C., Gerds, T. A., Jacqmin-Gadda, H., Joly, P. (2020). Regression analysis in an illness-death model with interval-censored data: a pseudo-value approach. *Statist. Meth. Med. Res.*, 29:752–764.

Scheike, T. H., Zhang, M.-J. (2007). Direct modelling of regression effects for transition probabilities in multistate models. *Scand. J. Statist.*, 34:17–32.

Scheike, T. H., Zhang, M.-J., Gerds, T. A. (2008). Predicting cumulative incidence probability by direct binomial regression. *Biometrika*, 95:205–220.

Self, S. G., Prentice, R. L. (1988). Asymptotic distribution theory and efficiency results for case-cohort studies. *Ann. Statist.*, 16:64–81.

Shih, J. H., Louis, T. A. (1995). Inferences on association parameter in copula models for bivariate survival data. *Biometrics*, 51:1384–1399.

Shu, Y., Klein, J. P., Zhang, M.-J. (2007). Asymptotic theory for the Cox semi-Markov illness-death model. *Lifetime Data Analysis*, 13:91–117.

Spikerman, C. F., Lin, D. Y. (1999). Marginal regression models for multivariate failure time data. *J. Amer. Statist. Assoc.*, 93:1164–1175.

Støer, N., Samuelsen, S. O. (2012). Comparison of estimators in nested case-control studies with multiple outcomes. *Lifetime Data Analysis*, 18:261–283.

Suissa, S. (2007). Immortal time bias in pharmacoepidemiology. *Amer. J. Epidemiol.*, 167:492–499.

Sun, J. (2006). *The Statistical Analysis of Interval-censored Failure Time Data*. New York: Springer.

Sverdrup, E. (1965). Estimates and test procedures in connection with stochastic models for deaths, recoveries and transfers between different states of health. *Skand. Aktuarietidskr.*, 48:184–211.

Szklo, M., Nieto, F. J. (2014). *Epidemiology. Beyond the Basics*. Burlington: Jones and Bartlett.

Thomas, D. C. (1977). Addendum to 'Methods of cohort analysis: appraisal by application to asbestos mining' by F. D. K. Liddell, J. C. McDonald, D. C. Thomas. *J. Roy. Statist. Soc., ser. B*, 140:469–491.

Tian, L., Zhao, L., Wei, L. J. (2014). Predicting the restricted mean event time with the subject's baseline covariates in survival analysis. *Biostatistics*, 15:222–233.

Titman, A. C. (2015). Transition probability estimates for non-Markov multi-state models. *Biometrics*, 71:1034–1041.

Titman, A. C., Putter, H. (2022). General tests of the Markov property in multi-state models. *Biostatistics*, 23:380–396.

Tsiatis, A. A. (1975). A nonidentifiability aspect of the problem of competing risks. *Proc. Nat. Acad. Sci. USA*, 72:20–22.

Tsiatis, A. A., Davidian, M. (2004). Joint modeling of longitudinal and time-to-event data: An overview. *Statistica Sinica*, 14:809–834.

Turnbull, B. W. (1976). The empirical distribution with arbitrarily grouped, censored and truncated data. *J. Roy. Statist. Soc., ser. B*, 38:290–295.

Uña-Alvarez, J., Meira-Machado, L. (2015). Nonparametric estimation of transition probabilities in the non-Markov illness-death model: A comparative study. *Biometrics*, 71:364–375.

van den Hout, A. (2020). *Multi-State Survival Models for Interval-Censored Data*. Boca Raton: Chapman and Hall/CRC.

van der Laan, M. J., Rose, S. (2011). *Targeted Learning. Causal Inference for Observational and Experimental Data*. New York: Springer.

van Houwelingen, H. C. (2007). Dynamic prediction by landmarking in event history analysis. *Scand. J. Statist.*, 34:70–85.

van Houwelingen, H. C., Putter, H. (2012). *Dynamic Prediction in Clinical Survival Analysis*. Boca Raton: Chapman and Hall/CRC.

Wei, L. J., Glidden, D. V. (1997). An overview of statistical methods for multiple failure time data in clinical trials. *Statist. in Med.*, 16:833–839.

Wei, L. J., Lin, D. Y., Weissfeld, L. (1989). Regression analysis of multivariate incomplete failure time data by modeling marginal distributions. *J. Amer. Statist. Assoc.*, 84:1065–1073.

Westergaard, T., Andersen, P. K., Pedersen, J. B., Frisch, M., Olsen, J. H., Melbye, M. (1998). Testis cancer risk and maternal parity: a population-based cohort study. *Br. J. Cancer*, 77:1180–1185.

Xu, J., Kalbfleisch, J. D., Tai, B. (2010). Statistical analysis of illness-death processes and semicompeting risks data. *Biometrics*, 66:716–725.

Yashin, A., Arjas, E. (1988). A note on random intensities and conditional survival functions. *J. Appl. Prob.*, 25:630–635.

Zeng, D., Mao, L., Lin, D. Y. (2016). Maximum likelihood estimation for semiparametric transformation models with interval-censored data. *Biometrika*, 103:253–271.

Zheng, M., Klein, J. P. (1995). Estimates of marginal survival for dependent competing risks based on an assumed copula. *Biometrika*, 82:127–138.

Zhou, B., Fine, J. P., Latouche, A., Labopin, M. (2012). Competing risks regression for clustered data. *Biostatistics*, 13:371–383.

Subject Index

Page numbers followed by * refer to a (*)-marked section, and page numbers followed by an italic *b* refer to a summary box.

Printed in the United States
by Baker & Taylor Publisher Services